MW01518034

ENERGY RESEARCH DEVELOPMENTS: TIDAL ENERGY, ENERGY EFFICIENCY AND SOLAR ENERGY

ENERGY RESEARCH DEVELOPMENTS: TIDAL ENERGY, ENERGY EFFICIENCY AND SOLAR ENERGY

KENNETH F. JOHNSON
AND
THOMAS R. VELIOTTI
EDITORS

Nova Science Publishers, Inc.
New York

LIBRARY OF CONGRESS CATALOGING-IN-PUBLICATION DATA
Energy research developments : tidal energy, energy efficiency, and solar energy / Kenneth F. Johnson and Thomas R. Veliotti.
    p. cm.
 Includes index.
 ISBN 978-1-60692-680-2 (hardcover)
 1. Renewable energy sources--Research. 2. Ocean wave power. 3. Energy conservation. 4. Solar power. I. Johnson, Kenneth F., 1965- II. Veliotti, Thomas R.
 TJ808.6.E54 2009
 621.042--dc22
                        2009017659

Published by Nova Science Publishers, Inc. ✦ *New York*

CONTENTS

PREFACE

Energy fuels modern society, though it is often taken for granted. The issues surrounding energy supply are of paramount importance to governments the world over. While calls for "clean" energy technologies are rampant and loud, conventional sources like coal and oil remain the most feasible. However, research continues to plow ahead to find better and more efficient ways to keep the demand for energy met. Nuclear power is a perennially controversial but enticing possibility, although many nations seek to phase-out nuclear power, citing safety and environmental concerns. The energy industry and production major economic components in any country, so all energy and electricity issues hold sway over us all. This new book presents leading research on energy from around the world with an emphasis on tidal energy, energy efficiency and solar energy.

Chapter 1 describes the solar desalination test plant in Abu Dhabi, UAE and gives a summary of its first year performance and economics. The plant has been operating successfully for 18 years supplying fresh water to the City of Abu Dhabi. The plant was commissioned in September 1984 and was running until the year 2002 when it was dismantled after fulfilling its objectives. The aim of the plant is to investigate the technical and economic feasibility of using solar desalination of seawater in providing fresh water to remote communities in the Middle East and to obtain long-term performance and reliability data on the operation of the plant. The plant has proved its technical feasibility and proved to be reliable in operation with few minor maintenance problems that required slight plant modification. Maintenance routines were established to maintain high plant performance. The economic feasibility of the plant was established by comparing the cost of water from a solar MED plant with a conventional MED plant using fossil fuel for plant capacity ranging from 100 m^3/day to 1000 m^3/day. It was found that the cost of water from solar MED plants is competitive with that from a conventional MED plant if the cost fuel continues to rise.

In Chapter 2, simulation results for the M_2 surface and internal tides in the Arctic Ocean are presented. The model results for the surface tide are close to those obtained by other authors. A comparison of the predicted amplitudes and phases of tidal sea surface level elevations with ground-based gauge measurement data shows that our estimates are better consistent with the data than those derived from the Fairbanks University model and worse than those derived from the Oregon State University model, assimilating all available empirical information. The internal tide waves (ITW) in the Arctic Ocean are of nature of trapped waves, localized near large-scale topographic irregularities. Their amplitudes are maximum (~ 4 m) near the irregularities and degenerate as the distance moves away from them. The ITW generation site occurs at a small part of the continental slope to the north-west of the New Siberian islands. The ITW decay scale at the section going across the above site is ~ 300 km and is not beyond the range of its values (100 – 1000 km) for other oceans. Meanwhile, the local baroclinic tidal energy dissipation rate at specific sites of Amundsen Trough and Lomonosov Ridge are significantly less than in other oceans. The same may be said about the integrated (over area and in depth) total (barotropic + baroclinic) tidal energy dissipation in the Arctic Ocean as a whole.

The idea of extracting power from the tides is not new, and some tidal barrages have been built across estuaries with large tidal ranges. However, in recent years interest has also emerged in smaller potentially less invasive impoundment options, known as tidal lagoons. In addition, a new concept has emerged, free-standing tidal current turbines, operating on tidal flows.

Chapter 3 reviews the state of play with tidal energy developments around the world, looking at barrages and lagoons, but covering tidal current turbine projects in more detail. It draws out some of the key design issues and assesses prospects for the future of tidal energy.

In Chapter 4, a tight and elastic polysaccharide solid containing excess water was proposed as a solid medium for electrochemistry. Conventional electrochemical measurements could be performed in this solid the same as in liquid water. Diffusion coefficients (D_{app}) of small molecules and ions in this solid were almost the same as in a liquid. Charge transfer resistance (R_{ct}) as well as double layer capacitance (C_{dl}) at the interface of the electrode/polysaccharide solid was similar as that of electrode/water interface. Ionic conductivity in this solid was the same as in liquid water. It was found that bulk convection does not take place in this solid, which allows discrimination of molecular diffusion from bulk convection. This solid was successfully applied to solidify the redox electrolyte solution of a dye-sensitized solar cell giving almost similar light-to-current conversion efficiency as that of a liquid type cell.

Chapter 5 concerns the experimental study of heat and mass transfer in a distillation cell. The cell is a parallelepiped cavity, of high form factor and with vertical active walls. Pure water is evaporated from a film of salted water which falls along a heated wall while the opposite wall is maintained at a relatively low temperature and is used as a condensation wall.

The experimental results show that that the mass transfer inside the distillation cell is dominated by the latent heat transfer associated to evaporation. These results allowed us to perform a parametric study of the operating parameters of the distillation cell. A convenient choice of these parameters is necessary to optimize the distillation yield. The performance of a three-stage distiller derived from the characteristics of a single-stage distiller has been studied.

As explained in Chapter 6, the bio-mechanization of organic waste offers energetic, economic, environmental and social advantages which harmoniously integrate it in the concept of continuous development. Among the principal recovered products, we are interested in agricultural biogas whose production was carried out successfully through the mechanization of cows dungs in an experimental digester of 800 liters. Following these first experiments, we present a feasibility study for the realization of a family digester of 15 m³ which can produce between 68.4 and 185.4 m³ of biogas according to the digestion temperature. In order to improve the efficiency of biogas production, we heat the digestion substrate up to a temperature of 40°C by means of an air low temperature heating system using 20 m² of solar collectors. The proposed digester will produce monthly the energy equivalent of three standard cans of butane gas (13 kg/can) and avoid the use of an equivalent of 927 kg of heating wood.

Fluid inclusions in transparent minerals, which trapped different hydrothermal fluids at various stages after precipitation of the minerals till now, offer us information about the physicochemical properties of the fluid such as temperature, salinity and chemical composition. From the point of view, fluid inclusion studies have been widely used to estimate reservoir temperature during drilling of a well and interpret the physicochemical evolution of reservoir fluid at geothermal field.

As discussed in Chapter 7, the minimum Th value of fluid inclusion in quartz, calcite and anhydrite from a geothermal well can be used to estimate the reservoir temperature with real time at a development site in most liquid-dominated and vapor- dominated geothermal fields, Japan. In numerous methods developed for fluid inclusion analysis, semi-quantitative convenient gas analysis of liquid-rich inclusion using a crushing experiment and microthermometry is available to estimate CO_2 content in inclusion fluid lower than about 2.0 mol%. To the contrary, a quadrupole mass spectrometer (QMS) is one of most useful quantitative analytical methods to clarify gas chemistry of tiny fluid inclusion. Based on the analytical results using the QMS on fluid inclusions in anhydrite and quartz from wells in the Japanese geothermal fields, the main component is H_2O with small amounts of CO_2, N_2 and CH_4 and Ar. The CO_2 and CH_4 contents in the inclusion fluids are nearly equal to or slightly higher than those in the present reservoir fluids. In contrast, the N_2 contents in the former are generally about one to three orders of magnitude higher than those in the latter. The differences in the CO_2 content and the CO_2/CH_4 and CO_2/N_2 ratios between the inclusion fluid and present reservoir fluid could be explained by degassing and/or dilution. In the degassing case, early stage fluid was trapped in the fluid inclusions before considerable boiling and vapor-loss. It appears that the CO_2 content in the reservoir fluid decreased more or less during progressing of the exploitation of geothermal field.

As size of low dimensional materials decreases, which leads to dramatic increase of surface-to-volume ratio, their properties are essentially controlled by related interface energetic terms, such as interface energy and interface stress. Although such changes in behavior can be dominant effects in nanoscale structures, we still have remarkably little experience or intuition for the expected phenomena, especially the size dependent properties and their practical implication, except for electronic systems. In Chapter 8, the classic thermodynamics as a powerful traditional theoretical tool is used to model these energetic terms including different bulk interface energies and the corresponding size dependences. Among the above modeling a special emphasis on the size dependence of interface energy is carried out. The predictions of the established models without free parameters are in agreement with the experimental or other theoretical results of different kinds of low dimensional materials with different chemical bond natures.

The use of natural resources is necessary for human life and activities. Finding strategies that can satisfy energy and material demands, be sustainable for the environment, and preserve natural resources from depletion, seems to be a crucial point for the modern society.

Chapter 9 deals with the role of Decision Support Systems (DSS) for environmental systems planning and management, with specific attention to the use of forest biomasses for energy production, and energy and material recovery from urban solid waste. The two mentioned issues are treated in two different sections to highlight the legislative and technological differences beyond the "treatment" and the "possibilities of use" of biomasses and solid wastes. Then, after a general introduction about how to build Environmental DSSs, decision models (characterized by decision variables, objectives, and constraints), and about the state of the art, the chapter is divided in two main sections: energy production from renewable resources, and material and energy recovery from solid urban waste.

The first one is focused on forest biomass exploitation and includes a detailed mathematical description of the formalized decision model and its solution applied to a real case study. The DSS described in the present study is organized in three modules (GIS, data management system, optimization) and is innovative, with respect to the current literature, because of the design of the whole architecture and of the formalization of the different decision models. In particular, the optimization module includes three sub-modules: strategic planning, tactical planning, and operational management. In this chapter, an example related to strategic and tactical planning is given.

The second one regards a more qualitative discussion about multi-objective decision problems for solid waste management in urban areas. In this case, the structure of DSSs able to help decision makers of a municipality in the development of incineration, disposal, treatment and recycling integrated programs, taking into account all possible economic costs, technical, normative, and environmental issues, is described. Finally, conclusions and future developments about EDSS for material recovery and energy production are reported.

When it comes to future reliable oil supplies, Canada's oil sands will likely account for a greater share of U.S. oil imports. As presented in Chapter 10, oil sands account for about 46% of Canada's total oil production and oil sands production is increasing as conventional oil production declines. Since 2004, when a substantial portion of Canada's oil sands were deemed economic, Canada, with about 175 billion barrels of proved oil sands reserves, has ranked second behind Saudi Arabia in oil reserves. Canadian crude oil exports were about 1.82 million barrels per day (mbd) in 2006, of which 1.8 mbd or 99% went to the United States. Canadian crude oil accounts for about 18% of U.S. net imports and about 12% of all U.S. crude oil supply.

Oil sands, a mixture of sand, bitumen (a heavy crude that does not flow naturally), and water, can be mined or the oil can be extracted in-situ using thermal recovery techniques. Typically, oil sands contain about 75% inorganic matter, 10% bitumen, 10% silt and clay, and 5% water. Oil sand is sold in two forms: (1) as a raw bitumen that must be blended with a diluent for transport and (2) as a synthetic crude oil (SCO) after being upgraded to constitute a light crude. Bitumen is a thick tar-like substance that must be upgraded by adding hydrogen or removing some of the carbon.

Exploitation of oil sands in Canada began in 1967, after decades of research and development that began in the early 1900s. The Alberta Research Council (ARC), established by the provincial government in 1921, supported early research on separating bitumen from the sand and other materials. Demonstration projects continued through the 1940s and 1950s. The Great Canadian Oil Sands company (GCOS), established by U.S.-based Sunoco, later renamed Suncor, began commercial production in 1967 at 12,000 barrels per day.

The U.S. experience with oil sands has been much different. The U.S. government collaborated with several major oil companies as early as the 1930s to demonstrate mining of and in-situ production from U.S. oil sand deposits. However, a number of obstacles, including the remote and difficult topography, scattered deposits, and lack of water, have resulted in an uneconomic oil resource base. Only modest amounts are being produced in Utah and California. U.S. oil sands would likely require significant R&D and capital investment over many years to be commercially viable. An issue for Congress might be the level of R&D investment in oil sands over the long-term.

As oil sands production in Canada is predicted to increase to 2.8 million barrels per day by 2015, environmental issues are a cause for concern. Air quality, land use, and water availability are all impacted. Socio-economic issues such as housing, skilled labor, traffic, and aboriginal concerns may also become a constraint on growth. Additionally, a royalty regime favorable to the industry has recently been modified to increase revenue to the Alberta government. However, despite these issues and potential constraints, investment in Canadian oil sands will likely continue to be an energy supply strategy for the major oil companies.

Tidal interactions between neighboring objects span across the whole admissible range of lengths in nature: from, say, atoms to clusters of galaxies i.e. from micro to macrocosms. According to current cosmological theories, galaxies are embedded within massive non-baryonic dark matter (DM) halos, which affects their formation and evolution. It is therefore highly rewarding to understand the role of tidal interaction between the dark and luminous matter in galaxies. The investigation in Chapter 11 is devoted to Early-Type Galaxies (ETGs), looking in particular at the possibility of establishing whether the tidal interaction of the DM halo with the luminous baryonic component may be at the origin of the so-called "tilt" of the Fundamental Plane (FP). The extension of the tensor virial theorem to two-component matter distributions implies the calculation of the self potential energy due to a selected subsystem, and the tidal potential energy induced by the other one. The additional assumption of homeoidally striated density profiles allows analytical expressions of the results for some cases of astrophysical interest. The current investigation raises from the fact that the profile of the (self + tidal) potential energy of the inner component shows maxima and minima, suggesting the possible existence of preferential scales for the virialized structure, i.e. a viable explanation of the so called "tilt" of the FP. It is found that configurations related to the maxima do not suffice, by themselves, to interpret the FP tilt, and some other relation has to be looked for.

The use of ensemble techniques for wind power forecasting aids in the integration of large scale wind energy into the future energy mix and offers various possibilities for optimisation of reserve allocation and operating costs. In Chapter 12 we will describe and discuss recent advances in the optimisation of wind power forecasts to minimise operating costs by using a multi-scheme ensemble prediction technique to demonstrate our theoretical investigations. In recent years a number of optimisation schemes to balance wind power with pumped hydro power have been investigated. Hereby the focus of the optimisation was on compensating the fluctuations of wind power generation. These studies assumed that the hydro plant was dedicated to the wind plant, which would be both expensive and energy inefficient in most of today's and expected future electricity markets, unless the wind generation is correlated and has a very strong variability. Instead a pooling strategy is introduced that also includes other sources of energy suitable to balance and remove the peaks of wind energy, such as biogas or a combined heat and power (CHP) plant. The importance of such pools of energy is that power plants with storage capacity are included to enable the pool to diminish speculations on the market against wind power in windy periods, when the price is below the marginal cost and when the competitiveness of wind power as well as the incentives to investments in wind power become inefficient and unattractive.

It will also be shown that the correlation of the produced wind power diminishes and the predictability of wind power increases as the wind generation capacity grows. Then it becomes beneficial to optimise a system by defining and applying cost functions rather than optimising forecasts on the mean absolute error (MAE) or the root mean square error. This is because the marginal costs of up and down regulation are asymmetric and dependent on the competition level of the reserve market. The advantages of optimising wind power forecasts using cost functions rather than minimum absolute error increase with extended interconnectivity, because this serves as an important buffer not only from a security point of view, but also for energy pricing.

In: Energy Research Developments
Editors: K.F. Johnson and T.R. Veliotti

ISBN: 978-1-60692-680-2
© 2009 Nova Science Publishers, Inc.

Chapter 1

MULTIPLE EFFECT DISTILLATION OF SEAWATER WATER USING SOLAR ENERGY - THE CASE OF ABU DHABI SOLAR DESALINATION PLANT

Ali El-Nashar

Zizinia, Alexandria, Egypt

Abstract

This report describes the solar desalination test plant in Abu Dhabi, UAE and gives a summary of its first year performance and economics. The plant has been operating successfully for 18 years supplying fresh water to the City of Abu Dhabi. The plant was commissioned in September 1984 and was running until the year 2002 when it was dismantled after fulfilling its objectives. The aim of the plant is to investigate the technical and economic feasibility of using solar desalination of seawater in providing fresh water to remote communities in the Middle East and to obtain long-term performance and reliability data on the operation of the plant. The plant has proved its technical feasibility and proved to be reliable in operation with few minor maintenance problems that required slight plant modification. Maintenance routines were established to maintain high plant performance. The economic feasibility of the plant was established by comparing the cost of water from a solar MED plant with a conventional MED plant using fossil fuel for plant capacity ranging from 100 m^3/day to 1000 m^3/day. It was found that the cost of water from solar MED plants is competitive with that from a conventional MED plant if the cost fuel continues to rise.

Keywords: desalination, solar energy, solar desalination, economic feasibility, operating performance, solar distillation.

1. Introduction

Many remote areas of the world such as coastal desert areas in the Middle East or some Mediterranean and Caribbean islands are suffering from acute shortage of drinking water particularly during the summer season. Drinking water for these locations are normally hauled in by tankers or barges or produced by small desalination units using the available saline water. The transportation of water by tankers or barges involves a lot of expense and is fraught with logistical problems which can make fresh water not only very expensive when available but also its supply being very susceptible to frequent interruptions. The use of small conventional desalination units using a fossil fuel such as diesel oil as the energy supply can suffer from the same procurement problems that are encountered with transporting fresh water, namely transportation expenses and supply reliability.

Some of the remote areas are blessed with abundant solar radiation which can be used as an energy source for small desalination units to provide a reliable drinking water supply for the inhabitants of the remote areas. Recently, considerable attention has been given to the use of solar energy as an energy source for desalination because of the high cost of fossil fuel in remote areas, difficulties in obtaining it, interest in reducing air pollution and the lack of electrical power source in remote areas. Desalination of seawater and brackish water is one of the ways for meeting future fresh water demand. Conventional desalination technology is fairly well established, and some of the processes may be considered quite mature although there is still considerable scope for improvement and innovation. Conventional desalination processes are energy intensive, and one of the major cost items in operating expenses of any conventional desalination plant is the energy cost. Thus, one of the major concerns about using desalination as a means of supplying fresh water to remote communities is the cost of energy. Apart from energy cost implications, there are environmental concerns with regard to the effects of using conventional energy sources. In recent years it has become clear that environmental pollution caused by the release of green house gases resulting from burning fossil fuels is responsible for ozone depletion and atmospheric warming. The need to control atmospheric emissions of greenhouse and other gases and substances will increasingly need to be based on growing reliance on renewable sources of energy.

Figure 1. Picture of Abu Dhabi solar desalination plant.

A solar-assisted desalination plant was designed, constructed and put into operation on September 1984 as part of a cooperative research program between Japan and the United Arab Emirates (UAE) to test the technical and economic feasibility of using solar energy for desalination of seawater[1,2,3,4]. The plant (see figure 1) has been in operation in a Umm Al Nar near Abu Dhabi City until the year 2002 when it was dismantled. This report describes the main features of the first year of operation and compares its economics with conventional systems using the same desalination technology.

2. History of Abu Dhabi Solar Desalination Plant

In July 1979, when Mr. Ezaki, the then Japanese Minister of International Trade and Industry, visited the United Arab Emirates (UAE) and discussed the utilization of solar energy utilization in the UAE with Dr. Mana Saeed Al-Otaiba, the UAE Minister of Petroleum and Mineral Resources, they agreed on a joint project between the two countries to develop solar energy utilization for desalination of seawater.

Under this agreement, several discussions were held at various levels. On January 22, 1983, the Record of Discussion (ROD) was finally signed for the joint implementation of a Research and Development Cooperation on Solar Energy Desalination Project by the New Energy Development Organization (NEDO) in Japan, and the Water and Electricity Department in the Abu Dhabi Emirate of the UAE[5].

An outline of the ROD is as follows:

- Execusion period of the project is 3 years starting January 22, 1983
- Location of the project is Umm Al Nar in the suburbs of Abu Dhabi City
- Product water capacity of the test plant has a yearly average value of 80 m^3/day
- Research operation period of the test plant is one year
- Japanese project executor: Engineering Advancement Association of Japan (ENAA)

The design, procurement and fabrication of the test plant started in February 1983 and the test plant was completed in October 1984. For the following year, research operation on the test plant was jointly conducted by ENAA and WED and was concluded in October 1985.

Upon completion of the cooperative research project, the test plant was put in operation and was used as a research tool for a number of research projects carried out by WED. The plant was decommissioned in June 2002 after successfully operating for 18 years producing fresh water to Abu Dhabi City.

3. Description of Abu Dhabi Solar Desalination Plant

The solar desalination plant is designed for an expected yearly average fresh water production of 80 m^3/day. A simplified schematic of the plant is shown in figure 2.A bank of evacuated tube solar collectors, whose orientation with respect to the sun has been optimized to collect the maximum amount of solar radiation, is used to heat the collector fluid to a maximum temperature of about 99°C. The effective collector area of this bank is 1862 m^2.

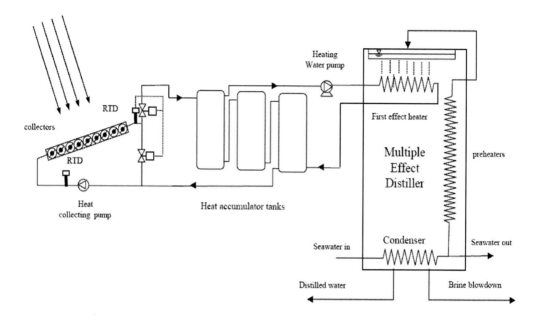

Figure 2. A simplified schematic of the solar desalination plant.

The heat collecting water leaving the collector bank flows into the top of the heat accumulator which has a total capacity of 300 m³. The heat accumulator is of the thermally stratified liquid type where, by virtue of density variation between the top and bottom layers, the higher temperature water is located in the upper region of the accumulator tank while the lower temperature water occupies the lower region. The lower temperature water is drawn from the tank bottom and pumped through the collectors by the heat collecting pump which has a capacity of 80m³/hr at 26m discharge head. The heat collecting water is drawn from the top of the accumulator tank by the heating water circulating pump and is forced to flow into the heating tubes of the first effect of the MED evaporator. This evaporator is designed for a maximum distillate production of 120m³. By transferring heat to the cooler brine flowing on the outside of the tubes, the heating water is cooled down and is then discharged into the accumulator.

The MED evaporator has 18 effects stacked one on top of the other with the highest temperature effect (No. 1) located at the top of the stack and the lowest temperature effect (No. 18) located at the bottom. The 18 effects are actually arranged in a double-stack configuration where effects 1, 3, 5,....17 are in one stack and effects 2, 4, 6...18 in the second. The double-stack arrangement is incorporated into one evaporator vessel as will be shown in detail later.

In addition to the 18 effects, the evaporator has a final condenser designed to condense the vapor generated in the bottom (last) stage (No. 18). Heat input supplied to the first effect by the heating water is repeatedly used by evaporating a portion of the brine flowing into each effect. The evaporator operates under vacuum that is effected by a positive displacement pump connected to the final condenser. The absolute pressure to be maintained in the final condenser is designed to be 50 mmHg. The pressure to be maintained in each effect varies from slightly below atmospheric in the first effect to about 50 mmHg in the 18th effect.

Seawater is used to condense the vapor generated in the 18th effect. Part of the discharged warm seawater leaving the final condenser returns to the sea, while the other part constitutes the evaporator feedwater. The feedwater flow rate amounts to 17.3 m³/hr; it flows through 17 preheaters before reaching the first effect, one preheater for each effect except the 18th effect. These preheaters are designed to raise the feedwater temperature incrementally by flowing from the bottom effect (No. 18) to the top effect (No.1).

3.1. Plant Description

3.1.1. The Solar Heat Collector Subsystem

The solar energy collecting system (SECS) has the function of collecting the solar energy when it is available during the day using the collector bank and storing this energy in the heat accumulator which supplies thermal energy to the evaporator with minimum fluctuations in the supply temperature. This is desirable since steady state operation of the evaporator near its optimum operating condition is highly recommendable.

The basic unit in the collector bank is the Sanyo evacuated tube solar collector which is shown in isometric in figure 3. This is a flat plate-type collector that employs selective coating absorber plates enclosed in glass tubes maintained under high vacuum of 10^{-4} mm Hg. Ten glass tubes with their absorber plates are incorporated in each collector. Along the centerline of each glass tube is located a single copper tube which is attached to the middle of the absorber plate. The heat collecting water flows through this center pipe and absorbs the solar energy collected.

The ends of each glass tube are sealed to a special stainless steel end cap using a ceramic glass material having a coefficient of thermal expansion approximately the same as that of the glass tube. The difference in the thermal expansion between the copper tube and the glass tube is taken up by bellows installed between the end cap and the copper tube.

Each collector consists of 10 individual tubes arranged in parallel. The heat collecting water moves inside the center tubes in a parallel/series arrangement whereby in five of the tubes the flow is in one direction and in the other five it is in the opposite direction.

Table 1. Specifications of a single collector

Item	Specification
Selective coating	Absorptivity $\alpha \geq 0.91$
	Emissivity $\varepsilon \leq 0.12$
Absorber area	1.75 m²
External dimensions	2860 mm x 985 mm x 115 mm
Net weight	64 kg
Flow rate	700 – 1,800 lit/hr
Max. operating pressure	6 bar

Attached to one end of the center tubes is a header tube with an orifice located in the middle of the header tube. The other ends are connected to return bends which are used to connect pairs of center tubes in series. Several collectors (14 in number) are connected in series by coupling the different header tubes. Each collector has an absorber area of 1.75 m^2 and is coated with a black selective coating having an absorptivity, $\alpha \geq 0.91$ and an emittance, $\varepsilon \leq 0.12$. The specifications of a single collector as provided by the manufacturer are shown in table 1.

The collector bank consists of 1064 collector units making up a total collector area of $1064 \times 1.75 = 1862$ m2. 28 collectors are combined to form a single array pair of collectors with its own support structure as shown in figure 4. Each array pair consists of two parallel stacks of collector with each stack consisting of 14 collectors in series. The array pair is 14.5 meters long and 6.0 meters wide and is oriented in the north/south direction at a slope of 1/50. Water is supplied from the main pipe on the south side and passes through the 14 collectors connected in series and exit into the main pipe on the north side.

Figure 3. Isometric view of a collector.

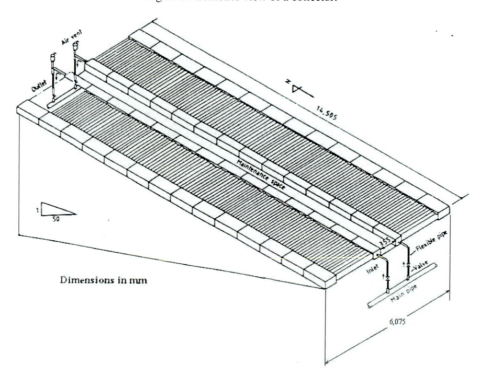

Figure 4. One array pair.

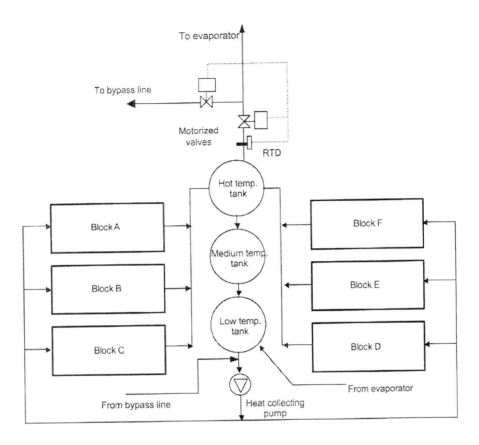

Figure 5. Collector bank consisting of six blocks A, B, C, D, E and F.

76 array pairs are arranged in a U-shape to form the whole collector bank. All array pairs are connected in parallel and each is provided with two isolating valves- at inlet and exit-, a drain valve, and an air vent. The bank is divided into six blocks designated A, B, C, D, E and F. Blocks A and F consists of 12 array pairs while the other blocks each consists of 13 array pairs. Figure 5 is a block diagram of the collector field.

3.1.2. The Heat Accumulator Subsystem

The heat accumulator subsystem (see figure 6) is designed to provide thermal energy to the evaporator during its 24 hours per day operation. It consists of three carbon steel tanks having a total capacity of 300 m³ and contains hot water at a temperature ranging from 74°C to 99°C and at atmospheric pressure. The tanks are insulated with a 100 mm layer of fiberglass to minimize heat loss to the ambient air. All three cylindrical tanks have the same internal diameter (3.8m) and wall thickness (9mm). However, the tank heights are not identical with tank No. 1 having an effective height of 10m while tanks No. 2 and 3 having an effective height of 7.6m. The heat collecting water from the collector bank is introduced at the top of tank No. 1. The heat collecting water to the collector bank is taken from the bottom of tank No. 3. Heating water to the evaporator is drawn from the top of tank No. 1 and returns to the bottom of tank No. 3. The water is therefore stratified in such a way that the top water layers of tank No. 1 are always at the highest temperature and the bottom layer of tank No. 3 at the lowest temperature.

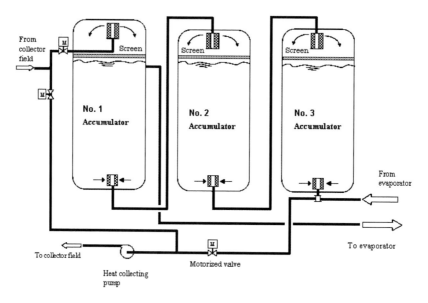

Figure 6. Heat accumulator subsystem.

The heat accumulator tanks have enough capacity to be able to provide the required thermal load (for the evaporator) for about 16 hours after sunset provided that the tanks are fully charged just before sunset. This feature makes it possible to operate the desalination unit during night time. Only during extended overcast or hazy days when sandstorms prevail we expect plant shut-down to occur due to insufficient energy collection.

3.1.3. MED Evaporator Subsystem

A horizontal tube, thin film multi-effect distiller (MED) is used for desalination of seawater. The distiller is manufactured by Sasakura Engineering Co., Ltd. It was chosen because of its capability to accommodate large load fluctuations and its small consumption of electrical power. The maximum capacity of the distiller is 120 m^3/day.

A flow diagram of the MED desalination process is shown in figure 7. Preheated feedwater is sprayed into the top of the first effect and descends down the evaporator stack, flowing as a thin film over the tube bundle in each effect. The feedwater flashes and thereby cooled by several degrees as it passes from one effect to the next. It is rejected at the bottom of the plant as cool, concentrated brine. In the top effect, heating water from the accumulator is used to partially evaporate the thin seawater film on the outside of the tubes. The generated vapor passes through demisters to the inside of the tubes in the second effect where it condenses to form part of the product.

It simultaneously causes further evaporation from the external seawater film and the process is repeated from effect to effect down the plant. The heat input from the accumulator is thus used over and over again in successive evaporation/condensation heat exchangers in each effect to produce more product and new vapor, thereby obtaining a maximum quantity of fresh water with minimum heat input. The vapor generated in the last effect (18th) is condensed in a seawater-cooled condenser and part of the seawater is used as feedwater to the stack. The remaining seawater is rejected to the sea and carries most of the heat away from the process.

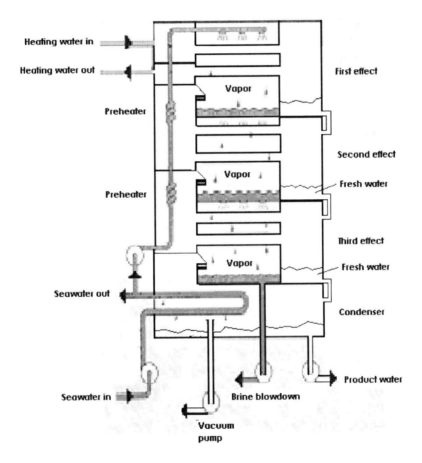

Figure 7. The MED evaporator.

3.2. Design Features

Table 2 shows the design conditions and specifications at the time of the original planning of the test plant. At the planning stage, no detailed solar radiation data was available for Abu Dhabi city and the only data available was that for nearby Kuwait. Therefore the data for Kuwait was used with the annual mean daily solar radiation on horizontal surface taken as 5000 kcal/m^2 day. Based on the measurements at the test plant made during 1985 and subsequent years, the annual average daily values were found to be slightly higher than this value (5270 kcal/m^2 day for 1985).

Table 2. Plant design conditions assumed for design of test plant

Design parameter	Assumed value/range
Solar radiation	5,000 kcal/m^2 day (annual mean value on horizontal surface)
Ambient temperature	30°C (daytime mean temperature)
Rainfall	18.1 – 390.1 mm/year
Wind speed	5 m/s (for collector design) 30 m/s (for structures)
Relative humidity	Max. 100%, min. 10%, normal 25-90%
Seawater temperature	55,000 ppm TDS (design base)
System capacity	80 m^3/day as expected yearly average
Solar collector type	Evacuated glass tube collector 1862 m^2 (effective absorbing area)
Heat accumulator	Thermally stratified vertical cylinder Capacity 300 m^3
Evaporator	Horizontal-tube, multiple effect stack type Evaporator, capacity 120 m^3/day, specific heat consumption 43.8 kcal/kg- product water

4. Measurements and Data Acquisition System

Table 3 lists the plant parameters measured every 15 minutes and sent to the data acquisition system (DAS). The DAS is shown in figure 8 and consists of two separate subsystems: one is the on-line control room subsystem and the other is the data analysis subsystem. The data analysis subsystem consists of a Thermodac 32 data logger manufactured by Eto Denki, Co., PC (model PC-8001 mkII manufactured by NEC company) and a PC printer (model PC-8023C manufactured by NEC company). The data analysis subsystem consists of a data logger (model Thermodac 3 manufactured by Eto Denki, Co.), PC (model PC-8801mk II by NEC company) and PC printer (model PC-8024 by NEC company).

The on-line control room subsystem has the following functions:

- Sampling of data at 15 minute intervals
- Calculate hourly average values once per hour
- Record data on CD at even hours (i.e. 8:00, 10:00, 12:00,...)
- Print a summary report every 12 hours.

The data analysis subsystem has the following functions:

- Make daily, weekly and monthly reports.
- Format new data disks
- Copy data disks for backup
- Edit hourly or daily data on data disks.

**Table 3. Items of data acquired every 15 minutes
from data loggers Thermodac 32 and Thermodac 3**

Channel #	Measuring survice	Tag #	Signal output	Unit
THERMODAC 32				
1	Ambient temp.	TE-111	DC 1-5 V	^{o}C
2	Collector field outlet temp.	TE-102-1	DC 1-5 V	^{o}C
3	Accumulator inlet temp.	TE-102-2	DC 1-5 V	^{o}C
4	Accumulator outlet temp.	TE-102-3	DC 1-5 V	^{o}C
5	Heating water inlet temp.	TE-104	DC 1-5 V	^{o}C
6	Heating water outlet temp.	TE-105	DC 1-5 V	^{o}C
7	No. 1 effect temp.	TE-206-1	DC 1-5 V	^{o}C
8	Preheater No. 1 outlet temp.	TE-203	DC 1-5 V	^{o}C
9	No. 18 effect temp.	TE-206-2	DC 1-5 V	^{o}C
10	Seawater temp.	TE-202	DC 1-5 V	^{o}C
11	Empty collector temp. #1	TE-301	Thermocouple	mV
12	Empty collector temp. #2	TE-302	Thermocouple	mV
13	Relative humidity	HUE-111	DC 1-5 V	%
14	Heat collecting water flow	FIT-101	DC 1-5 V	m^3/hr
15	Heating water flow	FIT-105	DC 1-5 V	m^3/hr
16	Product water flow	FQ-205	DC 24V pulse	m^3
17	Solar radiation	SQ-111	DC 24V pulse	kcal/hr m2
18	Heat collected from field	CAQ-102	DC 24V pulse	kcal
19	Heat used by evaporator	CAQ-105	DC 24V pulse	kcal
20	Heat collected by block F	CAQ-101-1	DC 24V pulse	kcal
21	Heat collected by block A	CAQ-101-2	DC 24V pulse	Kcal
22	Pump P-101 running hours (heat collecting pump)	N/A	On/off pulse	hr
23	Pump P-205 running hours (product water pump)	N/A	On/off pulse	hr
24	Electrical energy consumption	N/A	DC 24V pulse	kWh
THERMODAC 3				
1	No. 1 effect temp.	TEW-206-1	Pt 100	^{o}C
2	No. 4 effect temp.	TEW-206-4	Pt 100	^{o}C
3	No. 7 effect temp.	TEW-206-7	Pt 100	^{o}C
4	No. 10 effect temp.	TEW-206-10	Pt 100	^{o}C
5	No. 13 effect temp.	TEW-206-13	Pt 100	^{o}C
6	No. 16 effect temp.	TEW-206-16	Pt 100	^{o}C
7	No. 18 effect temp.	TEW-206-18	Pt 100	^{o}C

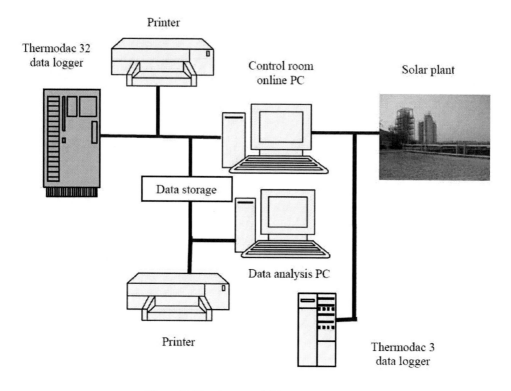

Figure 8. The data acquisition system (DAS).

The items of data shown in table 3 are transferred to the computer through an RS-232C interface to the control room computer to be processed. The results of the mean of four 15-minute data values are displayed every hour on the CRT in the control room. The following hourly values are displayed on the CRT:

- Climatic temperature (°C)
- Solar radiation (kcal/m^{2}hr)
- Heat consumption (kcal/kg dist.)
- OHTC of heater (first effect) (kcal/m^{2}hr°C)
- Seawater flow (m^3/hr)
- Seawater TDS (ppm)
- Seawater inlet temp. (°C)
- Seawater outlet temp. (°C)
- Heat collected by block F (kcal/hr)
- Heat collected by block A (kcal/hr)
- Heat collecting pump flow rate (m^3/hr)
- Heat accumulator inlet temp. (°C)
- Heat accumulator outlet temp. (°C)
- Heating water inlet temp. to evaporator (°C)
- Heating water return temp. from evaporator (°C)

- Heat supplied to accumulator (kcal/hr)
- Heat supplied to evaporator (kcal/hr)
- Heat supplied to evaporator (kcal/hr)
- Evaporator feedwater flow rate (m^3/hr)
- Preheater #1 outlet temp. ($^{\circ}$C)
- First effect temp. ($^{\circ}$C)
- 18th effect temp. ($^{\circ}$C)
- Product water flow rate (m^3/hr)
- Absorber plate temp. of empty collector ($^{\circ}$C)
- Header tube temp. of empty collector ($^{\circ}$C)

4.1. Measuring the Heat Collected in Block F

Measuring the heat loss in the piping system in such a large collector bank is obviously laborious and will require the accurate measurement of collector fluid temperature at many locations within the field. This was deemed impractical and was ruled out from the outset. The solution which was found practical is to isolate a single block of collectors and use it for test measurements in order to find the piping heat loss in this block, then estimate the heat loss in the piping system of the whole collector bank based on the results obtained from the measurements carried out on the selected block. Block F was selected for this purpose since this block was already provided with resistance temperature detectors at inlet and outlet of the block, as well as a vortex flowmeter for measuring the water flow rate through the block.

Figure 9 is a schematic diagram of block F showing the location of the temperature measuring probes. Two RTDs are attached at block inlet (location E) and block outlet (location F). The RTDs, which are three-wire sheathed platinum resistance elements, are model RN33-AMAS and are manufactured by Yokogawa Hokushin Corporation. They are connected to Thermodac 32 via resistance-to-voltage converters which produce 1-5 volt DC signal outputs that are fed to temperature recorders.

The two output signals from the resistance-to-voltage converters which are attached to the RTDs at locations E and F are connected to a programmable computing unit (PCU), model SPLR-100A manufactured by Yokogawa (see Yokogawa Instruction Manual for Model YF100 vortex flowmeters (1993)), in which the 4-20 mA signal from the vortex flow meter is also connected to the PCU. A variety of arithmetical computational functions can be performed by the PCU. Programs can be developed and written to ROM (Read Only Memory) using a dedicated programming language connecting the PCU to an SPRG Programmer. The pulse output signal from the PCU represents the heat collected between E and F and was measured by subtracting the two input temperature signals, T_F - T_E, and multiplying by the flow rate signal and the specific heat of the heat collecting fluid (water) at the operating temperature to obtain the heat collected between E and F (block heat collected).

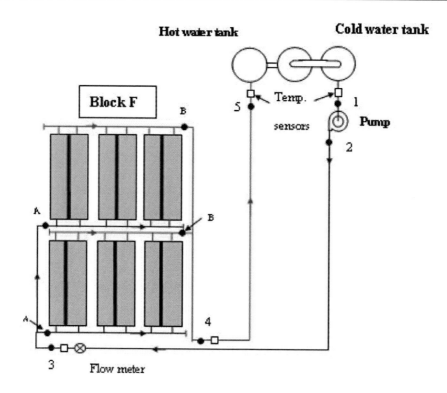

Figure 9. Location of measuring sensors in block F.

The analog signals from the resistance-to-voltage converters, as well as the pulse output signal from the PCU were connected to the DAS for recording and printing and for transfer to a PC computer system via an RS-232 interface cable. The pulse output signal was integrated over a period of one hour and scanning of the data was made every 15 minutes.

The inlet and outlet temperature measurements (T_A and T_B) of each of the 12 arrays making up block F were measured using copper-constantan thermocouples which were attached to the array header pipes as shown in figure 9. The assembly of these thermocouples into the array supply and return pipes, as well as the locations of the 24 (2 x 12 arrays) thermocouples used are also shown in this figure. The mV output signals from these thermocouples were recorded on an hourly basis. The heat collected by the 12 arrays was calculated by multiplying the temperature difference, $T_A - T_B$, by the flow rate and the specific heat in a manner similar to that used for estimating the block heat collected.

A remote converter-type vortex flowmeter (model YF105, Yewflo by Yokogawa) is used with a model YFA11 vortex flow converter are used for flow measurement. This vortex flow meter measures the flow rates and converts the measurements to a 4 to 20 mA DC output signal. The accuracy of the instrument is ± 1.0% of reading plus ± 0.1% of full scale.

A solar radiation sensor (pyranometer)- model H 201 manufactured by Nakaasa Instrument Co.- is used to measure the global insolation on the collector absorber plate. The sensor has a measuring range of 0 to 2 kW/m^2 and has an accuracy of ± 0.5% of full scale. The mV output signal from the sensor is first amplified before being converted into a pulse signal for connection to Thermodac 32 data logger. This signal is integrated over hourly intervals in order to obtain the hourly values of the solar radiation. The ambient temperature was measured by a three-wire RTD connected to the DAS. Figure 10 shows a block diagram of the data acquisition and analysis for the heat collection system. The estimated percentage error in the heat measurement was estimated as 1.5 – 3% while the error in the collector efficiency measurement 2.0 – 3.5%.

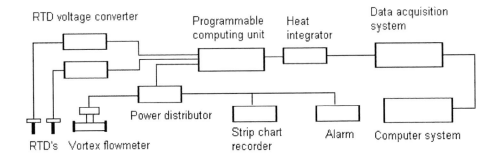

Figure 10. Block diagram of data acquisition system for heat collection.

5. Data Analysis

5.1. Calculating the Solar Radiation on Absorber Plate

The global solar radiation intensity on a tilted surface can be expressed by Eq. 1.

$$I_t = I_d + I_s \tag{1}$$

where I_d is the direct component of solar radiation and I_s is the diffuse component. According to [5], the components can be expressed as:

$$I_d = I_o \times P^{\frac{1}{\sin(h)}} \times \cos(\theta) \tag{2}$$

$$I_s = \frac{1}{2} \times I_o \times \sin(h) \times \frac{1 - P^{\frac{1}{\sin(h)}}}{1 - 1.4\ln(P)} \times \frac{1 + \cos(\alpha)}{2} \tag{3}$$

where P is the atmospheric transmittance, h is the solar altitude, α is the tilt angle to the ground, θ is the incidence angle of direct radiation on the tilted surface. The solar angles are shown in figure 11. The transmittance P is defined as the ratio between the normal solar radiation at the ground level to the corresponding value at the outer limit of the atmosphere:

$$P = \frac{I_N}{I_{oN}} \tag{4}$$

where I_N is direct normal radiation at the ground level and I_{oN} is the corresponding value at the edge of the atmosphere. Since the hourly global radiation is measured at a tilted surface having the same tilt angle as the collector absorber plates, I_t, it is possible to solve the following equation for the hourly values of P:

$$I_t = I_o \times P^{\frac{1}{\sin(h)}} \times \cos(\theta) + \frac{1}{2} \times I_o \times \sin(h) \times \frac{1 - P^{\frac{1}{\sin(h)}}}{1 - 1.4\ln(P)} \times \frac{1 + \cos(\alpha)}{2} \qquad (5)$$

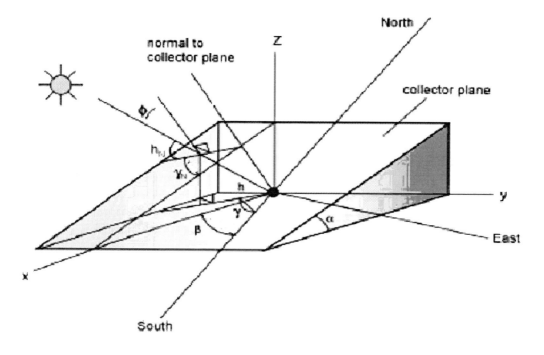

Figure 11. Solar angles.

With the hourly values of P estimated, it is possible to calculate the direct and diffuse components from Eqs. 2 and 3.

5.2. Calculating the Amount of Heat Collected and Collector Outlet Water Temperature

The heat collection amount, Q_c, was calculated from the measured collector inlet and outlet water temperatures and the water flow rate

$$Q_c = m_c C_p (T_{c2} - T_{c1}) \qquad (6)$$

where m_c is the water flow rate through the collector bank (or block) and T_{c1} and T_{c2} are the inlet and outlet water temperatures, respectively.

The heat collection efficiency is expressed by the following polynomial equation

$$\eta_c \equiv \frac{Q_c}{A_c I_t} = a + b\,x + c\,x^2 \tag{7}$$

a, b, and c are constants and x is a parameter defined as

$$x = \frac{\dfrac{T_{c1} + T_{c2}}{2} - T_a}{I_t} \tag{8}$$

where T_a is the ambient air temperature and I_t is the solar radiation *on* a tilted surface. Susbstituting the expression for Q_c from Eq. 6 and the expression for x from Eq. 8 into Eq. 7 we get a relationship for the collector outlet temperature, T_{c2} in terms of the collector inlet temperature, ambient temperature, solar radiation and collector absorber area

$$C_1 T_{c2}^2 + 2 C_2 T_{c2} + C_3 = 0 \tag{9}$$

where C_1, C_2 and C_3 are given by the following expressions

$$C_1 = c\,A_c$$

$$C_2 = c\,A_c\,(T_{c1} - 2\,T_a) + b\,A_c\,I_t - 2\,m_c\,I_t$$

$$C_3 = c\,A_c\,(T_{c1}\,2\,T_a)^2 + 2\,b\,A_c\,I_t\,(T_{c1} - 2\,T_a) + 4\,a\,A_c\,I_t^2 + 4\,m_c\,C_p\,I_t\,T_{c1}$$

Using the constants a, b, and c and for given values for T_{c1}, T_a and I_t, eqn 9 can be solved for T_{c2} to give

$$T_{c2} = \frac{-C_2 - \sqrt{C_2^2 - C_1 C_3}}{C_1}$$

with the above equations, the hourly values of the outlet water temperature T_{c2} can be derived if the hourly inlet water temperature, T_{c1}, is given.

5.3. Calculating the Performance of the Evaporator

The performance of the evaporator consists of estimating the overall heat transfer coefficients (OHTC) of the first effect (heater), the other evaporator effects (2nd – 18th effects), the 17 preheaters and the condenser as well as evaluating the economy (or specific heat consumption) of the evaporator. The list of measurements carried out for the evaporator are shown in table 4.

Table 4. Measurements made at evaporator

Measurement	Location	Symbol	Unit
Flow rate	Heating water flow	m_{hw}	m³/h
	Seawater flow	m_{sw}	m³/h
	Feedwater flow	m_{fw}	m³/h
	Product water flow	m_d	m³/h
Temperture	Heating water inlet	T_{hw1}	°C
	Heating water outlet	T_{hw2}	°C
	First effect	$T_{ev}(1)$	°C
	18th effect	$T_{ev}(18)$	°C
	1st preheater outlet	$T_{pr}(1)$	°C
	Seawater inlet to condenser	T_{con1}	°C
	Seawater outlet from condenser	T_{con2}	°C
Heat amount	Heat supplied by heating water	Q_{hw}	kcal/h
Total dissolved solids	Seawater	$C_b(0)$	Kg salt/kg water

5.3.1. Calculating the Brine Concentration for Each Effect

The brine total dissolved solids (TDS) in each effect is estimated from a mass balance for each effect considering the fact that the heater (first effect) receives all the feedwater while all the other effects receive only half of the amount of brine leaving the first effect. This stems from the fact that the evaporator consists of one effect at the top followed by two stacks in parallel with all odd-numbered effects in one stack and even-numbered ones in the other. The brine concentration in the first effect is estimated from:

$$C_b(1) = \frac{m_{fw}}{(m_{fw} - m_d/18)} \times C_b(0) \tag{10}$$

where m_{fw} is the feedwater flow rate, m_d is the distillate flow rate and $C_b(0)$ is the seawater concentration. The concentration of the even-numbered effects is obtained from

$$C_b(2N) = \frac{0.5 \times (m_{fw} - m_d/18)}{[0.5 \times (m_{fw} - m_d/18) - N \times m_d/18]} \times C_b(1) \tag{11}$$

where N varies from 1 (2nd effect) to 9 (18th effect)

The concentration of the odd-numbered effects is obtained from

$$C_b(2N + 1) = C_b(2N) \tag{12}$$

where N varies from 1 (3rd effect) to 8 (17th effect).

5.3.2. OHTC of Heater (First Effect)

In the heater (first effect) hot water from the accumulator flows through the tubes in a horizontal tube bundle while a relatively cold seawater is flowing as a thin boiling film on the outside of the tubes. Heat is therefore transmitted from the hot water inside the tubes to seawater on the outside. The overall heat transfer coefficient for the heater can be expressed as

$$U_h = \frac{Q_h}{A_h (\Delta T)_h} \tag{13}$$

where Q_h is the rate of heat transfer (kcal/h), A_h is the heat transfer area ($A_h = 24.9$ m^2) and $(\Delta T)_h$ is the log-mean-temperature difference obtained from the equation:

$$(\Delta T)_h = \frac{\{T_{hw1} - T_{ev}(1)\} - \{T_{hw2} - T_{pr}(1)\}}{\ln \dfrac{\{T_{hw1} - T_{ev}(1)\}}{\{T_{hw2} - T_{pr}(1)\}}} \tag{14}$$

The rate of heat transfer can be estimated from

$$Q_h = m_{hw} C_p (T_{hw1} - T_{hw2}) \tag{15}$$

The specific heat C_p is calculated at the average temperature $(T_{hw1} + T_{hw2})/2$.

5.3.3. Average OHTC of Other Evaporator Effects

In the 2nd –18th effects the mechanism of heat transfer is different from that in the first effect. In these effects, boiling takes place on the outside of the horizontal tubes while condensation occurs on the tube inside. The average overall heat transfer coefficient for the 17 evaporators was estimated from the equation

$$U_{ev} = \frac{Q_{ev}}{A_{ev} (\Delta T)_{ev}} \tag{16}$$

where Q_{ev} is the average heat transfer rate for each evaporator effect, $(\Delta T)_{ev}$ is the average log-mean-temperature difference for each evaporator and A_{ev} is the heat transfer area of each evaporator ($A_{ev} = 63.1$ m^2). The heat transfer rate Q_{ev} is obtained from the fact that the distillate production consists of two components: production by the 17 (2nd –18th) effects and production by the 17 (1st – 17th) preheaters:

$$m_d = (m_d)_{ev} + (m_d)_{pr} \tag{17}$$

where $(m_d)_{ev}$ is the distillate production by the 17 evaporator effects and $(m_d)_{pr}$ is the production by the 17 preheaters. $(m_d)_{pr}$ is calculated from a heat balance equation over all the preheaters:

$$(m_d)_{pr} = \frac{m_{fw} C_p (T_{pr}(1) - T_{con2})}{L_{av}} \qquad (18)$$

The latent heat of vaporization L_{av} is estimated at an average temperature of $[T_{ev}(1) + T_{ev}(18)]/2$. The average heat transfer by each evaporator effect is obtained from:

$$Q_{ev} = \frac{m_{ev}}{17} \times L_{av}$$

$$= \frac{m_d L_{av} - m_{fw} C_p (T_{pr}(1) - T_{con2})}{17} \qquad (19)$$

The average log-mean-temperature difference $(\Delta T)_{ev}$ is calculated as the average temperature difference between the heating steam inside the evaporator tube bundles and the boiling brine on the outside. Noting that the heating steam temperature in a particular effect i is slightly lower than the brine temperature in the preceding effect (i.e., effect i -1) due to:

- the boiling point elevation (*BPE*),
- the temperature drop across the demister inside $(\Delta p)_{demis}$,

$(\Delta T)_{ev}$ was estimated from the following relation

$$(\Delta T)_{ev} = \frac{T_{ev}(1) - T_{ev}(18) - \sum_{i=1}^{18} BPE - \sum_{i=1}^{17} (\Delta p)_{demis}}{17} \qquad (20)$$

The values of the boiling point elevation (*BPE*) and the demister pressure drop $(\Delta p)_{demis}$ are calculated using the correlations given in the appendix.

5.3.4. Average OHTC of Preheaters

In the 17 preheaters seawater flowing inside the tube bundles of heat exchangers is heated up by steam condensing on the tube outside. For any preheater, the average OHTC (kcal/hr m^{2}°C) can be calculated using the equation:

$$U_{ph} = \frac{Q_{ph}}{A_{ph} \times (\Delta T)_{ph} \times 17} \qquad (21)$$

where Q_{ph} is the heat transfer rate for all preheaters , kcal/hr and A_{ph} is the heat transfer area of each preheater, m^2 (A_{ph} = 19.5 m^2). Q_{ph} is calculated from the measured feedwater flow rate and the temperature difference between the outlet of preheater #1 (top preheater) and the outlet of the condenser:

$$Q_{ph} = M_{fw} \times C_p \times (T_{ph}(1) - T_{cond2}) \qquad (22)$$

The log-mean temperature difference $(\Delta T)_{ph}$ is assumed to be identical for each preheater.

5.3.5. OHTC of Condenser

The OHTC of the condenser is calculated from the measured seawater flow rate and condenser inlet and outlet temperatures according to the equation:

$$U_{cond} = \frac{Q_{cond}}{A_{cond} \times (\Delta T)_{cond}} \qquad (23)$$

where Q_{cond} is condenser heat flow rate, kcal/hr, A_{cond} is the condenser heat transfer area, m^2 ($A_{cond} = 35.3$ m^2). The log-mean temperature difference for the condenser is calculated from the equation:

$$(\Delta T)_{cond} = \frac{(T_{ev}(18) - T_{cond1}) - (T_{ev}(18) - T_{cond2})}{\ln \dfrac{(T_{ev}(18) - T_{cond1})}{(T_{ev}(18) - T_{cond2})}} \qquad (24)$$

5.3.6. Evaporator Economy

The performance ratio (PR) of the evaporator is calculated as the ratio between the product flow rate, M_d (kg/hr) and the equivalent amount of heating steam to the heater (first effect):

$$PR = \frac{m_d}{(Q_h / L)} \qquad (25)$$

where Q_h is amount of heat supplied to the heater (first effect) and L is the latent heat of vaporization (kcal/kg).

6. Weather Condition in Abu Dhabi

For a solar desalination plant, the solar radiation and ambient temperature have a great effect on the heat collection efficiency of the collectors while the temperature and salinity of the seawater influence the capacity of the distiller. Consequently, these data which have been yielded by the research operation constitute valuable basic material for analyzing the performance of the operation of the plant.The total solar radiation on a horizontal surface for 1985 (see table 5) came to 1,923,000 kcal/m^2 year which gave a mean daily value of 5,270 kcal/m2 day. When the monthly average total solar radiation an a horizontal surface is considered, the maximum value was achieved in June which was 1.95 times as high as the minimum value recorded in December.

Table 5. Daily total radiation on tilted surface at 21° 09' (data of 1985)

JAN. 1985 ~ DEC. 1985 [kcal/m²day]

DAY	JAN.	FEB.	MAR.	APR.	MAY	JUN.	JUL.	AUG.	SEP.	OCT.	NOV.	DEC.	ANN.
1	5280	5660	5910	6240	6400	6170	5810	5610	5640	5950	5540	5200	
2	4960	5760	5740	6490	6330	6260	6200	5060	5940	5870	5470	5100	
3	4390	5590	5710	6280	6160	6280	6040	5560	6240	5450	5480	5070	
4	3720	6130	6070	6380	5980	6150	6090	4990	6060	5930	5550	4940	
5	4800	5610	4830	4750	5060	6260	5620	5650	6120	6060	5540	2650	
6	2770	5620	4740	5840	4750	6100	5990	5730	6190	5910	4830	2550	
7	4140	6050	6000	6160	4920	6110	5460	5940	6100	5900	5390	5080	
8	5480	6040	5500	4510	4080	6130	4370	6070	6070	6090	5460	5370	
9	4720	5520	5570	6820	6220	6400	5140	5890	6090	5950	5450	5300	
10	5290	6160	6280	5500	6330	6340	5050	5720	6250	5890	5420	5430	
11	4550	6220	6250	6580	5660	6280	5030	5760	5100	5860	5380	5310	
12	5190	6380	6450	6790	6370	6000	5130	5600	6140	5730	5330	4110	
13	5070	6340	6330	6410	6220	6310	5300	2820	6180	5740	5260	5080	
14	4160	6240	6250	7000	6290	6080	5760	5900	6220	5460	4820	4590	
15	4360	6300	6590	6930	6290	6200	5470	5960	6140	5720	5120	4690	
16	3250	6450	6130	6620	6160	5960	5410	4470	6170	5780	5110	4740	
17	5300	6040	6020	6360	5230	6140	5320	5810	6210	5830	5300	4880	
18	4120	6400	6290	6500	5060	6310	4980	5890	6150	5950	5310	4540	
19	4550	6320	6370	6780	6360	6340	2800	5910	6190	5980	5280	3640	
20	5330	6620	6320	6610	6450	6370	4700	6130	6050	5820	4210	4100	
21	4530	6290	5600	6340	6060	5950	5320	6110	6870	5560	4870	2870	
22	4900	6670	6170	6490	6230	5920	5170	6130	6050	5310	4990	4090	
23	5260	5870	6230	6590	6440	5940	5491	6010	5980	5330	5230	4580	
24	5290	6020	5720	5560	6280	6000	5630	6070	5820	5430	5420	4570	
25	4780	6167	4250	6240	6160	6010	5670	6080	5770	5850	5170	4520	
26	2750	6380	5600	6880	6120	6150	5820	5940	5960	5490	5110	4820	
27	4800	5640	2330	5710	6150	6170	5710	5940	5970	5630	4990	4920	
28	5140	5540	5280	6270	6170	6000	5770	5940	5950	5600	3950	4650	
29	5660		5920	6070	6180	5760	5940	6020	6080	5650	5020	4330	
30	5600		5010	5570	5930	5800	5630	5950	6100	5580	5180	3730	
31	5640		6360		6290		5370	5890		5610		4500	
TOTAL	145800	170000	178800	187300	184400	183900	167200	176400	181800	177900	155200	139900	2050000
AVERAGE	4700	6070	5770	6240	5950	6130	5390	5690	6060	5740	5170	4510	5610

Table 6. Atmospheric transmittance at noon

Atmospheric Transmissivity at Noon

JAN. 1985 ~ DEC. 1985

DAY	JAN.	FEB.	MAR.	APR.	MAY	JUN.	JUL.	AUG.	SEP.	OCT.	NOV.	DEC.
1	0.7253	0.6914	0.6705	0.6638	0.6427	0.6241	0.6035	0.5381	0.5784	0.6322	0.6711	0.7062
2	0.7062	0.7135	0.6444	0.6634	0.6358	0.6282	0.6181	0.5130	0.6089	0.6363	0.6743	0.7038
3	0.6341	0.7064	0.6401	0.6430	0.6144	0.6351	0.5987	0.5423	0.6359	0.6058	0.6770	0.6633
4	0.5597	0.7446	0.6798	0.6461	0.5954	0.6510	0.6215	0.5517	0.6175	0.6440	0.6805	0.6785
5	0.6190	0.7144	0.2698	0.5657	0.5501	0.6429	0.5607	0.5616	0.6164	0.6561	0.6872	0.5513
6	0.3270	0.6900	0.5537	0.5954	0.5356	0.6248	0.5925	0.5538	0.6311	0.6330	0.6378	0.3724
7	0.4817	0.7398	0.6706	0.6473	0.5105	0.6223	0.5232	0.5897	0.6237	0.6415	0.6660	0.6956
8	0.7427	0.7304	0.5977	0.1246	0.3022	0.6418	0.4304	0.6091	0.6194	0.6610	0.6767	0.7247
9	0.6230	0.6685	0.6671	0.6960	0.6121	0.6812	0.5093	0.5887	0.6244	0.6474	0.6798	0.7223
10	0.7768	0.7281	0.6925	0.6176	0.6282	0.6687	0.4932	0.5611	0.6343	0.6426	0.6851	0.7293
11	0.7513	0.7299	0.6554	0.6965	0.6229	0.6579	0.4620	0.5510	0.6315	0.6452	0.6785	0.7211
12	0.6542	0.7471	0.6889	0.6834	0.6409	0.6288	0.4760	0.5663	0.6342	0.6377	0.6758	0.6329
13	0.6968	0.7389	0.6845	0.6985	0.6239	0.6506	0.5150	0.1261	0.6325	0.6400	0.6692	0.6995
14	0.6016	0.7467	0.6878	0.7119	0.6260	0.6105	0.5843	0.5902	0.6309	0.6454	0.5802	0.6728
15	0.6939	0.7591	0.7239	0.7123	0.6302	0.6314	0.4636	0.6094	0.6351	0.6459	0.6661	0.6588
16	0.3685	0.7592	0.6530	0.6769	0.6175	0.6219	0.5397	0.4358	0.6384	0.6464	0.6659	0.6706
17	0.7046	0.7368	0.6201	0.6387	0.4713	0.6140	0.5428	0.5752	0.6452	0.6629	0.5988	0.6870
18	0.6541	0.7345	0.6711	0.6436	0.4681	0.6489	0.4901	0.5766	0.6365	0.6753	0.6925	0.7020
19	0.6433	0.7223	0.6694	0.6767	0.6442	0.6527	0.1921	0.5826	0.6381	0.6773	0.6862	0.4378
20	0.7009	0.7541	0.6711	0.6669	0.6546	0.6587	0.4527	0.6111	0.6320	0.6665	0.6021	0.6122
21	0.6497	0.7321	0.6496	0.6408	0.6013	0.6015	0.5207	0.6156	0.6223	0.6581	0.6554	0.4925
22	0.6622	0.7615	0.6475	0.6475	0.6317	0.6044	0.5116	0.6231	0.6311	0.6636	0.6765	0.5501
23	0.6889	0.6929	0.6527	0.6744	0.6475	0.5986	0.5430	0.6049	0.6328	0.6479	0.6805	0.6541
24	0.6927	0.6650	0.6210	0.4791	0.6300	0.6074	0.5583	0.6064	0.6167	0.6406	0.7041	0.6614
25	0.6933	0.7189	0.4596	0.6239	0.6186	0.6130	0.5555	0.6135	0.6247	0.6778	0.6884	0.6592
26	0.3338	0.7233	0.6404	0.6926	0.6046	0.6275	0.5787	0.5951	0.6233	0.6530	0.6882	0.6758
27	0.6553	0.6275	0.2227	0.5396	0.6146	0.6319	0.5665	0.5868	0.6351	0.6659	0.6925	0.6884
28	0.7188	0.6133	0.5584	0.6496	0.6102	0.6125	0.5648	0.5919	0.6320	0.6631	0.5262	0.6632
29	0.7145		0.6347	0.6682	0.6192	0.5912	0.5770	0.6099	0.6467	0.6770	0.6832	0.6904
30	0.7041		0.5748	0.5484	0.6426	0.5808	0.5644	0.6022	0.6530	0.6689	0.6943	0.5175
31	0.6993		0.6609		0.6424		0.5117	0.5723		0.6653		0.6946
AVERAGE	0.7430	0.7562	0.6955	0.7024	0.6453	0.6638	0.6069	0.6146	0.6443	0.6753	0.6964	0.7207

Table 6 shows the atmospheric transmittance and the figures in this table were determined by calculations from the total solar radiation values on a tilted surface using the Bouger's and Berlage's formulae[5]. Generally, there is a tendency on any given clear day for the atmospheric transmittance at dawn and dusk when the solar altitude is low to work out on the high side and for it to be on the low side around midday when the solar altitude is high. Table 6 gives the values for the atmospheric transmittance at noon everyday and consequently the air mass (air mass= sin(h) where h is the solar altitude) is ranging from 1.0 to 1.5. Since values yielded for cloudy days are assumed to be meaningless, the same table gives the mean atmospheric transmittance values for the above five days in each month as the monthly averages. This indicates that the closer the atmospheric transmittance is to 1.0, the purer is the air, and seasonal changes are apparent in that it is high during winter and low during summer.

The monthly average daily mean, maximum and minimum ambient temperatures are shown in table 7. Table 8 shows the monthly average daily maximum relative humidity in Abu Dhabi and table 9 shows the seawater salinity. The results of the TDS analysis of seawater are indicated in table 9. Variations are visible in the measured values with the salinity being slightly higher in summer months compared to winter months.

Table 7. Monthly average daily mean, max. and min. ambient temperatures (1985)

Month	Jan.	Feb.	Mar.	Apr.	May	Jun.	Jul.	Aug.	Sep.	Oct.	Nov.	Dec.
Mean	20.2	19.4	22.8	26.1	30.5	32.1	34.0	35.2	32.3	29.4	25.5	21.0
Maximum	24.8	23.9	28.2	31.8	36.3	37.8	39.9	42.6	39.0	34.5	29.9	25.6
Minimum	16.7	15.7	18.7	21.3	25.9	27.2	30.0	30.7	27.3	24.8	21.5	17.1

Table 8. Monthly average daily maximum relative humidity in Abu Dhabi (1985)

Month	Jan.	Feb.	Mar.	Apr.	May	Jun.	Jul.	Aug.	Sep.	Oct.	Nov.	Dec.
Humidity	90.9	82.6	82.0	81.5	77.6	80.4	78.0	77.6	87.8	85.2	87.3	83.6

Table 9. Seawater salinity

Sampling date	25 Sep. 84	2 Jan. 85	2 Mar. 85	17 Jul. 85	22 Oct. 85
TDS (ppm)	52,100	51,200	51,900	53,500	53,000

7. Operating Characteristics

In this section the results of plant characteristics during the first year of operation is presented. Typical measured performance for each of the main plant subsystems, e.g. heat collecting subsystem, heat accumulator subsystem and evaporator subsystem, are presented. The major performance parameters for the whole plant are also shown.

7.1. Heat Collecting Subsystem

7.1.1. Heat Collector Efficiency

Instantaneous Heat Collection Efficiency

Figure 12 shows the measured collector efficiency of the whole collector bank for a typical month (June 1985) and the ideal efficiency of a single collector measured under controlled conditions at the manufacturer's laboratory. The ideal efficiency can be correlated to the x-parameter by the following polynomial equation:

$$\eta_c^0 = 0.913 - 2.46x - 1.92x^2 \tag{26}$$

The measured collector efficiency is seen to be lower than the ideal efficiency due to heat losses from the piping system as well as losses due to attenuation of solar radiation received by the absorber plates because of dust deposition on the glass tubes of the collectors. In order to exclude the data during the warm-up and cool down periods in the early morning and before sunset, only the data for the period 10:00 am to 5:00 pm were plotted. The amount of heat collected was estimated from the measured inlet and outlet water temperature to the collector field and the flow rate of water. Therefore, all the heat loss from the internal and external piping system was included in the instantaneous efficiency shown in figure 12.

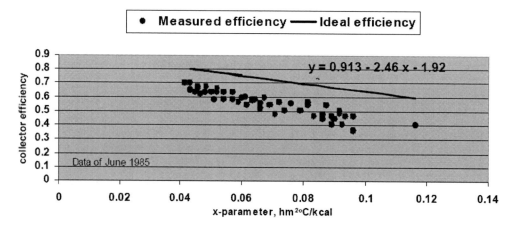

Figure 12. Measured efficiency of collector bank and the ideal efficiency of a single collector for a typical day.

Figure 13 shows the instantaneous efficiency of the collector bank at mid-day during the months of January and June 1985. In a clear day, the efficiency at mid-day (12:00 noon) is usually close to the highest value for that day. It can be seen that, for the month of January where some days are usually overcast, the mid-day efficiency drops for those overcast days. June is normally a sunny month with rare overcast periods, the mid-day efficiency fluctuates only slightly. The mid-day efficiency can drop slightly during periods of sand storms where the air is laden with small dust particles that reduce the solar radiation falling on the absorber plates of the collectors.

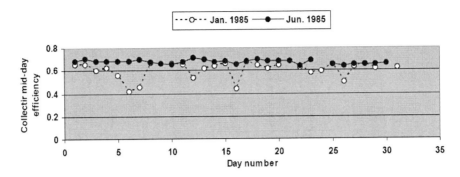

Figure 13. Daily instantaneous efficiency at mid-day during January and June 1985.

7.1.2. Daily Heat Collection Efficiency

The daily heat collection efficiency is defined as the ratio of the amount of heat produced by the collector bank divided by the amount of solar radiation falling on the absorber plates. The daily amount of heat produced and the daily amount of incident solar radiation are estimated from the summation of their hourly values during a day which are measured and recorded by the data acquisition system. Figure 14 shows the daily heat collection efficiency for the months of January and June 1985. It can be seen that the daily efficiency for sunny days is normally above 50% except for days with prolonged overcast periods, such as in January, during which the daily efficiency can drop below 40%.

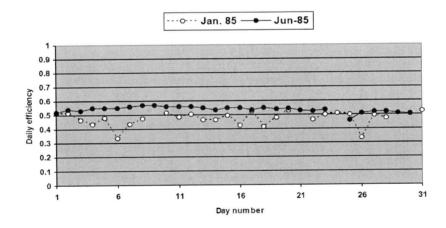

Figure 14. Daily heat collection efficiency for January and June 1985.

7.2. Heat Accumulator System

7.2.1. Heat Loss from the Heat Accumulator

Table 10 gives the monthly heat loss from the heat accumulator as a percentage of the incident solar radiation on the collector field for several months during 1985. As can be seen, the heat loss varies from 4.6% (for June) to 6.6% (for December). The percentage heat loss is increased during winter months compared to summer months and also increases during the month where the plant experiences several shutdowns either emergency shutdown or automatic shutdown due to insufficient accumulator charge. Months where the plant has been in emergency shutdown for long periods of pump maintenance, for example, can have accumulator loss exceeding 8%.

Table 10. Monthly heat loss from the heat accumulator as a percentage of incident solar radiation (data of 1985)

Month	Jan.	Mar.	May	Jun.	Sep.	Nov.	Dec.
Heat loss %	5.6	6.3	5.0	4.6	5.6	5.3	6.6

7.2.2. Thermal Stratification Ratio

The thermal stratification ratio is the ratio of the mass of the strata of water in the heat accumulator where a temperature gradient exists to the total mass of water inside the accumulator. The total mass of water inside the 3 accumulator tanks is essentially constant at 300 m^3. The thickness of the temperature gradient strata is measured using the temperature sensor (RTD) located in the middle of tank # 2 (mid-temperature tank). The heat accumulator operates in two main modes: the simultaneous heat collection and discharge mode during day periods and the discharge mode during night periods. Table 11 shows the percentage thermal stratification ratio for each of these two modes for a number of days. During the heat discharge mode (at night), the temperature gradient strata averages 19.5 m^3 which corresponds to a stratification ratio of 6.5%. During the simultaneous heat collection and discharge mode (during day time) this strata averages 21.9 m^3 which corresponds to a stratification ratio of 7.3%.

Table 11. Thermal stratification ratio of heat accumulator (data of 1985)

Date	Heat discharge	Simultaneous heat collection and discharge
Jan. 14	4.6%	6.0%
Jan. 15	3.8%	7.9%
Jan. 19	4.6%	4.8%
Feb. 2	6.7%	8.1
Feb. 13	6.6%	6.1%
Feb. 20	6.1%	8.3%
Mar. 10	10.6%	6.3%
Mar. 20	8.8%	3.9%
Mar. 29	6.5%	5.8%
Apr. 4	9.4%	5.7%
Apr. 10	7.2%	7.6%
Apr. 20	1.7%	11.5%
May 3	6.4%	8.9%
May 11	5.6%	9.2%
May 20	8.3%	9.3%
Average	6.5%	7.3%

7.3. Evaporating System

7.3.1. Evaporator Performance

The performance ratio (*PR*) is defined here as the amount of product water produced by the evaporator per 526 kcal of heat supplied by the heating water. Table 12 shows the average *PR* values and average specific heat consumption for several months during the first year of the test plant operation. The effect of the product water flow rate on the *PR* is shown in figure 15 which is based on actual tests carried out during plant commissioning in November 1984.

Table 12. Performance ratio of the evaporator

Month	Product water flow m³/hr	Specific heat consumption kcal/kg	Performance Ratio
Jan. 1985	4.6	40.71	12.9
Feb. 1985	5.3	39.04	13.5
Mar. 1985	5.0	39.31	13.4
Apr. 1985	5.2	38.83	13.5
May 1985	5.0	39.04	13.5
Jun. 1985	5.1	39.98	13.2
Sep. 1985	4.9	40.31	13.0

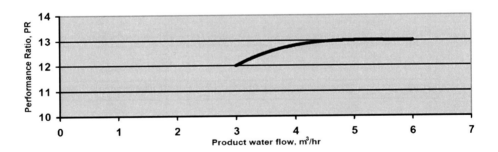

Figure 15. Performance Ratio versus product water flow.

Overall Heat Transfer Coefficients

The overall heat transfer coefficients (OHTC) shown in figure 16 are estimated from the measured temperatures and flow rates. Shown in this figure are the average HTC for the evaporators, the average OHTC of the preheaters and the OHTC of the heater (first effect) and condenser. The evaporators are heat exchangers in which vapor is condensed inside tubes while seawater brine boils on the outside of the tubes. These heat exchangers have the highest OHTC compared with the other heat exchangers as shown in the figure. The data shown in this figure represent typical values of the OHTC,s obtained during the first year of plant operation. Some deterioration in the OHTC's has occurred during the subsequent years which necessitated acid cleaning to remove scale deposited on the heat exchanger tubes.

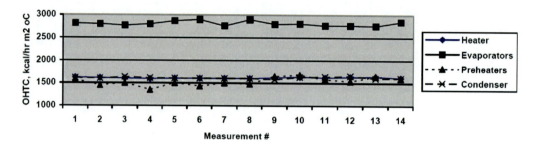

Figure 16. Measured OHTC during the period June 1-7, 1985.

7.4. Performance of the Plant

Figure 17 and figure 18 show pie chart plots depicting the January and June split of the incident solar radiation falling on the collector bank among collector loss (including piping loss and loss due to dust effect), accumulator heat loss, heat loss by the evaporator and the heat going for desalination. It can be seen that for January, 37% of the incident solar radiation

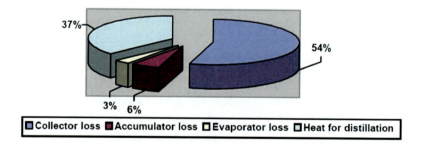

Figure 17. Split of the incident solar radiation for January.

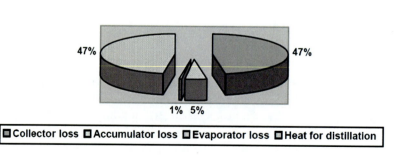

Figure 18. Split of the incident solar radiation for June.

is converted into thermal energy for desalination while for June it is 47%. This is due to the fact that for winter months heat losses from the collector bank, the accumulator tanks and the evaporator are larger than during the summer period because ambient winter temperatures are substantially lower than summer temperatures.

Figure 19 shows the monthly values of three major plant performance parameters: the specific water production in kg/m^2 day, the specific heat consumption in kcal/kg dist. and the specific power consumption in kWh/m^3. The specific water production is defined as the rate of water production per unit collector area. It varies between 43.9 kg/m^2 day and 78.1 kg/m^2 day with the lower values for the winter months and the higher values for the summer months. The specific heat consumptions is defined as the rate of heat supplied to the evaporator per unit of water production. It is essentially constant at about 39 kcal/m^3 of product water. The specific power consumption, defined as the electrical energy required per cubic meter of product water, varies between 7.08 kWh/m^3 and 8.15 kWh/m^3 with the higher values typical for winter months and lower values for summer months.

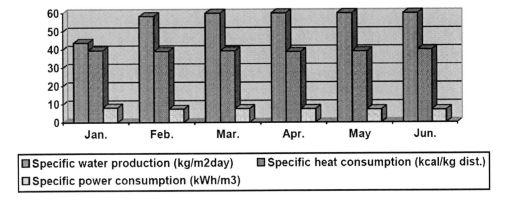

Figure 19. Plant performance parameters for the first six months of 1985.

8. Plant Maintenance and Modifications

In this section, the major maintenance and plant modification activities will be described. Among the major maintenance activities that had to be carried out regularly are the following:

- Cleaning the collectors to remove dust and dirt deposited.
- Inspection for corrosion replacing corroded components
- Inspect vacuum inside collector glass tubes
- Monitor scale formation on evaporator heat exchanger tubes and carry out acid cleaning if necessary
- Monitor level of water in heat accumulator and add makeup water and anticorrosive chemical as required
- Evaporator pump maintenance
- Monitor pressure difference across seawater intake filters and clean when necessary

Plant modifications were necessary in order to avoid the harmful effects of emergency plant shutdown due to power failure.

8.1. Heat Collecting System

8.1.1. Cleaning the Solar Collector Field

The performance of the solar collector field is affected by the extent of dust deposition on the glass tube which influences the transmittance of the glass tube to solar radiation. It is therefore important to clean the collector field at regular intervals to maintain good performance. The cleaning was carried out using a high-pressure water jet spray device. Since fresh water is used for cleaning, it is important to economize on the use of fresh water for collector cleaning without adversely affecting the performance of the collectors.

The amount of water required for each cleaning session depends on the extent of dust deposition on the collectors; more water is required when more dust has accumulated on the collectors. Since sandstorms are seasonal in character, the amount of dust deposited in a particular period depends on the month of the year. Several tests were carried out to determine the required amount of fresh water needed for each collector block. Figure 20 shows the measured water quantity used for each cleaning session for one block for different months. As can be seen, there is wide variation in the quantity of cleaning water required which may be attributed to variation in the personal skills of the different cleaners as well as variation in the amount of dust deposition on the glass tubes. The average quantity of water for each cleaning session is about 1000 liters.

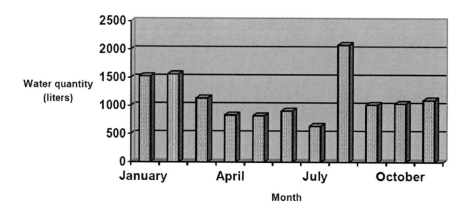

Figure 20. Water quantity required for cleaning a single block for different months.

8.1.2. Corrosion of the Collector Air Vent Valves

A total of 76 air vent valves were installed on the arrays of solar collectors to vent the air out during the water filling up process. Water started to leak from these valves after few months of operation and several valves have to be replaced with spare ones. Because of a limited supply of vent valves, few of the leaking valves have to be plugged as a makeshift measure until new valves are procured, a procedure usually taking several months due to administrative delays. The problem with the leaky vent valves occurred soon after a power failure happened during a sunny day causing the heat collecting pump to shutdown while the water is still inside the collectors which caused partial evaporation of the water and frequent opening and closing of the vent valves. To prevent further damage to the vent valves, it was decided that plant re-start after the restoration of power following a power failure is made either early in the morning or at dusk.

8.1.3. Vacuum Loss Inside Glass Tubes

As previously stated, the glass tubes operate under a high vacuum of 10^{-4} mmHg. High collector efficiency depends on maintaining this high vacuum level inside the collector. The vacuum level is monitored every year by checking the condition of the getter. At the end of the first year of operation, the condition of the vacuum on almost all glass tube was essentially as new.

8.1.4. Scale Prevention

The most important item for the evaporator is scale prevention. There are two kinds of scale that could form in seawater distillation: a hard scale and a soft scale. The hard scale consists mainly of calcium sulfate ($CaSO_4$) and very difficult to remove from the heat transfer surfaces once formed. The only way to avoid the formation of this scale is to operate the evaporator within the solubility limits of Ca^{+2} and SO_4^{2-} ions in the brine. The evaporator was designed in such a way that in normal operation, the concentration of these ions are not allowed to reach saturation. On the other hand, soft scale which consists mainly of calcium carbonate (Ca CO_3) and magnesium hydroxide ($Mg(OH)_2$) can be avoided by injecting a chemical inhibitor such as Belgard EV into seawater feed. In our evaporator 10 ppm of Belgard EV is injected and was found satisfactory with the high-salinity seawater of the Gulf (52,000 ppm). Scale formation was monitored on an hourly basis during plant operation by observing the heat transfer coefficient of the heater (first effect) since it has the most likelihood of scale formation since it operates at the highest temperature. For an emergency, an acid injection system is used to carry out an acid cleaning procedure to remove any soft scale formed.

8.1.5. Anti-corrosion Chemical for Use in the Heat Collecting Water

An anticorrosive chemical must be used in the heat collecting water, which circulates through a closed system, to protect the equipment from corrosion. "High Clean CL-100" was selected in view of the heat collector tube material being of copper and the heat accumulator and piping material being of iron. "High Clean CL-100" was found to be effective in protecting both materials from corrosion. This chemical is a solution of alkanol-amino salt of nitrogen-containing condensate having the chemical symbol: [R-$SO_2NH(CN_2)n$ COOH]   where $R = C_6H_5$ and $n = 1$ to 3. The solution is a light brown transparent liquid having a pH of 8 – 8.5. A concentration of 5,000 ppm was recommended by the manufacturer. However, the concentration of the chemical as poured into the makeup water tank after draining and leakage was gradually reduced as the stock of the chemical decreased because the heat collecting water was forced to be drained more frequently than was planned due to unexpected power failures and pump troubles. The average concentration of the chemical in the collector water during the first year of operation was approximately 2,8000 ppm. Table 13 shows the results of analysis of the heat collecting water of four samples taken during the first year of operation. It can be seen that there was virtually no change in the total Fe content indicating an adequate anticorrosive effect for Fe. On the other hand, the Cu content tended to increase gradually as the concentration of the anticorrosive chemical drops. The thickness of the copper tubes corroded in the solar collector field was estimated at 0.001 mm/year[5].

Table 13. Chemical analysis of heat collecting water

Sampling data	31 Jan. 85	27 Mar. 85	30Jul. 85	21 Oct. 85
pH	8.6	8.5	7.9	8.2
Conductivity (μS/cm)	209	210	373	242
Temperature (^{o}C)	-	20	28	18
M Alkalinity (ppm as $CaCO_3$)	526	486	468	452
Total hardness (ppm as $C(ppm)aCO_3$)	22	8	20	4
Chloride (ppm as Cl^-)	37	21	44	33
Silica (ppm as SiO_2)	1.3	2.0	1.9	2.8
Total iron (ppm as Fe)	1.6	1.82	1.1	1.4
Copper (ppm as Cu^{++})	0.55	0.62	1.8	2.83
Total Nitrogen (ppm as N)	217	177	156	161
High-Clean CL-100	5,167	4,214	3,714	3,843

8.1.6. Measures against Power Failure

The original plant design was such that in the case of a power failure during the day, the heat collecting pump stops and the water in the collectors, subject to solar radiation, evaporates resulting in dangerous water hammer in the header pipes. In order to protect against such power failure, the solar field piping system was modified by installing three motorized valves (normally closed) on three 25 mmϕ lines, one motorized valve (normally open) on the 125 mmϕ on the outlet piping of the heat collecting pump and a check valve on the 125 mmϕ on the return header pipe from the collector field to the accumulator. The modifications are shown in figure 21. In the modified system, following a power failure, the three 25 mmϕ motorized values will be open thus draining all the water in the collector field and the 125 mmϕ motorized valve in the pump outlet will close thus preventing the collector water from flowing by gravity from the heat accumulator to the collector field. The modified system proved effective in protecting against the hazard of evaporation in the collector field. The collector field so not automatically restart following the restoration of power and the valves has to be rest manually by the operator.

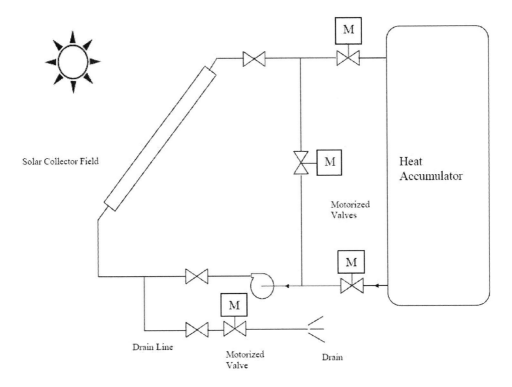

Figure 21. Modifications to the solar collector field to protect against power failure.

8.2. Evaporating System

8.2.1. Evaporator Pump Maintenance

Pump problems were one of the main causes of plant downtime. The plant had 12 pumps: seawater intake pump, seawater pump, seawater feed pump, brine blow-down pump, product water pump, vacuum pump, drain pump, heat collecting pump, heating water pump, drinking water pump, plant water pump and priming vacuum pump. All these pumps are motor-operated centrifugal pumps pumping seawater, brine or product water except the vacuum pump which is an oil ring-seal vacuum pump drawing in a mixture of vapor and non-condensable gas from the evaporator. Most of the pump problems were started by a high-pitch noise and vibration which gets worse as time goes on. The pumps dealing with seawater or its brine were found to be more amenable to failure than the other pumps due to seawater corrosion problems that affect the pump bearings.

Continuous operation of the evaporator had to be interrupted by pump trouble several times. The pumps which gave most of the problems were the drain pump, the seawater intake pump. Most of the problems were resolved by replacing the pump/motor bearings or replacing the mechanical seal (or packing material). Many emergency plant shutdowns happened because of "intake pit level low" signal due to air leakage through the shaft packing of the seawater intake pump because of a worn out packing material. The suction line of this pump operates at a vacuum due to the fact that the seawater suction point is below the pump level. The air leakage into the pump would break the vacuum and drastically reduces the pump flow.

The drain pump gave a lot of trouble. All the drains from the evaporator was collected into a pit in which the pump suction line is immersed. A level switch is used to start and stop the operation of the pump. This pump was found to be easily clogged with debris which accumulates inside the pit and caused a relay trip due to motor over-current. To solve this problem, a fine screen was installed on the tip of the suction line inside the pit to prevent the trash from entering the pump. The screen has to be cleaned regularly and trash removed from it. The product water pump was another source of problems and was responsible for plant shutdown particularly in the first year of plant operation. Faulty bearings and a pump shaft were replaced.

8.2.2. Inspection of the Evaporator

An overhaul was carried out at the end of one year of plant operation which consisted of:

- The first effect, which is subject to high temperature and hence high possibility of scaling was inspected after the first year of operation. An accumulation of silt (fine sand) approximately 1 cm thick was found on the bottom of the 1st effect and was removed.
- Blackened spots were seen on the outer surface of the first effect tube bundle but they remained the same as those at the start of plant operation and should cause no problems.
- Scale marks were seen on parts of the tube bundle of the first effect near the water chambers at the heater inlet and outlet, but they did not represent any advanced state. A re-inspection few months later revealed no advanced state.
- No abnormal conditions were seen on the 18th effect.

8.2.3. Change in Operating Sequence

A modification was carried out in a part of the operating sequence so that the test plant could be safely operated. This change brought about the following improvements in plant operation:

- If the intake strainer is clogged with foreign matter like seaweed, the seawater intake pump will be stopped temporarily.
- If the concentration of the discharged brine becomes too high, an alarm will be triggered and the plant shut down. This is to prevent scale formation on the heat transfer tubes if the flow of heating water increases more than necessary or if the feedwater quantity is reduced.
- If the product water quantity becomes too small, an alarm will be triggered and the plant shutdown. This is to prevent the product water pump fro running at no load if the temperature of the heating water drops enough to make water production too low.
- The manner of emergency plant shutdown will be either of the two types given below depending on the degree of emergency:

    ○   Immediate plant shutdown for motor relay tripping, too low seawater flow, or the like.

    ○   Gradual plant shutdown for too high first effect evaporator temperature, too high brine concentration, or the like.

8.2.4. Modification of the System for Injecting Anti-scale Chemical

The original plan called for a single system for injecting the scale preventing chemical "Belgard EV" into seawater feed. However, because the injection pump has failed during unmanned operation of the plant and it was feared that the plant could have been subjected to scaling, it was decided to modify the injection system to include the standby system as a secondary injection system. In case the primary injection system fails, the secondary system automatically turns on.

8.2.5. Modification of the Method of Feeding Sealing

Water to the Priming Vacuum Pump

Because the seawater level is lower than the level of the seawater intake pump, when the plant is not in operation, the seawater in the suction line of this pump drains back by gravity to the sea and the line is full of air. The priming vacuum pump, which is a water ring sealed vacuum pump, is used to prime (start) the seawater intake pump by insuring that this pump and its suction line is full of water. Seawater was initially used for the sealing water, but since the primary vacuum pump is operated at plant startup only, seawater inside the pump and sealing tank has caused considerable amount of rust and made the water brown. It was therefore decided that fresh water be used for the sealing water instead of seawater.

9. Simulation Program and its Validation

9.1. Simulation Program

9.1.1. Outline

The performance of a solar desalination plant that uses solar energy as the heating source for a seawater evaporator is affected by the amount of solar radiation, ambient temperature and other climatic conditions. These conditions are extremely unstable and are subject to constant change. This means that computer-based simulation is indispensable if, to cope with these changes, the system operation states are to be grasped and the amount of product water annually is to be forecast.

As shown with the basic design, computer simulation was used to investigate the operating characteristics even at the planning stage of the test plant. However, it was not possible to grasp with any accuracy the climatic conditions when design was commenced and forecasted values based on experience to date were used even for the performance of the various components.

Subsequently, research operation was begun, actual climatic data were accumulated and the characteristics of the equipment started to become clear through the winter and summer component tests and the continuous operation tests. Thus, a new look was taken at the simulation program. As a result of comparing and examining the calculated values and th actually measured values, it was concluded that the program must be revised in order that the operation of the test plant was reproduced as faithfully as possible and more accurate forecasts were made to cope with different operating conditions and plants of different scales. The program was revised and manuals were prepared to use the program between October 1985 and January 1986.

9.1.2. Flow Chart of the SOLDES Program

Evacuated glass tube-type of collectors are used and the absorber area can be varied from 500 m^2 to 20,000 m^2. The heat collecting system uses a bypass circuit. When the temperature of the heat collecting system drops below the set value, the bypass operation is performed, and once the temperature rises above the set value, operation is switched over to the accumulator side. Solar cell type control is excercised to start and stop the heat collecting water pump. The accumulator is treated as a thermal stratified type, and the temperature distribution inside the tank is determined while bearing in mind the inflow and outflow of the heat collecting water and heating water. The number of collectors used and the heating water flow rate are varied so that the maximum operating capacity of the evaporator is not exceeded. However, for the simulation, consideration was given to maximize the effective use of the collectors and a bypass circuit was installed between the accumulator and the evaporator. By this means, a system of control was adopted where some of the heating water returning from the evaporator is bypassed and forwarded to the evaporator so that the temperature of the heating water entering the evaporator is kept below the rating. This is particularly useful when the accumulator water temperature is excessively high.

The evaporator capacity can be varied over the range of 100 to 2,000 m^3/day, the maximum brine temperature can be varied from 60 – 80°C and the number of effects of the evaporator can be varied fro 13 to 32. The simulation program takes into consideration the influence of the heating water temperature, heating water flow rate, seawater temperature and seawater flow rate.

Figure 21 is an abbreviated flow chart of the simulation program. The equipment specifications, calculation conditions and other data are input to the program and are divided into two groups: System data No. 1 and System data No. 2. The data input with the System data No. 1 group serve to output error messages and suspend the execusion of the calculations in cases where there are errors in the input data or where allowable ranges have been exceeded. Any calculation period ranging from one day to one year can be designated and a balance in the system is determined every 30 minutes in relation to the calculation loop. The calculation results can be output either by the day or by the month. The "SOLDES" simulation program for the solar desalination plant is composed of 22 subroutine programs, two sets of system data and four types of climatic data.

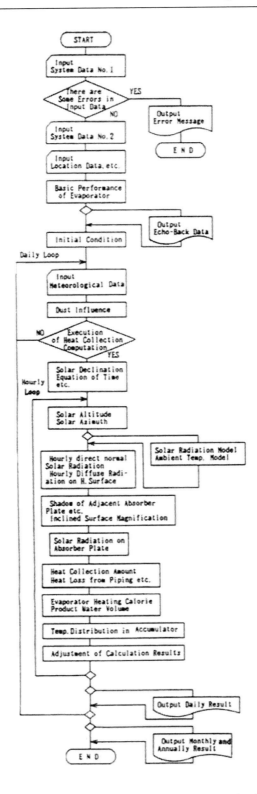

Figure 22. Flow chart of Solar Desalination Plant computer simulation program.

9.1.3. Program Input and Output Data

In conducting a computer simulation of a solar desalination plant, input data need to be revised according to changes in plant specifications such as absorber area, accumulator capacity, or evaporator capacity. The items of data input that can be revised using program "SYDT1" are as follows:

- Title of simulation run
- Starting and finishing date of simulation
- Specifications of daily data print out for each month
- Correction coefficient of solar radiation by month
- Correction coefficient of ambient temperature by month
- Collector absorption area of field
- Azimuth angle of collector
- Collector support angle
- Tilt angle of absorber plate
- Heat collecting water flow rate
- Heat collection pump rated power consumption
- Heat accumulator capacity
- Heat accumulator initial temperature distribution
- Evaporator capacity
- Maximum brine temperature
- Number of effects of evaporator
- Heating water flow rate
- Seawater flow rate
- Power consumption of vacuum pump and other evaporator pumps
- By-pass valve open/close temperatures
- Evaporator start/stop temperatures
- Correction coefficient of dust influence
- Specification of collector cleaning days

Table 14. Types of meteorological data which can be used in "SOLDES"

Data name	Solar radiation data	Ambient temperature data
MEDT1	Hourly total solar radiation on tilt surface Input of data obtained at Abu Dhabi solar desalination plant	Hourly ambient temperature Input of data obtained at the Abu Dhabi solar desalination plant
MEDT2	Instantaneous total solar radiation on horizontal surface by hour	Instantaneous ambient temperature by hour
MEDT3	Daily total solar radiation on tilt surface	Daily mean, daily maximum and daily minimum ambient temperature
MEDT4	Daily total solar radiation on horizontal surface	Daily mean, daily maximum and daily minimum ambient temperature

Four types of meteorological data can be accommodated in the program (see table 14). The type MEDT1 is used in the solar desalination plant where the hourly global radiation and hourly ambient temperature were measured. The other types of meteorological data can accommodate hourly or daily global solar radiation on horizontal surface instead of tilt surface as well as hourly temperature or mean-max-min daily temperatures.

9.1.4. Mathematical Models

Mathematical models were developed for the different components in preparation of the simulation program "SOLDES". The following models were developed :

1. *Numerical method to estimate solar radiation on a tilted surface*
   This model calculates the beam and diffuse components of solar ration on a tilted surface having the same angle as the absorber plate of the solar collectors given as input the following information:

   o Measured hourly solar radiation on a tilted surface and measured hourly ambient temperature (MEDT1) as the case in the solar desalination plant.
   o Measured hourly solar radiation on a horizontal surface and measured hourly ambient temperature (MEDT2),
   o Estimated daily total solar radiation on tilted surface and estimated mean, maximum and minimum daily ambient temperatures (MEDT3)
   o Estimated daily total solar radiation on horizontal surface and estimated mean, maximum and minimum daily ambient temperatures (MEDT4)

The solar radiation on tilted surface, I_t is estimated from the following equation:

$$I_t = I_b + I_d = I_0 \times P^{\frac{1}{\sin(h)}} \times \cos(\theta_0) + \frac{1}{2} \times I_0 \times \sin(h) \times \frac{1 - P^{\frac{1}{\sin(h)}}}{1 - 1.4 \times \ln(P)} \times \frac{1 + \cos(\alpha_0)}{2} \quad (27)$$

If the atmospheric transmittance, P, is given as known data, hourly solar radiation on a tilted surface can be calculated from the above equation. However, the only data available is the hourly global radiation or the daily global radiation. Therefore, to proceed with the computer simulation, it is necessary to convert global radiation to hourly bean and diffuse components. This is achieved by estimating the hourly P values from Eq. 23 knowing the global radiation, then calculating the beam and diffuse components from the following the following equations:

$$I_b = I_0 \times P^{\frac{1}{\sin(h)}} \times \cos(\theta_0) \quad (28)$$

$$I_d = \frac{1}{2} \times_0 \times \sin(h) \times \frac{1 - P^{\frac{1}{\sin(h)}}}{1 - 1.4 \times \ln(P)} \times \frac{1 + \cos(\alpha_0)}{2} \quad (29)$$

when daily global radiation is available instead of the measured hourly values, the hourly solar radiation is first estimated using an iterative procedure then the beam and diffuse components are estimated from Eqs. 28 and 29[5].

2.   *Effect of shade on solar radiation on absorber plate*

The amount of solar radiation on the absorber plate is less than the amount falling on a tilted surface having the same angle as the absorber plate but situated in the open. This is due to the shading cast on the absorber plate by several collector components:

- Shade by the neighboring absorber plates (shade length = s_1)
- Shade by the neighboring glass tubes (shade length = s_2)
- Shade by the header box of the collector (shade length = s_3)

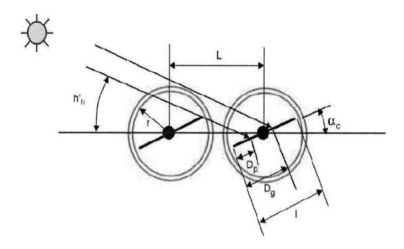

Figure 23. Shadow effect due to adjacent plate and adjacent glass tube.

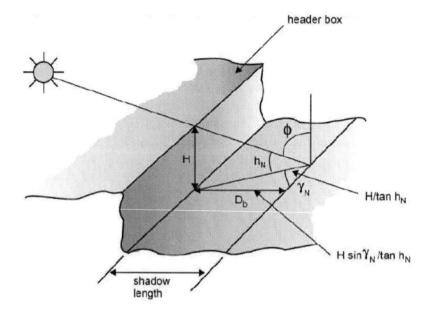

Figure 24. Shadow effect due collector header box.

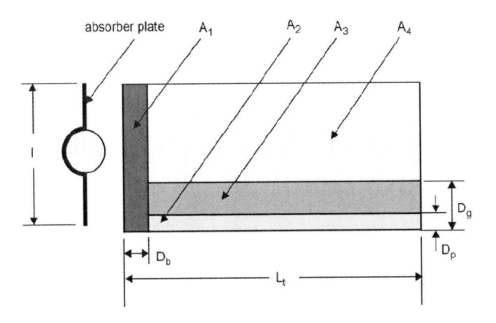

Figure 25. Different shadow areas in a plan view.

Figure 23 shows the shadow length due to adjacent absorber plate (s_1) and adjacent glass tube (s_2). Figure 24 shows the shadow length due to collector header box (s_3). The three shadow areas are shown in a plan view in figure 25.

These shadow effects are particularly evident in early morning and late afternoon but have a minimal effect throughout the rest of the time. Based on the solar angles (solar altitude and solar azimuth), absorber plate dimensions and tilt angle, pitch of absorber plates, glass tube diameter and collector header box height, a model was developed to estimate the hourly value of the length of the shade cast by each of the above three effects. We present here the results of this model.

$$\text{Morning:} \quad s_1 = \frac{l \times [\cos(\alpha_c) \times \tan(h_N^{'}) + \sin(\alpha_c)] - L \times \tan(h_N^{'})}{\cos(\alpha_c) \times \tan(h_N^{'}) + \sin(\alpha_c)} \tag{30}$$

$$\text{Afternoon:} \quad s_1 = \frac{l \times [\cos(\alpha_c) \times \tan(h_N^{'}) - \sin(\alpha_c)] + L \times \tan(h_N^{'})}{\cos(\alpha_c) \times \tan(h_N^{'}) - \sin(\alpha_c)} \tag{31}$$

$$\text{Morning:} \quad s_2 = \frac{\dfrac{l}{2} \times \sin(\alpha_c) + \dfrac{\gamma}{\cos(h_N^{'})} - \tan(h_N^{'}) \times [L - \dfrac{l}{2} \times \cos(\alpha_c)]}{\cos(\alpha_c) \times \tan(h_N^{'}) + \sin(\alpha_c)} \tag{32}$$

$$\text{Afternoon: } s_2 = \cfrac{-\dfrac{l}{2} \times \sin(\alpha_c) + \dfrac{\gamma}{\cos(h_N')} - \tan(h_N') \times [(L - \dfrac{l}{2} \times \cos(\alpha_c)]}{\cos(\alpha_c) \times \tan(h_N') - \sin(\alpha_c)} \tag{33}$$

$$s_3 = \frac{H \times \sin(\gamma_N)}{\tan(h_N)} \tag{34}$$

with reference to figure 25 we can write:

Total area of absorber plate............$A = L_T \times l$ (35)

Shadow area #1........................$A_1 = s_3 \times l$ (36)

Shadow area #2........................$A_2 = s_1 \times (L_T - s_3)$ (37)

Shadow area #3........................$A_3 = (s_2 - s_1) \times (L_T - s_3)$ (38)

The area exposed to direct solar radiation is therefore the difference between the total area of the absorber plate, A, and the three shadow area, i.e.

$$A_4 = A - (A_1 + A_2 + A_3) \tag{39}$$

It is convenient to introduce the ratios R_1, R_2, R_3 and R_4 such that

$$\begin{aligned}
R_1 &= \frac{s_3 \times l}{A} \\
R_2 &= \frac{s_1 \times (L_T - s_3)}{A} \\
R_3 &= \frac{(s_2 - s_1) \times (L_T - s_3)}{A} \\
R_4 &= \frac{(l - s_2) \times (L_T - s_3)}{A}
\end{aligned} \tag{40}$$

It is to be noted that areas #1 and #2 are completely shadowed by solid obstacles which does not transmit any radiation whereas area #3 (shadow due to adjacent glass tube) has an attenuated solar radiation due to the transmittance of three laters of glass through which each solar ray has to travel (see figure 23). It is assumed that the daily average transmittance of the three glass layers is equal to 0.7. This assumption is suggested by the manufacturer (Sanyo) and is based on detailed hourly computer simulation. Consequently, the net beam radiation on the absorber plate can be expressed by the equation:

$$I'_b = I_b \times (R_4 + 0.7 \times R_3) \tag{41}$$

The net total radiation on the absorber plate is obtained by adding the diffuse component to the beam component

$$I'_t = I'_b + I_d \tag{42}$$

3. Ambient temperature model

When hourly ambient temperature data is not available, hourly data can obtained from knowledge of the daily mean, maximum and minimum values using a model developed for this purpose.

4. Effect of dust on transmittance of glass tubes

The effect of dust on the transmittance of the glass tubes varies with the season. The heat collection amount drops sharply especially when there is a sandstorm and dust accumulates rapidly. This causes the transmittance of the glass tubes to drop. When it rains, on the other hand, the transmittance is restored because the rainfall washes away the dust. Climatic conditions in the UAE are such that there is little rainfall, but nevertheless, the cleaning effect of rainfall appears in the collector measurements. In the UAE, moreover, the temperature differences between day and night are relatively large; in the mornings, dew accumulates on glass tube surface and they become damp, so mucu so that water droplets sometimes fall from them. If the level of dew accumulation is slight, dust can easily adhere to the tube surface, but if the extent of dew is great enough, it will serve to wash away dust from the tube.

The relationship between the cumulative level of dust affecting the transmittance of the glass tube and the cleansing effect of various natural climatic conditions is extremely complex. In order to develop a model of the effect of dust accumulation on the transmittance of the glass tube, it was assumed that the clean glass tube transmittance of 98% is restored after each tube cleaning and that following a tube cleaning the transmittance drops exponentially due to dust effect according to the equation (see Sayigh et al. [..]):

$$\frac{\tau - \tau_m}{\tau_0 - \tau_m} = \exp(-0.055 \times N) \tag{43}$$

where τ is the transmittance after N days has elapsed since cleaning, τ_m is the transmittance after one month has elapsed since cleaning, τ_0 is the transmittance immediately measured after cleaning (= 98%). The monthly drop in transmittance $(\tau_0 - \tau_m)$ is dependent on the month with summer months experiencing larger drops than winter months due to the sandstorm prevailing mainly in summer months.

5.   Control of heat collecting operation

The solar collector field is operated by the solar controller which uses a solar cell to switch the heat collecting pump on and off depending on the solar radiation. The solar radiation condition for pump startup, I_{on} (kcal/hm^2), and shutdown, I_{off} (kcal/hm^2), are determined from the measured values of the heat accumulator outlet temperature, T_{w1} ($^{\circ}$C) and the ambient temperature, T_a ($^{\circ}$C) according to the relations:

$$I_{on} \geq 5 \times (T_{w1} - T_a) - 25$$
$$I_{off} < 5 \times (T_{w1} - T_a) - 10 \tag{44}$$

6.   Heat collection amount from solar collector field

The heat collection efficiency is defined as the amount of heat collected divided by the amount of solar radiation falling on the absorber plates of the solar collector field and is expressed by the following formula:

$$\eta_c = \alpha + \beta.x + \gamma.x^2 \tag{45}$$

where η_c is the collector efficiency, α, β, γ are constant parameters and x is a variable defined as:

$$x = \frac{-\frac{1}{2}(T_{c1} + T_{c2}) - T_a}{I_t'} , \quad T_{c1}$$

and T_{c2} are the collector field inlet and outlet water temperature and I_t' is the solar radiation on the absorber plate.

The rate of heat collected, Q_c, from the solar field is obtained from the equation:

$$Q_c = A_c.\eta_c.I_t' = m_c.C_p.(T_{c2} - T_{c1}) \tag{46}$$

Using Eqs. 34 and 35 the collector outlet temperature can be obtained as follows:

$$T_{c2} = \frac{-C_2 - \sqrt{C_2^2 - C_1 C_3}}{C_1} \tag{47}$$

where C_1, C_2 and C_3 are given by the following equations:

$$
\begin{aligned}
C_1 &= \gamma.A_c \\
C_2 &= \gamma.A_c(T_{c1} - 2T_a) + \beta.A_c.I_t' - 2m_c.C_p.I_t' \\
C_3 &= \gamma.A_c.(T_{c1} - 2T_a)^2 + 2\beta A_c.I_t'(T_{c1} - 2T_a) + 4\alpha.A_c.I_t'^2 + 4m_c.C_p.I_t'.T_{c1}
\end{aligned}
\tag{48}
$$

with this equation the collector outlet water temperature can be obtained at each hour of the day given the inlet water temperature, the solar radiation, water flow rate, and the collector parameters $\alpha i, \beta$ and γ.

7. *Evaporator performance*

A simple model is used for predicting the performance of the MED evaporator at part load given the evaporator's capacity, maximum brine temperature, number of effects, heating water flow rate, heating water temperature and seawater temperature. The aim is to calculate the hourly distillate production, the hourly heating water outlet temperature and the hourly pumping power requirement.

The performance ratio and specific heat consumption of the evaporator depends on the number of effect and was estimated from the following equations provided by the manufacturer (Sasakura):
Number of effects $N \leq 13$:

$$
PR = -1.875 \times 10^{-2} \times N^2 + 1.15 \times N - 1.625
\tag{49}
$$

Number of effects $N > 13$:

$$
PR = -2.500 \times 10^{-3} \times N^2 + 0.625 \times N + 2.56
\tag{50}
$$

Specific heat consumption, SPC (kcal/kg):

$$
SHC = \frac{526}{PR}
\tag{51}
$$

The effective temperature difference per single effect δT at the design condition was estimated from the equation:

$$
\delta T = \frac{(T_{2d} - T_{1d} - BPE_{av} \times N)}{N}
\tag{52}
$$

where T_{2d} is the design brine maximum temperature in the first effect (top effect), T_{1d} is the design brine temperature in the last effect (bottom effect), BPE_{av} is the average boiling point elevation in the N effects. For the evaporator in the solar desalination plant the following values are used: $T_{2d} = 68^{\circ}C$, $T_{1d} = 40.7^{\circ}C$, $BPE_{av} = 0.71^{\circ}C$. The average heat transfer coefficient for the N effects, U (kcal/h m^{2o}C) was estimated from:

$$U = (1888 \times L + 1313) \times \frac{C(T_1) + C(T_2)}{2} \tag{53}$$

where C is a correction coefficient which is dependent on the brine temperature, L is the load. The design value is that when $L = 1.0$, i.,.

$$U_d = (1888 + 1313) \times \frac{C(T_1) + C(T_2)}{2} \tag{54}$$

The correction coefficient can be expressed by the equation:

$$C(T_b) = -0.4678 + 0.050 \times T_b - 0.0005 \times T_b^2 + 0.17 \times 10^{-5} \times T_b^3 \tag{55}$$

The operating condition of the evaporator depends on the load parameter $L = (M_d/M_{d100\%})$. The operating temperatures for each load are evaluated as follows:

• Calculate the brine temperature in the last effect from knowledge of the seawater temperature and the condenser load:

$$\begin{aligned} T_2 &= T_{c2} + 1.2 \\ &= T_{c1} + \frac{Q_c}{m_{sw} \times C_p} + 1.2 \\ &= T_{c1} + \frac{0.9 Q_h}{m_{sw} \times C_p} + 1.2 \end{aligned} \tag{56}$$

In this equation the assumption is made that the condenser load, Qc, is 90% of the heating load, Q_h due to heat loss to the environment. It is also assumed that the last effect brine temperature is smaller than the condenser outlet temperature by 1.2oC. These assumptions are verified by actual measurements at the evaporator of the solar plant.

Calculate an approximate value for the first effect temperature, T1,

$$T_1 = T_2 + (L \times \delta T + BPE) \times N \tag{57}$$

Calculate an average overall heat transfer coefficient for the N effects,

$$U = (1888 \times L + 1313) \times \frac{C(T_1) + C(T_2)}{2} \tag{58}$$

- Calculate an improved value for the first effect temperature, T_1,

$$T_1 = T_2 + (L \times \delta T \times \frac{U}{U_d} + BPE) \times N$$

- Iterate to get improved values for T_1.

The heating water outlet temperature was assumed to be equal to the first effect temperature: $T_{h2} = T_1$ and the heating water inlet temperature is calculated from a heat balance equation:

$$T_{h1} = T_{h2} + \frac{Q_h}{m_h \times C_p} \tag{59}$$

The power consumption (in kW) of the evaporator was assumed to depend mainly on its capacity and was estimated from the following equations provided by the manufacturer (Sasakura):

Evaporator capacity $m_d \leq 500$ m³/day:

$$P = -1.25 \times 10^{-5} \times m_d^2 + 9.0 \times 10^{-2} \times m_d + 5.125 + Z \tag{60}$$

Evaporator capacity $m_d \geq 1000$ m³/day:

$$P = -6.00 \times 10^{-6} \times m_d^2 + 5.1 \times 10^{-2} \times m_d + 32.0 + Z \tag{61}$$

Evaporator capacity 500 m³/day $< m_d < 1000$ m³/day

$$P = -4.48 \times 10^{-5} \times m_d^2 + 0.1272 \times m_d - 5.4 + Z \tag{62}$$

where Z is the pumping power (kW) of the vacuum pump and is given by:

$$Z = 0.050535 \times m_d + 0.93045 \tag{63}$$

9.2. Comparison of Simulation and Actually Measured Values

In order to determine the accuracy of the simulation program, a comparison between the simulation results and the results from actual plant operation was conducted for January and

June 1985. The operating condition of the test plant underwent various changes over the test period and are not constant. The plant had also experienced several power failures that caused complete shutdown and the plant had to be restarted manually after the restoration of power. The power failures that occurred during the test period (January and June 1985) are listed in table 15.

Table 15. Power failures causing emergency shutdown

Month	Date	Time
January 1985	21	11:10
	28	16:30
	29	14:30
June 1985	16	0:41
	22	9:35
	24	13:15

In order to account for the variation in the operating parameters occurring during the month as well as the emergency shutdowns that occurs during each month, each monthly period is divided into a number of smaller periods with the operating parameters maintained constant during each small period. The computer program is run for each small period using the prevailing operating condition of the test plant. The operating parameters which were input to the program are:

- Collector absorber area in service (January: 1543.5 m^2 – 1837.5 m^2, June: 1543.5 m^2)
- Simulation start and end dates (January: 1st – 20th, June: 1st – 15th)
- Temperature of heat accumulator at the start of simulation period (bottom 64.0°C, top 76.5°C)
- Heating water flow rate (16.5 m^3/hr)
- Monthly bypass valve open and close temperatures (open: 70.0°C, close: 73.0°C)
- Monthly evaporator start and stop temperatures (start: 72.0°C, stop: 66.0°C)
- Specification of collector cleaning days (January: 1st, 30th, June: 8th)

Figure 26 shows the measured and simulation results of the daily net amount of heat collected by the solar collector field and delivered to the heat accumulator during the month of January 1985. This amount is equivalent to the total heat collected minus the heat loss due to the collector piping system as well the heat loss due to dust effect on the glass tubes of the collectors. January is usually characterized by severe variations in the daily solar radiation and this variation is reflected on the daily amount of heat collected as shown in the figure. Figure 27 shows the measured and simulation results of the daily water production for January 1985. The simulation results appear to be in reasonable agreement with the measured values with the exception of January 6 for which there is a large discrepancy between measured and simulation values. This is because, while the operation of the evaporator was stopped at 11:30 due a drop in the temperature of the heating water in the actual plant, operation of the evaporator was continued in the simulation until 19:30 when it was stopped. This, in turn, was due to the fact that in the simulation the temperature of the heating water was slightly higher than the set level at which operation was to stop.

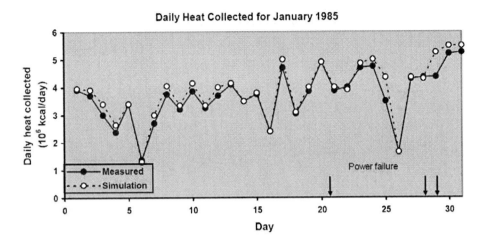

Figure 26. Daily heat collected for January 1985.

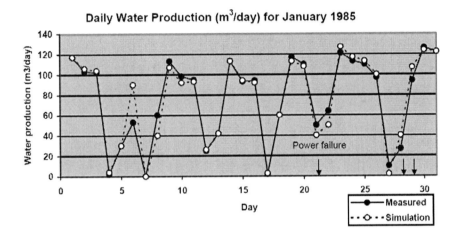

Figure 27. Daily water production for January 1985.

Figure 28 shows the simulation and measured values of the net amount of heat collected for June. The agreement seems to be quite good. Figure 29 shows the results of the daily water production for June. It is observed that the simulation result for the water production on June 1 is low. It appears that this happened because the initial temperature setting of the accumulator at the start of the calculations was lower than the actual temperatures. On June 24, the actual operation data are missing starting from 11:30 because of a power failure. To remove the dust on the glass tubes in June, blocks A and B were cleaned on June 4 and blocks D and E on June 8 while block F was cleaned regularly every 4 or 5 days. In the simulation, on the other hand, all blocks were presumed to have been cleaned on June 8. In June, the effect of the dust was significant and it appears that an error was produced due to the differences between the dust model and the actual dust accumulation. Nevertheless, the value for the amount of heat collected from June 10 through June 20 match relatively well since by this data the dust has been removed. As indicated here, there were areas where the simulation did not match the actual operation status exactly but, overall, the simulation was quite accurate even in cases where the evaporator was frequently started and stopped, as it was in January.

Daily Heat Collected (10⁶ kcal/day) for June 1985

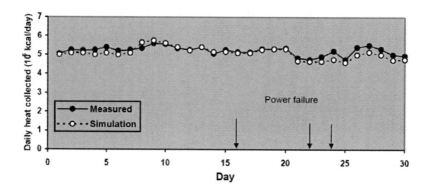

Figure 28. Daily heat collected for June 1985.

Daily Water Production (m³/day) for June 1985

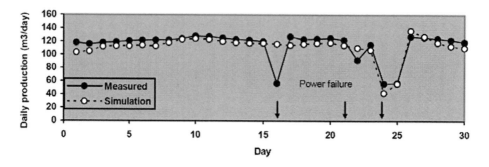

Figure 29. Daily water production for June 1985.

Table 16. Comparison between measured and simulation results for January and June 1985

		Jan. 1985	Jun. 1985
Heat collection amount (kcal/month)	Simulation	120,100,000	152,400,000
	Measured	115,100,000	153,100,000
	Error (%)	4.3	-0.45
Product water amount (m³/month)	Simulation	2,290	3,340
	Measured	2,340	3,430
	Error (%)	-2.1	-2.6

Table 16 shows a comparison between the actually measured values and the calculation values for January and June. Although the operation of the evaporator and the heat collection operation are greatly influenced by subtle changes in the set temperature conditions for operation and the ever-changing climatic conditions, it can be seen that the two sets of figures match relatively well. Consequently, it is assumed that this program will be quite serviceable for studying the optimum operating conditions of the test plant and also for designing a similar plant under different conditions.

10. Evaluation of the Test Plant

10.1. Optimum Operating Conditions

The simulation program of the test plant was run after inputting the values for the following:

- Performance of individual equipment of the plant obtained from the research operation (for example, the collector efficiency y= $0.84 - 2.46$ x $- 1.9$ x^2, and the maximum capacity of the evaporator = 6.0 m^3/h).
- Weather conditions for January through December 1985 (average solar radiation over inclined surface = 5,589 kcal/m^{2}day, average daily atmospheric temperature = 27.4 °C, and average seawater temperature = 27.5°C).
- Standard operating conditions of the test plant as shown in table 17.
- Heating water quantity (variables) = 12, 13.5, 15, 16.5, 18, 19.5 and 21 m^3/h.

Table 17. Operating condition of test plant for simulation calculations

Item	Value
Collector absorber area	1862 m^2
Heat collection water flow	83.6 m^3/h
Frequency of solar collector cleaning	Once a month
Maximum brine temperature	68°C
Seawater flow rate	36.7 m^3/h
Feedwater flow rate	17.5 m^3/h
Evaporator startup temperature	Heating water temp. corresponding to 80% load
Evaporator shutdown temp.	Heating water temp. corresponding to 60% load

A summary of the results of the simulation is shown in table 18. As seen from this table, a heating water quantity of 16.5 m^3/h results in the maximum water production: 43,400 m^3/year or 118.9 m^3/day. However, the plant should not be operated such that TME110 (i.e., the number of hours the plant is operated at above 110% load) exceeds 5% of any months total operating hours. The data in table 18 show the following in terms of TME110:

- TME110 for April is 7.6% even if the heating water flow rate is reduced to 12 m^3/h.
- TME110 for the other months can be below 5% when the heating water flow rate is 16.5 m^3/h or below.
- By increasing the heating water flow to 18.0 or 19.5 m^3/h without allowing TME110 to exceed 5%, the water production for the month concerned can be increased over that for the heating water flow of 16.5 m^3/h.

Since it was found that the quantity of heat collected in April exceeds the allowable capacity of the evaporator, the collector absorber area for April needs to be reduced in order to protect the evaporator from scaling. In order to find the number of arrays that need to be taken out of service during April, the simulation program was run for different number of arrays in operation. It was found that 4 arrays of the 76 arrays available need to be drained making the available absorber area for this month 1,764 m^2. The other operating conditions are as given in table 17. These results translate into the optimum operating conditions shown in table 19.

Table 18. Summary of simulation results for different heating water flow rates

Month	H/W flow m^3/h	Production m^3/month	TME110 hrs	Startups
Jan.	21	2360	11	28
	19.5	2530	10	20
	18	27600	0	11
	16.5	2890	0	5
	13.5	2830	0	4
	12	2740	0	7
Feb.	21	3680	267	1
	19.5	3710	207	0
	18	3700	109	0
	16.5	3680	68	0
	15	3640	46	0
	13.5	3600	23	0
	12	3560	5	0
Mar.	21	3680	291	8
	19.5	3860	233	2
	18	3870	190	1
	16.5	3850	126	1
	15	3820	85	1
	13.5	3780	41	1
	12	3740	14	1
Apr.	21	4030	428	1
	19.5	4050	385	0
	18	4040	355	0
	16.5	4010	317	0
	15	3990	227	0
	13.5	3950	140	0
	12	3900	55	0
May	21	3940	224	3
	19.5	3980	89	1
	18	3950	53	1
	16.5	3940	21	0
	15	3910	2	0
	13.5	3850	0	1
	12	3830	0	0

Table 18. Continued

Month	H/W flow m³/h	Production m³/month	TME110 hrs	Startups
Jun.	21	3940	212	0
	19.5	3930	163	0
	18	3900	72	0
	16.5	3870	33	0
	15	3840	18	0
	13.5	3800	4	0
	12	3750	0	0
Jul.	21	3350	16	6
	19.5	3440	10	1
	18	3420	4	1
	16.5	3400	2	1
	15	3350	0	2
	13.5	3340	0	1
	12	3310	0	1
Aug.	21	3850	70	3
	19.5	3860	60	1
	18	3860	19	1
	16.5	3830	8	1
	15	3790	2	1
	13.5	3750	0	1
	12	3700	0	1
Sep.	21	4050	350	0
	19.5	4050	291	0
	18	4020	131	0
	16.5	3990	74	0
	15	3960	42	0
	13.5	3920	12	0
	12	3870	0	0
Oct.	21	3800	149	9
	19.5	4010	85	0
	18	4000	58	0
	16.5	3970	30	0
	15	3930	11	0
	13.5	3890	4	0
	12	3850	1	0
Nov.	21	2830	3	24
	19.5	3180	0	9
	18	3300	0	4
	16.5	3340	0	1
	15	3320	0	1

Table 18. Continued

Month	H/W flow m³/h	Production m³/month	TME110 hrs	Startups
	13.5	3280	0	1
	12	3190	0	1
Dec.	21	2150	0	28
	19.5	2170	0	27
	18	2370	0	18
	16.5	2600	0	7
	15	2650	0	4
	13.5	2580	0	4
	12	2420	0	12
Total	21	41700	2020	111
	19.5	42800	1532	61
	18	43200	989	37
	16.5	43400	677	16
	15	43000	421	10
	13.5	42600	223	13
	12	41900	74	23

Table 19. Optimum operating conditions for the Abu Dhabi solar desalination plant

Month	Absorber area (m²)	H/C flow (m³/h)	Bypass valve Open (°C)	Bypass valve Close (°C)	H/W flow (m³/h)	F/W flow (m³/h)	Evaporator Start (°C)	Evaporator Stop (°C)
Jan.	1862	83.6	63	67	16.5	17.5	67	62
Feb.	1862	83.6	65	69	13.5	17.5	70	64
Mar.	1862	83.6	68	72	12.0	17.5	73	67
Apr.	1764	79.2	69	69	13.5	17.5	73	68
May.	1862	83.6	70	74	16.5	17.5	73	69
Jun.	1862	83.6	71	75	16.5	17.5	75	70
Jul.	1862	83.6	70	74	19.5	17.5	74	69
Aug.	1862	83.6	72	76	18.0	17.5	76	71
Sep.	1862	83.6	74	78	13.5	17.5	78	73
Oct.	1862	83.6	69	73	16.5	17.5	73	68
Nov.	1862	83.6	67	71	16.5	17.5	71	66
Dec.	1862	83.6	65	69	15.0	17.5	69	64

10.2. Simulation Results

Simulated operation of the test plant under the optimum operating conditions noted above was carried out using the solar radiation data of 1985. The results are shown in table 20. The effective annual water production is 42,900 m³, the annual operating rate of the evaporator is 97.6% (number of evaporator starts and stops = 13), and the specific power consumption is 5.1 kWh/m³ of product water.

Table 20. Simulation results at optimum operating condition

Item	Simulation output	Ratio
Total radiation on tilt surface	3,780,000,000 kcal/year	1.0
Drop in solar radiation due to dust	195,000,000 kcal/year	0.052
Heat collection amount	2,212,000,000 kcal/year	0.585
Amount of heat supplied to heat accumulator	1,950,000,000 kcal/year	0.516
Heat supplied to evaporator	1,808,000,000 kcal/year	0.478
Heat effectively used by evaporator	1,796,000,000 kcal/year	0.475
Product water	43,000 m^3/year	
Collector cleaning water	76 m^3/year	
Effective product water	42,900 m^3/year	
Heat collecting pump running hours	3,655 hr/year (10.0 hr/day)	
Evaporator operating hours	8,546 hr/year (yearly average rate 97.6%)	
Frequency of operator startup	13 times/year	
Power consumption per m^3	5.1 kWh/m^3	
Anti-scalant consumption (Belgard EV)	1,496 kg/year (0.035 kg/m^3 product)	
Sodium hypochlorite consumption (NaClO)	5,577 kg/year (0.13 kg/m^3 product)	

10.3. Evaluation of the Solar Plant

In this section we compare the planned values used in the design of the solar plant, the experimental results of the research operation obtained during the first year of plant operation and the values introduced in the simulation program which are based on the measured values. The results are summarized in table 21. Column one of this table shows the design values; column two the results of actual measured performance data and column three shows the introduced in the simulation program.

(a) The solar collector field

The heat collection efficiency of the solar collectors was about 9% lower than that for the basic design under the same weather conditions. The efficiency of a clean single collector was measured at the manufacturer's laboratory under ideal operating conditions and was reported as the catalog efficiency, η°_c, which can be represented by the following equation:

$$\eta^\circ_c = 0.913 - 2.46\,x - 1.92\,x^2 \qquad (64)$$

where x is a parameter defined as : $x = \{(T_o + T_i)/2 - T_a\}/I_t$ and the unit of $^\circ$C hr m^2/kcal. When a number of collectors are connected together to form a block in the solar collector field, the efficiency of such a clean block was measured and found to fit the following equation:

$$\eta_m = 0.84 - 2.46\,x - 1.92\,x^2 \qquad (65)$$

The measured efficiency is therefore lower than the catalog efficiency by about 8% for x = 0 and 9.3% for x = 0.05 °C hr m^2/kcal. The drop in collector efficiency of a group of interconnected collectors can be attributed to the heat loss by internal piping and connectors.

In the basic design, the influence of dust deposition on the solar collectors was assumed to cause a 10% loss of the incoming solar radiation. Measurements of the heat loss due to dust effect carried out during plant operation showed that the influence of dust deposition has a seasonal character with the loss in solar radiation varying from 4% during winter months to as much as 20% in the summer. To account for the monthly variation of the dust effect in the simulation program, a mathematical model was developed to estimate the dust effect from month to month.

Heat loss from the collector piping system is another cause of heat loss that has to be accounted f-cor. The piping system consists of insulated pipes varying in diameter from 30 mm to 125 mm as well as valves, pipe supports, expansion joints and safety valves. In the basic design, heat loss from the piping system was estimated based only on heat loss from all the pipes with heat loss from valves, supports, etc. neglected. Measurements were made to estimate the total piping loss which takes into consideration all components of the piping system. For a single block, the measured heat loss from the piping system was correlated by the following formula:

$$Q_{loss-m} = 66.6 \ (T_w - T_a)^{1.3} \quad \text{kcal/hr} \tag{66}$$

The theoretical (calculated) value of the piping heat loss excluding valves, supports, etc. was estimated as:

$$Q_{loss-c} = 93.5 \ (T_w - T_a) \quad \text{kcal/hr} \tag{67}$$

For a $(T_w - T_a) = 40$°C , the measured heat loss is 2.15 times the theoretical value.

(b) The heat accumulator

The heat accumulator (HA) in the solar plant consists of 3 series-connected thermally stratified vertical cylinders. Heat loss from the accumulator tanks was estimated in the basic design as 1.05°C per day for a water temperature of 99°C and an ambient temperature of 30°C. For a storage capacity of 300 m^3 of water, the estimated daily heat loss is 315,000 kcal/day. Heat loss measurements from plant heat balance indicates that daily temperature drop varies from month to month with the average yearly value of 2.61°C per day which is more than double the value used in basic design. This value is also based on a water temperature of 99°C and ambient temperature of 30°C.

(c) The MED evaporator.

As seen from table 21, the evaporator is the only piece of equipment in which the performance in the basic design was upgraded in the simulation program: 20% up in water production capacity, and 0.2 to 7.5% less in specific heat consumption. In the basic design, the maximum evaporator capacity was 120 m^3/day (5 m^3/hr) and specific heat consumption is 43.8 kcal/kg- product at 35°C seawater temperature and 55,000 ppm salt concentration. This design maximum production was assumed to be achieved at a heating water (HW) temperature of 99°C and a HW flow rate of 18.4 m³/hr. Measured values of maximum production was 130 m³/day (5.42 m³/hr) at a corresponding specific heat consumption of 42.4 kcal/kg- product.

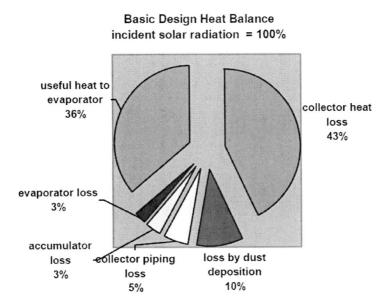

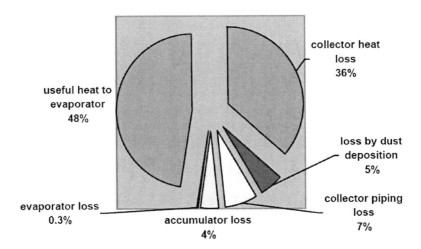

Figure 30. Total plant heat balance - comparison between design and simulation values.

Table 21. Comparison between performance values used in basic design, measured values and simulation program

Item	Basic Design	Measured value	Simulation
Heat collection efficiency	Catalog efficiency: $\eta = 0.913 - 2.46\,x - 1.92\,x^2$ $x = \{ (T_i + T_o)/2 - T_a \}/ I_t$ °C hr m²/kcal	Value of η when x=0 is: Winter = 0.83 ~ 0.84 Summer = 0.80 ~ 0.81 Average = 0.82 Slope of curve is: Winter = large Summer = small Efficiency of block F includes heat loss from internal piping and dust effect	* Efficiency used for simulation is: $\eta = 0.84 - 2.46 - 1.92\,x^2$ * For completely clean condition, the value of η for x=0 has been increased by 0.2 to 0.84 because of the fact that dust isn't completely removed after cleaning.
Heat loss from piping	Calculated value ignores valves and piping supports. For a single block: $Q_{loss} = 93.5\,(T_w - T_a)$ kcal/hr T_w = water temp., T_a=air temp.	Value measured for a single block is: $Q_{loss} = 66.6\,(T_w - T_a)^{1.3}$ kcal/hr	Value used is: $Q_{loss} = 66.6\,(T_w - T_a)^{1.3}$ kcal/hr
Dust influence	It is assumed that the solar radiation is reduced by 10% due to dust effect throughout the year	Dust effect varies seasonally. In winter about 4% and in summer reaches more than 20%.	Based on measured values, a model was developed to estimate the monthly dust influence.
Heat collecting system	Bypass operation is not considered. Water enters the accumulator immediately after pump start up.	A heat collecting system using bypass operation and control valves are used.	Heat collection amount takes into consideration bypass operation.
Heart accumulator	* When calculated heat loss from accumulator is converted to $\Delta T = T_w - T_a = 69°C$ is $\Delta T = T_w - T_a = 69°C$, the heat loss is $Q_{loss} = 312123$ kcal/day. * Heat loss is based on complete mixing in the tank and no stratification considered	* Based on measured values the heat loss at $\Delta T = 69°C$ is $Q_{loss} = 765000$ kcal/day. * Thermal stratification ratio was measured as 7.5% during daytime and 6.5% at night	* After heat loss is calculated, compensation is performed using $(Q_{loss})_{measured}/(Q_{loss})_{calc} = 2.45$ * Heat accumulator is divided into 1000 equal parts, mixing takes place at the top and bottom of tank. Mixing at top = 7.5% and mixing at bottom = 5.0%.

Table 21. Continued

Item	Basic Design	Measured value	Simulation
Operation of evaporator when excessive heat is collected	The evaporator can be operated up to a boiling temperature of 100°C.	Adjust the heat collecting area or the heating water flow rate so that the load of the evaporator does not exceed 110%.	Install a 3-way proportional control valve between the heat accumulator and the evaporator in order to control the heating water temperature so that the load of the evaporator does not exceed 110%.
Evaporator water production	Max. capacity = 5 m³/hr	Max. capacity = 6.46 m³/hr	Max. capacity = 6.0 m³/hr
Evaporator operating flow rates	* Heating water flow rate = 18.4 m³/hr * Feedwater to evaporator = 10.5 – 17.5 m³/hr * Seawater to evaporator = 39.4 m³/hr	* Heating water flow rate = 12 – 17 m³/hr * Feedwater flow rate = 17 – 21 m³/hr * Seawater flow rate = 36–40 m³/hr	* Heating water flow rate = 11.2 – 22.4 m³/hr * Feedwater flow rate = 17.5 m³/hr * Seawater flow rate = 20.2 –40.3 m³/hr
Evaporator operating temperatures	Operating temp. at 5 m³/day load: H/W inlet temp. = 99°C H/W outlet temp. = 87°C 1st effect temp. = 83°C 18th effect temp. = 43°C SW temp.(design value) = 35°C SW concentration = 55,000 ppm	Operating temp. at 5 m³/day : Jan. Jul. 71°C 79°C 58°C 66°C 58°C 66°C 28°C 39°C 21.8°C 32.4°C 51,200 53,500	Operating temp. at 5 m³/day load 78.5°C 67.5°C 67.5°C 40.8°C 35°C 52,000

The design and measured annual plant heat balance are shown in pie chart form in figure 26. In this figure it is assumed that the incident solar radiation represent 100% and the percentage of this energy going to each of the different losses are specified along with the net amount of useful energy for desalination. The effective heat input to the evaporator increased a remarkable 11.5% from 36.0% for the basic design to 47.5% in terms of the ratio to the total solar radiation quantity on the collector tilted surfaces. This sharp increase combined with the improved heat efficiency of the evaporator to produce a synergistic effect: The water production increased from an annual average of 80 m^3/day for the basic design to 117.8 m^3/day, and the effective water production (after subtracting the quantity of water used for cleaning the collectors) was 117.5 m^3/day, a dramatic 46.9% increase.

In addition to using performance data from actual plant measurement, the simulation program incorporates various improvements, which include:

- Adopting a heating water temperature control system based on a three-way proportional control valve for efficient use of the energy of the collected heat.
- Reducing power consumption by adopting pumps with appropriate capacity (several pumps used in the solar plant are oversized).

11. Economic Considerations and Comparison with Conventional MED Plants

On the basis of data obtained from the research operation of the test plant, an economic study was made to estimate the cost of water from solar MED plants with a capacity in the range 100 – 1000 m^3/day and compare the results with conventional MED plant. The pumping power of the solar and conventional MED plant is to be provided by a diesel generator of appropriate capacity. Steam for the conventional MED plant is to be provided by a steam generator. The product water costs of these practical plants were calculated and are shown below. Note that these trial designs incorporate improvements such as the inclusion of the three-way proportional control valve.

11.1. Basic Economic Parameters

The product water cost was calculated using the following economic parameters:

- Expected life of plant components

    o Evaporator.......................20 years
    o Heat accumulator...............20 years
    o Solar collectors..................20 years
    o Diesel generator.................10 years

- Interest rate................................8% per annum

11.2. Capital Equipment Cost

11.2.1. Capital Cost of MED Evaporator

The cost of MED evaporator is based on budget offers from different manufacturers and on cost information available in the open literature. The capital cost depends on the design capacity, number of effects and maximum brine temperature according to the correlation given by Fosselard et al.[6]:

$$C_{ev} = 5,005,000(\frac{m_d}{2,500})^{0.7}(\frac{PR}{8})^{0.5}(\frac{70}{T_{b\max}})^{0.47} \qquad 100 \leq m_d \leq 1000 \qquad (68)$$

where

C_{ev} = evaporator capital cost, \$
m_d = rated (design) capacity, m^3 day–1
PR = performance ratio
T_{bmax} = design maximum brine temperature, °C.

The performance ratio (PR) of the evaporator (defined as the distillate output in kg per 1055 kJ of heat input) is related to the number of effects by the following equation[1]:

$$PR = -0.809 + 0.932N - 0.0091N^2 \qquad (69)$$

11.2.2. Capital Cost of Solar Thermal Collectors

Solar collectors used for this application should be capable of producing hot water at a temperature ranging between 70 and 90°C. Evacuated tube collectors and high-efficiency flat plate collectors can be used to produce hot water at a temperature in excess of 80°C. The specific cost of the solar collectors are assumed to range 200 - 600 \$ m–2 (flat plate and evacuated tube collectors). The cost is assumed to include both the solar collector proper as well as the support structure, piping, valves, etc.

11.2.3. Capital Cost of Heat Accumulator

The heat accumulator is assumed to be a vertical cylindrical tank made of mild steel with a thick layer of fiberglass insulation to reduce heat loss. The tank is designed to operate at atmospheric pressure and is provided with a pressure relief valve as a safety measure. Hot water from the collector field is supplied to the tank at the top via a special water distribution grid that ensures that hot water diffuses slowly through the surrounding water with causing too much turbulence in order to enhance thermal stratification through the tank.

The capital cost of the heat accumulator as obtained from manufacturer's data is obtained from the following relation

$$C_{st} = 7803.9 \times m_{st}^{0.525} \qquad 100 \leq m_{st} \leq 600 \qquad (70)$$

where,

C_{st} = cost of heat accumulator, \$
m_{st} = storage capacity, m^3

11.2.4. Capital Cost of Steam Generator for Conventional MED Systems

Low pressure and low capacity steam generators are required to supply the MES evaporator with the low-temperature thermal energy necessary to drive the unit. The capacity of the steam generator depends on the capacity of the MES unit as well as its performance ratio. For a unit having a $PR = 13$ and having a capacity 200 m^3 day^{-1} at design conditions, requires an estimated 0.6 ton h^{-1} of low-pressure steam.

A fire-tube packaged steam generator producing steam at 10 bar and having an efficiency (LHV) of 86 per cent is considered appropriate. The capital cost of such unit, C_b (\$) is obtained from [7] and adjusted to the current cost level using the Marshall and Swift Equipment Cost Index. The resulting correlation is shown below:

$$C_b = 115,700 + 18,200 \times \frac{m_s}{12} \qquad \mathbf{0.15 \leq m_s \leq 12} \tag{71}$$

where,

C_b = capital cost of steam generator, \$
m_s = steam generating capacity, ton h^{-1}

11.2.5. Capital Cost of Diesel Generator

A diesel generator whose capacity will obviously depend on the capacity of the plant itself can supply the electrical demand of the desalination plant. The capital cost of such diesel generator is obtained from the following relation:

$$C_{dg} = 50\ 800\ (\frac{P}{40})^{0.5494} \tag{72}$$

where C_{dg} is the cost in \$ and P is the rated power in kW.

11.3. Operation and Maintenance Expenses

11.3.1. Consumable Chemical Expenses

The cost of consumable items is as follows:

- Cost of chemicals

  o Scale preventive (Belgard EV)   3.42 \$/kg

- o   Anti-corrosive (Nalco 2000)   1.9 $/kg
- o   Seawater disinfection (NaClO)   0.40 $/kg
- o   Caustic soda   1.78 $/kg
- o   Sodiun bisulfite   0.97 $/kg
- o   Coagulant   3.4 $/kg
- o   Calcium chloride   .0.81 $/kg

Chemical consumption for the solar MED plant were estimated from the following assumptions:

- o   Belgard EV   to be added to the feed water at 10 ppm
- o   Nalco 2000   to be added at 7,000 ppm to the makeup water, amounting to 30% of the accumulator capacity per year.
- o   Sodium hypochlorite (NaClO)   to be added to intake seawater at 18 ppm
- o   Sodium hypochlorite (NaClO)   to be added to product water at 2 ppm
- o   Sodium bicarbonate (NaHCO$_3$)   to be added to product water at 23 ppm
- o   Calcium chloride (CaCl$_2$)   to be added to product water at 18 ppm

11.3.2. Electrical Energy Consumption

Electrical energy consumption by the solar-MED and conventional MED systems are provided by a diesel-generator of appropriate capacity.

- •   Fuel consumption of diesel generator   3.0 kWh per liter of diesel oil
- •   Cost of one liter of diesel oil (c_f) $c_{barrel}/167$ $/liter
- •   One barrel of oil   = 167 liter
- •   Cost of one barrel of oil   50 –120 $/barrel
- •   Cost of electricity   = $c_f/3.0$ $/kWh
- •   Electricity consumption (solar MED)   kWh/m^3 product water
- •   Electricity consumption (conv. MED)   kWh/m^3 product water

The electrical power required by the solar MED plant consists of the following components:

- •   Pumping power for the evaporator which is given by:

$$P_{ev} = (4.805 + 0.094 \times m_d - 2.1 \times 10^{-5} m_d^2) \times (2.1 + 0.06 \times PR)/2.88 \quad (73)$$

- •   Power of the vacuum pump

$$P_{vac} = -1.866 + 0.057 \times m_d - 2.7 \times 10^{-6} \times m_d^2 \quad (74)$$

- •   Power of the heat collecting pump

$$P_c = [83.6 \times (\frac{A_c}{1862})]^{1.13} \times 2.691 \times 10^{-2} \quad (75)$$

- Power of the heating water pump

$$P_{hw} = 1.5 \times (\frac{m_d}{130}) \tag{76}$$

Thus, the total pumping power, kW, can be expressed as

$$P_1 = P_{ev} + P_{vac} + P_c + P_{hw} \tag{77}$$

The electrical power required by the conventional MED plant consists of the following components:

- Pumping power of the evaporator as before
- Pumping power for the steam generator

The pumping power of the steam generator was estimated from the relation:

$$P_b = 65.0 \times \frac{m_s}{15.0} \tag{78}$$

where m_s is the design capacity of the steam generator, ton/hr. The steam generator is capable of providing low-pressure heating steam for the MED evaporator and medium pressure steam to a steam ejector to create the operating vacuum inside the evaporator. The total electrical power required by the conventional MED plant is:

$$P_2 = P_{ev} + P_b \tag{79}$$

11.3.3. Spare Parts Cost

An amount of 2% of the direct capital investment has been estimated as yearly cost of spare parts.

11.3.4. Personnel Cost

The staff required for operation and preventive maintenance (for 3 shift operation) are assumed as follows:

Plant capacity	Supervisor	Mechanic	Electrician	Chemist	helper
100 – 300 m³/day	1	2	1	0.5	2
300 – 1000 m³/day	1	4	2	1	4
Monthly salary $	1000	600	600	600	400

The above estimates results in a cost of water due to personnel of 1.5 $/m³ for a 100 m³/day plant and 0.27 $/m³ for a 1000 m³/day plant.

11.4. Estimating the Cost of Water Produced

The estimates of the cost of water that are given below are based on the life-cycle cost analysis of the plant which includes capital, O&M and fuel costs. The total life-cycle cost, TLC, is the sum total of the capitals cost plus the present value of all future O&M annual expenses:

$$TLC = C_{\text{tot}} + PW\ (F) + PW\ (OM) \tag{80}$$

where,

C_{tot} = total capital cost including engineering, installation and management costs,
$PW\ (F)$ = present worth of all annual fuel costs incurred throughout the lifetime of the plant (for conventional plants),
$PW\ (OM)$ = present worth of all annual O&M expenses incurred throughout plant lifetime.

The present worth of the annual fuel and O&M expenses are calculated from the following expressions [5]

$$PW\ (F) = F_0 (\frac{1+g_f}{k-g_f})[1-(\frac{1+g_f}{1+k})^N]$$
$$PW\ (OM) = OM_0 (\frac{1+g_{om}}{k-g_{om}})[1-(\frac{1+g_{om}}{1+k})^N] \tag{81}$$

where

F_0 = fuel cost in the first year of operation, $
OM_o = O&M cost in the first year of operation, $
g_f = annual fuel escalation rate (assumed 0.03)
g_{om} = annual O&M cost escalation rate (assumed 0.03)
k = interest rate (assumed 0.08)
N = plant lifetime, years ($N = 20$ years)

The cost of water, c_w, ($/m^3$) was calculated as follows

$$c_w = \frac{TLC}{m_d (365) N (PF)} \tag{82}$$

where,

m_d = desalination plant rated capacity, m^3/day
PF = plant factor (assumed 0.85)

It should be noted that all water costs given in this section do not include seawater intake and outfall costs or cost of land. These additional costs are very much site dependent and has to be added to the cost estimates given here.

12. Results of the Economic Study

Figure 31 shows how the cost of water varies with the number of effects and cost of collector (in $/m^2$) for a solar MED plant having a capacity of 130 m^3/day which is identical to that of the test plant. The design maximum temperature T_b = 90°C and the fuel cost is assumed to be c_b = 50 $/barrel. The cost of water is seen to be quite sensitive to the number of effects and to a lesser degree on the cost of collector with the cost of water varying between 8 $/m^3$ and 4 $/m^3$. The water cost tends to decrease with increasing the number of effects and reducing the cost of collector. The increase in the number of effects results in an increase in the performance ratio and thus leads to a reduction in the heat demand of the evaporator for a given water production. Since this heat demand is produced by a field of solar collectors, the reduction in this demand is expected to cause a similar reduction in the area of the collector field and thus in its corresponding capital cost. On the other hand, the increase in the number of effects results also in an increase of the capital cost of the evaporator due to the increased structural complexity of the evaporator but this increase in capital cost is usually small compared to the benefits of larger number of effects.

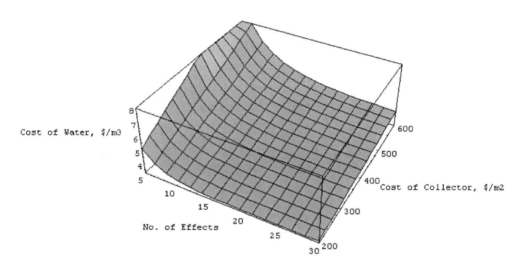

Figure 31. Effect of the number of effects and the collector cost on the resulting water cost – fuel cost, c_{barrel} = 50 $/barrel, max. brine temp. , T_b = 90°C (solar MED plant capacity = 130 m^3/day).

Figure 32 shows the influence of the number of effects and cost of fuel (expressed in $/barrel) on the resulting cost of water for a solar MED plant having a capacity of 130 m^3/day. The evaporator is assumed to have a maximum brine temperature, T_b = 90°C and the collector cost is assumed to be c_{col} = 300 $/m^2$. It can be seen that increasing the cost of fuel to the diesel generator results in an increase in the water cost due to the increase of the cost of electricity produced by the diesel generator.

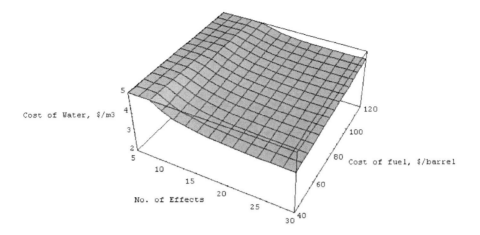

Figure 32. Effect of the number of effects and the fuel cost on the resulting cost of water – collector cost, c_{col} = 300 \$/m^2, max. brine temp., T_b = 90°C (solar MED plant capacity = 130 m^3/day).

Figure 33 demonstrates how the plant capacity and fuel cost affect the cost of water. As expected, higher plant capacity results in a lower water cost and higher fuel cost results in a higher water cost.

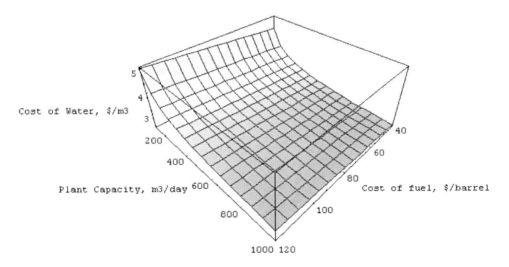

Figure 33. Cost of water as a function of plant capacity and fuel cost, N_{eff} = 25, T_b = 70°C, c_{col} =300 \$/m^2 (solar MED plant).

Figure 34 shows the effect of the cost of collector on the resulting water cost for solar MED plant capacity ranging from 100 m^3/day to 1000 m^3/day. It can be seen that the collector cost has a vital contribution to the cost of water. For a plant capacity of 1000 m^3/day, the cost of water is 2.24 \$/m^3 for a collector cost of 200 \$/m^2 and is 3.35 \$/m^3 for a collector cost of 600 \$/m^2.

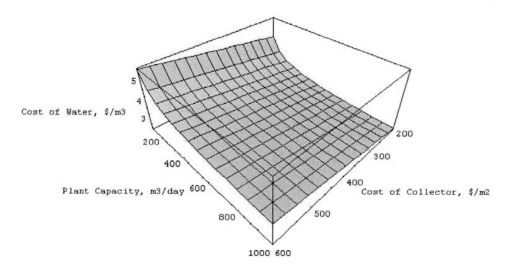

Figure 34. Cost of water as a function of plant capacity and collector cost, N_{eff} = 25, T_b = 90°C, c_{barrel} = 50 $/barrel (solar MED plant).

The effect of the cost of fuel on the water cost is displayed in figure 35 for different conventional MED plants of different capacities. For a plant capacity of 1000 m3/day and a fuel cost of 60 $/barrel (close to the current oil price), the cost of water is 2.68 $/m3

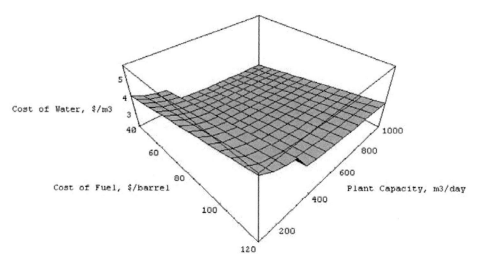

Figure 35. Cost of water as a function of plant capacity and cost of fuel, N_{eff} = 25, T_b = 70°C (conventional MED plant).

Figure 36 is 3-dimentional plot of the cost of water versus the number of effects and cost of fuel for a conventional MED plant having a capacity of 130 m^3/day. A conventional MED plant with this capacity produces water at a cost of 3.92 $/m^3 assuming an oil price of 60 $/barrel. As the oil price doubles to 120 $/barrel, the cost of water is expected to reach 4.99 $/m^3. The corresponding water cost from a solar MED plant is 4.58 $/m^3 at a collector cost of 300 $/m^2.

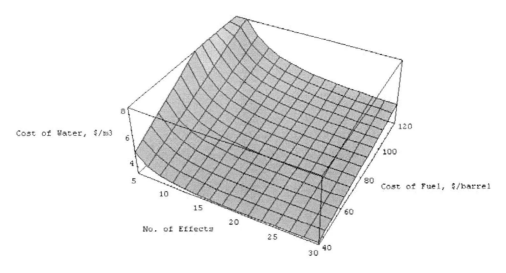

Figure 36. Effect of the number of effects and the cost of fuel on the resulting cost of water – max. brine temp., T_b = 90°C (conventional MED plant capacity = 130 m³/day).

The difference between the cost of water of a solar MED plant (c_{w1}) and the corresponding cost of a conventional MED plant (c_{w2}) is shown in figure 37 for a capacity of 130 m³/day and for different collector and fuel costs. It can be seen that for higher fuel costs and lower collector costs the difference ($c_{w1} - c_{w2}$) is negative indicating that the cost of water from a solar MED plant is cheaper than that of a conventional MED plant having the same capacity.

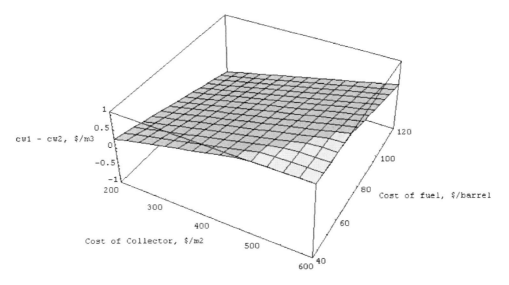

Figure 37. Difference in cost of water between solar MED plant, cw1, and conventional MED plant, cw2 , (cw1 - cw2) ,$/m³

Acknowledgement

The author would like to thank Dr. Darwish Al Qubaisi for his continued support and encouragement throughout this project.

Conclusion

The research operation of the test plant, conducted jointly between ENAA and WED, ended successfully at the end of October 1985, as scheduled, giving many useful results that are summarized below.

(1) System Reliability

The reliability of the automated continuous operation of the test plant for its first year of operation has been successfully demonstrated. However, the plant has suffered from pump trouble, etc. associated with plants of the same type in the early stage of development.

The plant's heat collecting subsystem, the heat accumulator subsystem and evaporator subsystem have also proved to have no particular problems:

- Corrosion of the heat collectors and heat accumulators was prevented by the use of corrosion prevention chemical additive;
- Scaling in the evaporator was prevented by using a scale inhibitor which maintained a good evaporator performance throughout the first year without any need for an acid cleaning;
- Experimental results showed that there was no problem with the vacuum in the evacuated glass tube collectors, despite early worries that the vacuum might deteriorate if the collectors were left at no load (drained condition) without operation under conditions of high solar radiation typical of the Middle East. The collector joints were found to pose no problems in the first year of operation.

The above results have led to estimating the life expectancy of the solar collectors, accumulators and evaporator at 20 years. The reliability of the plant has been further demonstrated during its succeeding years of operation.

(2) Response to varying weather condition

Weather conditions, including solar radiation, ambient temperature, seawater temperature and humidity varies widely according to the time of day and season. For example, the solar radiation on a horizontal surface varied from 2,150 kcal/m^2 day to 7,080 kcal/m^2 and the daily average temperature varied between 16.7°C and 38.0°C for 1985 in Abu Dhabi.

Despite such wide fluctuations, the results of the simulation program call for the evaporator of the test plant to be shutdown only 13 times a year for lack of thermal charge in the accumulator. Based on the simulation results also, the plant's annual operating time is

8,546 hours which translate into an availability of 97.6%. Thus, the test plant has a very good adaptability to weather conditions.

(3) Establishment of a method for cleaning of solar collectors

One of the early worries about the test plant was whether there was an easy and economical method for removing fouling material (dust) due to sandstorms and polluted air. In an effort to solve this problem, an investigation was made of solar installations in the Middle East, and based on the results of this investigation, a preliminary test was made for collector cleaning. As a result, high-pressure water spray gun was adopted as the method of cleaning the solar collectors. This method worked very well. A cleaning once a month reduced the heat loss from the solar collectors from the planned 10% to 5.2%. The annual amount of water used for this purpose was about 0.2% of the total water production. The applicability of this cleaning method to practical plants was therefore demonstrated.

(4) Comparison of the test plant results with that of the basic design

(a) Improved effective use of solar radiation
It was difficult to make an accurate comparison of the effective use of solar radiation between the test plant and the basic design because the weather conditions were different. Nevertheless, the table below shows that the ratio of solar radiation used for water production to the total solar radiation obtained using the simulation program (47.5%) increased markedly over the value that was predicted by the basic design (36.0%).

Table 22. Plant energy balance

Energy quantity		Basic design	Simulation
Total solar radiation on tilted surface		100%	100%
Heat loss due to fouling		10%	5.2%
Heat loss from solar collectors		42.7%	36.3%
Quantity of collected heat	Heat loss from collector piping	5.3%	6.9%
	Heat loss from heat accumulators	2.4%	3.8%
	Heat loss from evaporator	2.6%	0.3%
	Heat quantity used for desalination	36.0%	47.5%

The reasons for this marked increase are:

- The heat loss due to dust deposition was reduced by almost half, although the heat loss from the solar collectors was actually larger than was planned in the basic design.
- The quantity of collected heat was larger than was planned in the basic design because it was possible to collect heat at lower temperature.
- The number of evaporator starts and stops was considerably reduced over the planned number, which in turn reduced the heat loss from the evaporator.
- mprovements made on the test plant: Motor-operated valves, check valves, etc. were installed at the inlet and outlet of the solar collectors to make provision for power

failure during the day. If a power failure occurs, this installation automatically drains out the solar collector water, preventing water hammer due to overheated collector water.

For the test plant, there was no choice but to reduce the collector absorber area when dealing with peak radiation intensity during summer time. If all the collectors installed are to be fully used, however, it is suggested that a three-way proportional control valve is to be added between the evaporator and the heat accumulators and to mix some of the return heating water from the evaporator with the hot water from the accumulator to prevent the evaporator from overloading.

(b) Development of computer simulation program

A simulation program was prepared based on the results of the research operation of the test plant. It is a substantially improved version of the simulation program over that used in the basic design. The results of running this new program showed water productions of 102% and 97.4% of the test plant's actual performance for January and June 1985, respectively, and there was good agreement.

The simulation program is used to calculate the water production for data input into it, and involves the weather conditions at the plant concerned, such as solar radiation, ambient temperature and seawater temperature; and the specifications and capacities of the major individual plant components – capacity (100 to 2,000 m^3/day), maximum brine temperature (60 to 80°C), number of effects (13 to 32) for the evaporator, for example; and the operating conditions of the plant (eg. Flow rate and temperature of the heating water). If this simulation calculation is repeated for various sets of data, the optimum combination for the geographical area concerned, including the specifications and capacities of solar collectors, heat accumulator and evaporator, can be determined. Other data can also be obtained that may be useful in the selection of pump capacities, piping lengths, angles of absorber plates of solar collectors, etc. Therefore, the simulation program can be used to facilitate precise conceptual design of solar plants of similar design.

(c) Water production costs

- The collector cost has a vital contribution to the cost of water. For a plant capacity of 1000 m^3/day, the cost of water is 2.24 $/$m^3$ for a collector cost of 200 $/$m^2$ and is 3.35 $/$m^3$ for a collector cost of 600 $/$m^2$.
- For a plant capacity of 130 m^3/day which is identical to that of the test plant, the cost of water is quite sensitive to the number of effects and to a lesser degree on the cost of collector with the cost of water varying between 8 $/$m^3$ and 4 $/$m^3$.
- The difference between the cost of water from a solar MED plant and the corresponding cost from a conventional MED plant can be in favor for high fuel cost and low collector cost situation indicating that the cost of water from a solar MED plant can be cheaper than that of a conventional MED plant under these conditions.

Nomenclature

A_c	collector absorber area (m^2)
BPE	boiling point elevation, °C
C_b	capital cost of steam generator, $
c_{barrel}	cost of 1 barrel of oil, $
c_{col}	specific collector cost, $/m^2
C_{dg}	capital cost of diesel generator, $
C_{ev}	capital cost of evaporator, $
C_{st}	capital cost of heat storage tank, $
C_{tot}	total capital cost of plant, $
c_f	cost of 1 liter of fuel, $
c_{w1}	cost of water from a solar MED plant
c_{w2}	cost of water from a conventional MED plant
d_i	insulation thickness of pipe i
F_o	fuel cost in the first year of plant operation, $
g_f	fuel cost annual escalation rate
g_{om}	annual escalation rate for O&M expenses
H	height of collector header box (m)
h	solar altitude (rad)
h_N	solar altitude measured with respect to collector plane (rad)
I	solar radiation on horizontal surface (kcal/h m^2)
I_0	solar radiation at the outer limit of the atmosphere (kcal/h m^2)
I_b	beam component of solar radiation (kcal/h m^2)
I_d	diffuse component of solar radiation (kcal/h m^2)
I_t	solar radiation on tilted surface (kcal/h m^2)
k	interest rate
L	pitch of absorber plate (m), also latent heat of vaporization (kcal/kg)
l	width of absorber plate (m)
l_i	length of pipe i, m
m	mass flow rate (kg /s)
m_d	product flow rate ((kg/ s)
m_{st}	heat storage capacity, m^3
m_s	steam flow rate, ton/hr
N	No. of effects; also lifetime in years
OM_o	O&M cost in the first year of plant operation, $
P	pump power (kW), atmospheric transmittance
P_{ev}	pumping power of MED evaporator, kW
P_{vac}	pumping power of vacuum pump, kW
P_c	power of heat collecting pump, kW
P_{hw}	power of heating water pump, kW
P_b	power of steam generator pump, kW
P_1	total power of solar MED plant, kW
P_2	total power of conventional MED plant, kW
PF	plant factor

$PW(F_o)$	present worth of fuel cost, $
$PW(OM_o)$	present worth of O&M cost, $
Q_a	rate of heat supplied to accumulator, kcal/ h
Q_c	rate of heat collected by solar field and evaporator condenser, kcal/ h
Q_{ev}	rate of heat supplied to evaporator, kcal/ h
Q_{ph}	rate of heat transfer in preheaters, kcal/ h
r_i	radius of pipe i, m
s_1	shadow length by neighboring absorber plate, m
s_2	shadow length of neighboring glass tube, m
s_3	shadow length of collector header box, m
T	temperature, ^oC
T_b	brine temperature, ^oC
T_{bmax}	maximum brine temperature, ^oC
TLC	total life cycle cost, $
U	overall heat transfer coefficient, kcal/ h m^2 ^oC
V_a	wind speed, m/s
x	collector parameter, ^oC h m^2/kcal

Greek symbols

λ	heat conductivity of insulation material (kcal/h m^oC)
θ	incidence angle on collector plane (rad)
γ	solar azimuth angle (rad)
α	support angle of collector (rad)
τ	transmittance of glass tube
α_0	heat trasnsfer coefficient at air side, (kcal/h m^{2o}C)
η_c	collector efficiency
αc	tilt angle of absorber plate with respect to collector (rad)
γ_N	solar azimuth angle measured from tilted collector plane (rad)

Subscripts

a	ambient
av	average
c	collector or condenser
c1	inlet to condenser
c2	outlet from condenser
F	block F
in	inlet
m	measured
out	outlet
w	water

Appendix. Physical Properties of Seawater

Density (kg/m3)

$$\rho(T,C) = 1002.28 - 0.18302 \times T + 703.13 \times C - 0.32954 \times C \times T$$

Boiling point elevation (oC)

$$BPE(T,C) = -0.05 + \frac{(0.9576 + 0.8189 \times 10^{-2} \times T + 0.1647 \times 10^{-4} \times T^2) \times C}{0.114}$$

Latent heat of vaporization (kcal/kg)

$$L(T) = 0.5976 \times 10^3 - 0.565 \times T + 0.7828 \times 10^{-4} \times T^2 - 0.2859 \times 10^{-5} \times T^3$$

Specific heat at constant pressure (kcal/kgoC)

$$C_p = 1.0022 - 0.6405 \times 10^{-4} \times T - 1.14185 \times C + 6.0118 \times 10^{-3} \times T \times C +$$
$$2.109 \times 10^{-6} \times T^2 + 2.1753 \times C^2 + 1.5937 \times 10^{-2} \times T \times C^2 - 0.135 \times 10^{-2} \times T \times C$$

The temperature T is in $^{\circ}$C and the salt concentration C is in kg /kg water.

References

[1] El-Nashar, A.M., "Economics of small solar-assisted multiple-effect stack distillation plants", *Desalination.* **130** (2000) 201-215

[2] El-Nashar, A.M., "Predicting part-load performance of small MED evaporators- a simple simulation program and its experimental verification", *Desalination.* **130** (2000) 217-234

[3] El-Nashar, A.M., "The economic feasibility of small solar MED seawater desalination plants for remote arid areas", *Desalination.* **134** (2001) 173-186

[4] El-Nashar, A.M., "Validating the performance simulation program "SOLDES" using data from an operating solar desalination plant", *Desalination.* **130** (2000) 235-253

[5] ENAA & WED, Research and development cooperation on a solar energy desalination plant, Final Report, 1986.

[6] Fosselard, G. and Wangnick, K. "Comprehensive study on capital and operational expenditures for different types of seawater desalting plants (RO, MVC, ME, ME-TVC, MSF) rated between 200 m^3/day and 3000 m^3/day", Proceeding Fourth World Congress on Desalination and Water Reuse, Vol. IV, Kuwait 4-8, 1989

[7] Garcia-Rodriguez and Gomez-Camacho, C., "Design parameter selection for a distillation system coupled to a solar parabolic trough collector", *Desalination.* **122** (1999) 195-204

[8] Garcia-Rodriguez, L. and Gomez-Camacho, C., "Conditions for economical benefirs of the use of solar energy in multi-stage flash distillation", *Desalination.* **125** (1999) 133-138

[9] Garcia-Rodriguez, L. and Gomez-Camacho, C., "Thermo-economic analysis of a solar multi-effect plant installed at the Platforma Solare de Almeria (Spain)", *Desalination.* **122** (1999) 205-214

[10] Goosen, M. F.A.; Sablani, S.S.; Shayya, W.H.; Paton, C. and Al-Hinai, H., "Thermodynamic and economic considerations in solar desalination", *Desalination.* **129** (2000) 63-89

[11] Milow, B. and Zarza, E., "Advanced MED solar desalination plants. Configurations, costs, future – seven years of experience at the Platforma Solare de Almeria (Spain)", *Desalination.* **108** (1996) 51-58

[12] Sayigh, A. et al., "Dust effect on solar flat surface devices in Kuwait", Proceedings of the international symposium on Thermal Applications of Solar Energy, April 7-10 (1985), 95.

[13] Tsilingiris, P.T., "The analysis and performance of large-scale stand-alone solar desalination plants", *Desalination.* **100** (1995) 249-255

[14] Voivontas, D.; Misirlis, K.; Manoli, E.; Arampatzis, G. and Assimacopoulos, D., " A tool for the design of desalination plants powered by renewable energies", *Desalination.* **133** (2001) 175-198

[15] Voivontas, D.; Yannopoulos, K.; Rados, K.; Zervos, A. and Assimacopoulos, D., "Market potential of renewable energy powered desalination systems in Greece", *Desalination.* **121** (1999) 159-172

In: Energy Research Developments
Editors: K.F. Johnson and T.R. Veliotti

ISBN: 978-1-60692-680-2
© 2009 Nova Science Publishers, Inc.

Chapter 2

DYNAMICS AND ENERGETICS OF THE M_2 SURFACE AND INTERNAL TIDES IN THE ARCTIC OCEAN: SOME MODEL RESULTS

B.A. Kagan[*], A.A.Timofeev and E.V. Sofina

P.P. Shirshov Institute of Oceanology, Russian Academy of Sciences, St. Petersburg
Branch, 30, Pervaya Liniya, 199053, St. Petersburg, Russia

Abstract

Simulation results for the M_2 surface and internal tides in the Arctic Ocean are presented. The model results for the surface tide are close to those obtained by other authors. A comparison of the predicted amplitudes and phases of tidal sea surface level elevations with ground-based gauge measurement data shows that our estimates are better consistent with the data than those derived from the Fairbanks University model and worse than those derived from the Oregon State University model, assimilating all available empirical information. The internal tide waves (ITW) in the Arctic Ocean are of nature of trapped waves, localized near large-scale topographic irregularities. Their amplitudes are maximum ($\sim$ 4 m) near the irregularities and degenerate as the distance moves away from them. The ITW generation site occurs at a small part of the continental slope to the north-west of the New Siberian islands. The ITW decay scale at the section going across the above site is $\sim$ 300 km and is not beyond the range of its values (100 – 1000 km) for other oceans. Meanwhile, the local baroclinic tidal energy dissipation rate at specific sites of Amundsen Trough and Lomonosov Ridge are significantly less than in other oceans. The same may be said about the integrated (over area and in depth) total (barotropic + baroclinic) tidal energy dissipation in the Arctic Ocean as a whole.

Keywords: surface and internal tides; dynamics and energetics; the Arctic Ocean

[*] E-mail address: kgn@gk3103.spb.edu. tel.: +7(812)3282729; fax: +7(812)3285759. Corresponding author Prof. B.A. Kagan.

1. Introduction

Due to reasons of non-scientific character, the surface tides in the Arctic Ocean have been studied from a theoretical viewpoint better than the internal tides. Really, if about 10 tidal models were developed to simulate the surface tides [Schwiderski, 1979; Kowalik, 1981; Proshutinsky, 1993a, b; Gjevik and Straume, 1989; Polyakov and Dmitriev, 1994; Lyard, 1997; Padman and Erofeeva, 2004], then the all we know about the internal tides (otherwise, about the internal tidal waves (ITW)) was obtained using data of individual in-situ measurements of seawater temperature and conductivity in western adjacent seas of the Arctic Ocean (Zubov, 1950; Levine, 1983, 1990, Levine et al., 1985, 1987; D'Assaro and Marehead, 1991; Plüddemann, 1992; Pisarev, 1996; Kozubskaya et al., 1999; Konyaev, 2000, 2002; Konyaev et al., 2000; Serebryannii and Shapiro, 2001; Sabinin and Stanovoy, 2002; Smirnov et al., 2002; Serebryannii, 2002]. From these data and general considerations, it follows that the ITW and generally the internal waves (IW) in the Arctic Ocean differ from their images in other oceans by

- a low (more than an order of magnitude) energy level and, hence, smaller ITW amplitudes than in other oceans. For example, according to Levine et al. (1987), total energy, integrated over the internal wave frequency band, is $15 - 30$ times less than its observed values in low latitudes;
- a horizontal non-isotopic field, as opposed to its isotropicity in low latitudes. This implies that the IW in the Arctic Ocean differ in generation and propagation from those observed in other oceans; and
- the proximity of the critical latitude (this concerns the ITW with semidiurnal periods). Its result are the impossibility to propagate the ITW as free waves beyond the critical latitude, a rapid decay as the distance moves away from generation sites, when present, and a decrease for critical relative topographic slopes (the definitions appearing here may be found in Section 3).

The first two ITW dissimilarities of the Arctic Ocean are usually associated with the non-uniqueness of mechanisms of the ITW generation and dissipation due to the presence of ice cover [Levine et al., 1987] or different forcings in ice-covered and ice-free seas [D'Assaro and Morehead, 1991]. These explanations seem contrary to a solution of the problem on the ITW vertical structure in ice-covered seas [Savchenko and Zubkov, 1976] and modeling data for the stratified and homogeneous Arctic Ocean [Kowalik, 1981; Kowalik and Proshutinsky, 1991]. The articles outlined above testify that the effect of ice on low-frequency IW (in particular, on the ITW) can be neglected. A different situation arises for high-frequency IW, whose parameters depend on whether the ice cover is present or not [Muzylev and Oleinikova, 2007].

Theoretical studies of the ITW in the Arctic Ocean are more limited than experimental ones. We managed to find only 3 articles that were devoted to the problem being discussed. These are the articles of Polyakov et al. (1994), Morozov et al. (2002) and Morozov and Pisarev (2002). In the first of them, due to a low horizontal resolution, the adopted three-dimensional finite-difference model eliminates the possibility to simulate the ITW. Whereas, in the remaining two articles, the two-dimensional (in the vertical plane) nonlinear model of

Vlasenko (1991) was employed to simulate the ITW generation and propagation in western adjacent seas of the arctic continental shelf. The ITW were assumed to be plane and propagate to coast or continental slope at a right angle. This assumption is justified by the incommensurability of the ITW-induced variability in the normal and transversal directions: the variability in the first case is much more than in the second one. The mere fact of the incommensurability of the ITW-induced variability in the normal and transversal directions is true. This does not rule out the necessity of simulating the ITW three-dimensional structure, precisely which it is in the reality. Moreover, there is evidence [Craig, 1988, Holloway, 1996, 2000; Cummings and Oey, 1997; Katsumata, 2006] that the conversion of barotropic tidal energy into the baroclinic one and back depends radically on the dimension of the model in use (whether the model is two-dimensional (2D) or three-dimensional (3D)). We shall stress once again that the cited investigations are mainly concerns with western adjacent seas of the Arctic Ocean, whereas the Central and Kanadian Arctics and eastern adjacent seas of the Siberian continental shelf still remain unstudied in both experimental and theoretical respects. Otherwise, the issue as to the ITW climatology in the Arctic Ocean has not been solved.

There is an additional motivation for studies on the ITW in the Arctic Ocean. We bear in mind the necessity of evaluating a mixing rate and developing an adequate scheme of its parameterization. As known, observed values of the vertical eddy viscosity can vary from 0.1 cm^2/s in the thermocline [Ledwell et al., 1993, 1998] to 10 cm^2/s in the abyssal over rough topography [Polzin et al., 1977; Ledwell et al., 2000]. The maintenance of such an intensity of vertical eddy mixing requires a large supply of mechanical energy. One of sources of this energy is the interaction of barotropic tidal waves with bottom and/or coastline irregularities. Usually, the global supply of energy to the ITW is evaluated using measurements and tidal models [Egbert and Ray, 2000, 2001; Jaine and St. Laurent, 2001; Niwa and Hibiya, 2001]. However, because a direct relationship between forcing and the mixing rate in the deep ocean has not been established, we shall be forced to base on experimental data only.

From these data, it is evident that the mixing rate is enhanced in the ITW generation sites [Ledwell et al., 2000; Kunze and Toole, 1997; Lueck and Mudge, 1997; Polzin et al., 1997; Lien and Gregg, 2001; Pinkel et al., 2000; Kunze et al., 2002; Moum et al., 2002] and attenuated outside these sites. Another possibility to evaluate the vertical eddy mixing intensity in the deep ocean is related to an analysis of the global tidal energy dissipation budget. To maintain the observed density stratification, corresponding to the rate of thermohaline conveyor overturning, there is a need to dissipate $\sim 2 \times 10^{12}$ W [Munk and Wunsch, 1998]. An independent analysis of tides, performed using analytical methods, satellite altimetry and model results, showed that the consumption of barotropic tidal energy for the ITW generation was about $\sim 1 \times 10^{12}$ W [Egbert and Ray, 2000, 2001; Jaine and St. Laurent, 2001]. In the end, this energy should be dissipated and maintained the intensity of diapycnal mixing. Thus, the IW (in particular, the ITW) are one of basic sources of energy in oceans and, if this source is not considered, an understanding of the present-day state of oceans and its changes in the future, like the climatic system as a whole, is impossible in principle. Attempts at developing the global ocean circulation models with tidally induced mixing are available [Jaine and St. Laurent, 2001; Simmons et al., 2004].

The aim of the present article is to simulate the poorly known ITW dynamics and energetics in the Arctic Ocean for winter conditions and show distinctive features of the Arctic Ocean ITW. The tasks of the article are:

- to modify the 3D finite-element hydrothermodynamic model QUODDY-4, incorporating the rotated co-ordinate system, which allows us to obviate "the pole problem", and supplementing the model with the effects of the equilibrium tide;
- to simulate (in the frames of the modified model) the spatial structure of the M_2 ITW;
- to use the model results for revealing the ITW generation sites, if any;
- to find spatial distributions of the ITW energy budget components, including the conversion of barotropic tidal energy into baroclinic one and back; and
- to map the ITW-induced diapycnal mixing coefficient, determining changes in the mixing of tidal origin.

The article is organized as follows: in Section 2, a brief description of a modified version of the 3D finite-element hydrothermodynamic model QUODDY-4 and means of its implementaton are given; Section 3 is devoted to a discussion of the model results for the M_2 surface and internal tides. At last, Section 4 summarizes the work and setts off one of problems, which has a bearing on the surface and internal tides in the Arctic Ocean and, accordingly, on information relative to a contribution of the Arctic Ocean to the global tidal energy dissipation budget.

2. The Model

The basis for research on the surface and internal tides in the Arctic Ocean forms the modified 3D finite-element hydrothermodynamic model QUODDY-4, which differs from its original version [Ip and Lynch, 1995] by introducing the rotated coordinate system (Sein, a private communication), bypassing "the pole problem", and describing the effects of the equilibrium tide. The model contains the system of 3D hydrothermodynamic equations, namely, primitive equations of motion, written in hydrostatic and Boussinesq approximations, as well as evolutionary equations for temperature and salinity and equations of continuity and seawater state. The vertical eddy viscosity and diffusivity are considered to be unknown and determined using a 2½-level turbulence closure scheme (i.e. kinetic turbulence energy (TKE) and the turbulence scale are found from the relevant evolutionary equations and the Kolmogorov approximate similarity relationships). The horizontal eddy viscosity and diffusivity are computed using the well-known Smagorinsky formulae. The bottom stress is parameterized by a quadratic resistance law, suggesting that the velocity in the near-bottom layer is described by the logarithmic law. The heat and salt fluxes at the bottom and ocean surface are taken as zero, while the corresponding values of TKE and the turbulence scale are found from the condition of local balance between the TKE generation and dissipation, with allowance made (in accordance with the law of the wall) for a linear change in the turbulence scale within the near-bottom layer. It is also assumed that the ocean surface is ice-free. This assumption is justified by the fact that the influence of ice on surface and internal tides may be neglected, at least, as a first approximation.

The enumerated boundary conditions, together with the boundary conditions at open and solid boundaries, are presented and detailly described in [Ip and Lynch, 1995; see also Kagan and Timofeev, 2005]. Recall that the tidal sea surface level elevations at the open boundary are specified, and the normal derivation of the tangential (to the boundary) component of the integral transport is taken as zero. In addition, if the domain under steady is stratified, as with our case, the normal derivations of horizontal velocity, temperature, salinity, TKE and turbulence scale at the open boundary are also taken as zero. The latter condition follows from the Orlansky radiation condition if the rate of signal propagation is sufficiently large. The drag coefficient in the quadratic resistance law, parameterizing the bottom stress, is assumed to be equal to its standard value (0.003). At the solid boundary, the no-slip condition for the normal component and the slip condition for the tangential component of velocity and integral transport are specified. At the initial time instant, the condition of the rest for velocity, the condition of horizontal uniformity for temperature and salinity, and the condition of the lack of relict turbulence are prescribed.

The depths are taken from 1-minute database IBCAO (Figure 1); tidal sea surface level elevations at the open boundary, from the high-resolution Arctic tidal model [Padman and Erofeeva, 2004]. The horizontal grid resolution is assumed to be varying from 2.7 km near the coast and bottom uplifts to 25 km in the open ocean. The ocean is divided into 40 layers of the non-uniform vertical span, determining the vertical resolution. The time step is 11.178 s. The vertical profile of the Brunt-Väisälä frequency is computed using the data on temperature and salinity, that are contained in [Joint US-Russian Allas, 1997]. The remaining model parameters are specified the same as in the original version of the model.

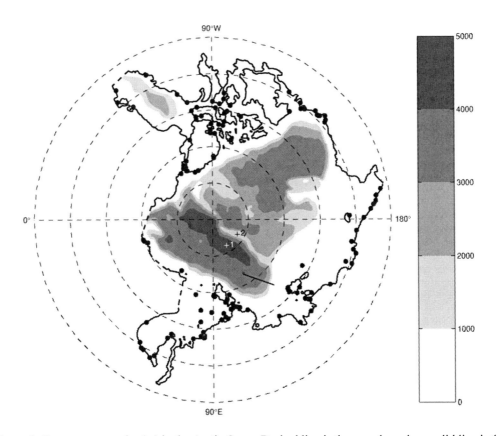

Figure 1. Bottom topography (m) in the Arctic Ocean.Dashed line is the open boundary; solid line is the section, going across the ITW generation sites; solid circles are the ground-based gauge stations where the sea surface level measurements are produced and which are used to verify the model parameters for tidal sea surface level elevations. Crosses (white and black) indicate the locations of sites at Amundsen Trough and Lomonosov Ridge, mentioned in the text.

The governing equations, which must satisfy the above initial-boundary conditions, are integrated with time using the simple prismatic finite elements, with the help of which basic and trial functions are discretized. Discretization in time is produced with a semi-implicit scheme, providing that all advective and diffusive terms in the equations for horizontal velocity, temperature, salinity and eddy characteristics are attributed to the previous time step. As a result, the evolutionary equations are solved at each grid node as a system of one-dimensional (in the vertical direction) inhomogeneous ordinary differential equations. This predetermines (to avoid numerical instability) choosing a small time step. Its value has already been pointed out.

The model equations are integrated before the quasi-periodic state is established, which is defined as the state when the relative changes in all components of the barotropic and baroclinic tidal energy budgets are 5 %. If the Väisälä frequency is taken as horizontally uniform, this condition is satisfied within 18 tidal cycles after the establishment of the quasi-periodic regime for the M_2 barotropic tidal flow, which in turn is established from the state of the rest within 12 tidal cycles. When the quasi-periodic state of the stratified ocean is established, the integration is yet applied to another tidal cycle; then, it is stopped, and the resulting solution is subjected to a harmonic analysis. In our case (the stratified ocean), the harmonic analysis is applied only to a time series of tidal sea surface level elevations. The time series for other dependent variables are not subjected to the harmonic analysis, so as not to exclude any manifestations of non-linearity.

3. Model Results

In the Introduction, we have already mentioned the articles in which model results for the surface tide in the Arctic Ocean are presented. These articles are mainly based on 2D (in a horizontal plane) tidal models disregarding the effects of density stratification and the phase difference between the bottom stress and the barotropic (depth-averaged) tidal velocity. From these articles, including our one, it follows that the M_2 surface tide in the Arctic Ocean is initiated by the Kelvin wave traveling from the North Atlantic. The interaction of this wave with the reflected Kelvin wave, generated by partial reflection of the ingoing wave in the Arctic Ocean, leads to the formation of the amphidrome with left (anti-clockwise) rotation of isophases in the Central Arctic (Figures 2,3). Two more amphidromes of left rotation are detected at the entrances to Amundsen Gulf and Davis Strait. The distributions of cotidal lines in the straits between islands of Kanadian Arctic Archipelago differ between various published tidal charts due to choosing different spatial resolutions and topographies. In the adjacent seas of the Siberian continental shelf, a chain of amphidromes of left rotation is found out. These amphidromes owe their origin to the interference of ingoing and outgoing Poincaré waves and the predominance of the eastward wave as compared to the westward one. The above feature conforms with the model results presented in [Androsov et al., 1998; Kagan et al., 2008]. In this respect, our results will not change the current view of the spatial structure of the M_2 surface tide in the Arctic Ocean.

It is interesting to compare our model results with the observational data obtained at coast and island stations, where sea surface level elevation measurements are available (see Figure1).

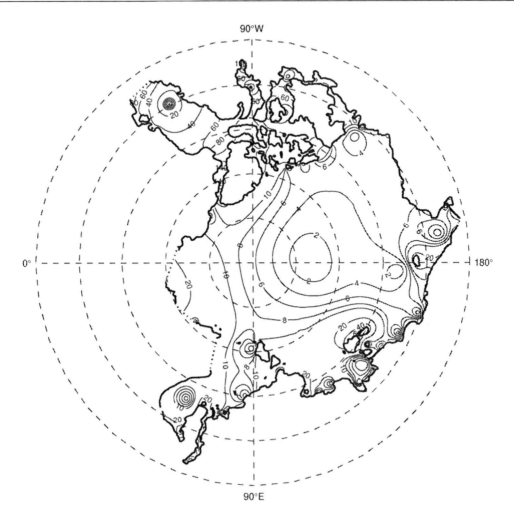

Figure 2. Chart of isoamplitude (cm) for the M2 wave in the Arctic Ocean.

This comparison shows that the root mean square absolute and relative errors ($\sum_n (D_n / N)$ and $\sum_n (D_n / A_n N)$, respectively) amount to 5.9 cm and 31.8 %. Here,

$$D_n = 2^{1/2} \left[A_{on}^2 + A_{mn}^2 - 2A_{on} A_{mn} cos(\varphi_{on} - \varphi_{mn}) \right]^{1/2}$$ is the rms (over a tidal cycle) error; A_{on}, ϕ_{on} and A_{mn}, ϕ_{mn} are observed and model values of the amplitude and the phase at the n-th site; N is the number of these sites (in our case, it is 119). As can be seen, the errors are better than those (9.6 cm and 43.7 %) derived from the Fairbanks University model [Kowalik and Proshutinsky, 1991] and worse than those (3.4 cm and 24.7 %) derived from Oregon State University model [Padman and Erofeeva, 2004]. This fact is not surprising: the Oregon State University model assimilates all available empirical information, including the TOPEX/POSEIDON and ERS satellite altimetric data, and uses the same (assimilated) data for comparing them with model results.

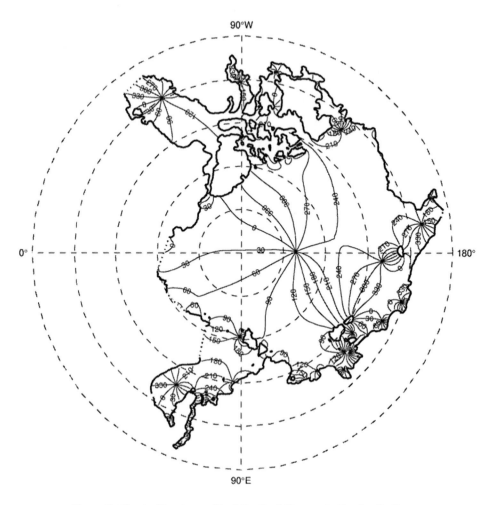

Figure 3. Chart of isophases (deg) for the M2 wave in the Arctic Ocean.

Similarly, for the same ground-based gauge observational data, the mean absolute error as applied to tidal sea surface level elevation amplitudes ($N^{-1}\sum_n |O_n - M_n|$, where O_n and M_n are observed and model values of the amplitude at the n-th measurement station), as well as the linear correlation ($s_{OM}/s_O s_M$, where $s_{OM} = N^{-1}\sum_n (O_n - \overline{O})(M_n - \overline{M})$, $s_O = \sqrt{N^{-1}\sum_n (O_n - \overline{O})}$, $s_M = \sqrt{N^{-1}\sum_n (M_n - \overline{M})}$, $\overline{O} = N^{-1}\sum_n O_n$, $\overline{M} = N^{-1}\sum_n M_n$), the bias $\left(N^{-1}\sum_n (M_n - O_n) \right)$ and the scatter index $\left(\sqrt{N^{-1}\sum_n (M_n - O_n)^2} / \overline{O} \right)$ are 4.8 cm, 0.9899, 2.03 cm and 0.2980 for our model, whereas for the Fairbanks University model and the Oregon State University model they are 6.55 cm, 0.9308, -0.02 cm, 0.4275 and 2.49 cm, 0.9922, -1.64 cm, 0.1426, respectively. We see that our estimates are again generally better than those derived from the Fairbanks University model and worse than those derived from the Oregon State University model. The reason for this has already been indicated.

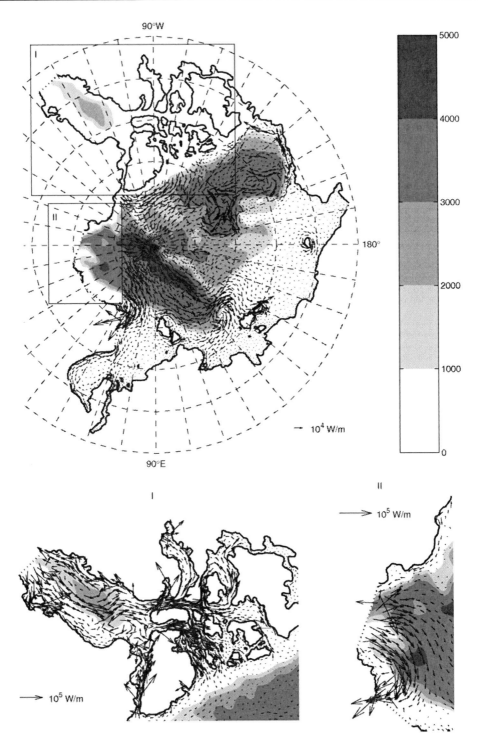

Figure 4. Spatial distribution of the horizontal flux per unit length of barotropic tidal energy (W/m) for the M2 wave in the Arctic Ocean. In the inserts are shown the horizontal fluxes in an increased scale.

We shall now discuss the model field of the averaged (over a tidal cycle) horizontal flux per unit length of barotropic tidal energy (Figure 4). Notice that the fluxes are mainly oriented along isobaths. This orientation makes the revealing of ITW generation sites difficult. An exception in the Arctic Ocean is a part of the continental slope to the north-west off the New Siberian islands, where the horizontal flux of barotropic tidal energy is approximately directed across isobaths. Accordingly, as noted in linear internal wave theory [Baines, 1973], an ITW generation site can occur in this region. However, if a local latitude is close to the critical latitude, where tidal and inertial frequencies are coincidence with one another, relative topographic slopes (topographic slopes normalized by slopes of ITW characteristics) are near zero or negative. Whence, the critical relative topographic slopes, at which the ITW generation is maximal, can reduce in the Arctic Ocean. On summing up, we may say that a commonly accepted view of the ITW generation site as a site where the averaged horizontal flux of barotropic tidal energy is orthogonal to isobaths and any one of relative topographic slopes or a part of the appropriate slope is critical (close to unity), fails in the Arctic Ocean.

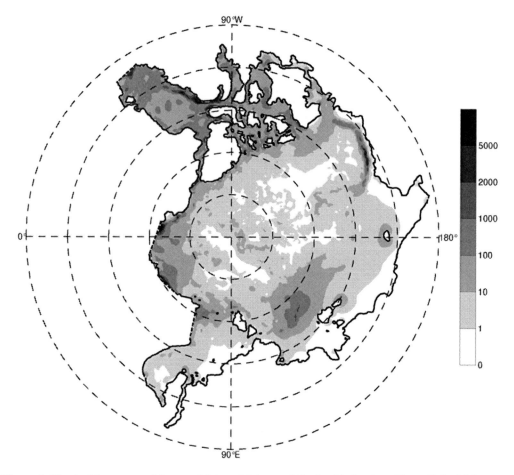

Figure 5. Chart of the averaged (over a tidal cycle) integral (in depth) density of baroclinic tidal energy (J/m^2) for the M2 wave in the Arctic Ocean.

For the ITW generation site to be revealed in the Arctic Ocean the use of the averaged integrated (in depth) density of baroclinic tidal energy seems to be preferable.

Judging from Figure 5, a local increase in baroclinic tidal energy density, exceeding $2000\ J/m^2$, indeed occurs to the north-west off New Siberian islands. In the remainder of the

Arctic Ocean, the baroclinic tidal energy density may be as much as the above value, only provided that departures of quasi-stationary residual tidal velocity from the barotropic one are large. Almost everywhere in the Arctic Ocean, the density of baroclinic tidal energy varies from 0 to 100 J/m^2. Figure 5 is interesting in one more respect. It turns out that the baroclinic tidal energy density depends on depth. It tends to rise on the continental shelf and diminish in the deep ocean. In so doing, its difference in value at troughs and ridges in the deep ocean is as much as one order of magnitude. This diference does not seem to be insignificant if characteristic changes in local depth (see Figure 1) are taken into account.

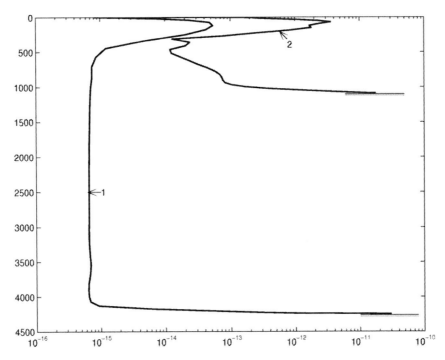

Figure 6. Vertical profiles of the local normalized (by mean seawater density) averaged rate of baroclinic tidal energy dissipation (W/kg) for the M2 wave at two sites of Amundsen Trough (1) and Lomonosov Ridge (2).Depths (m) is plotted along the ordinate axis, the values of $-\varepsilon/\rho_0$ (W/kg), along the abscissa axis. Horizontal lines correspond to local depths.

Further, it becomes pronounced most clearly when comparing vertical profiles of the local normalized (by ρ_0, where ρ_0 is the mean seawater density) averaged baroclinic tidal energy dissipation rate $-\varepsilon/\rho_0$. In Figure 6, the vertical profiles of $-\varepsilon/\rho_0$ are plotted at two locations of Amundsen Trough and Lomonosov Ridge. In both cases, the local baroclinic tidal energy dissipation rate is greater in the surface and near-bottom layers than at intermediate horizons. The fact of increasing $-\varepsilon/\rho_0$ as we approach the bottom is confirmed by experimental data [Ledwell et al., 2000; St. Laurent et al., 2001; Rudnick et al., 2003; Klymak et al., 2004; Alford et al., 2006]. The difference between the values of $-\varepsilon/\rho_0$ in the surface and near-bottom layers may be as one order of magnitude and more, the value of $-\varepsilon/\rho_0$ near the bottom at Lomonosov Ridge not exceeding 2×10^{-11} W/kg. Hence, it is about two-three orders of magnitude less than at Mid-Atlantic and Hawaiian Ridges [St. Laurent and Nash, 2004]. Generally, the local baroclinic tidal energy dissipation rate in the Arctic Ocean is less than its value in oceans of moderate and low latitudes. This property is one more distinctive feature of the ITW in the Arctic Ocean.

Now, if the depth-integrated horizontal flux of baroclinic tidal energy (not shown) and the depth-integrated baroclinic tidal energy dissipation (Figure 7) are known, we can determine the ITW decay scale, defining as the ratio between these quantities. It emerges that, for the section going across the ITW generation site, the ITW decay scale in the Arctic Ocean is ~ 300 km, i.e. it is not beyond the range of its values (100 – 1000 km) for Mid-Atlantic and Hawaiian Ridges [St. Laurent and Nash, 2004]. However, since the integrated baroclinic tidal energy dissipation in the Arctic Ocean is small (Figure 7), and the ITW decay scale is identical in value, whence it follows that the integrated horizontal flux of baroclinic tidal energy in the Arctic Ocean is also small as compared to its values in oceans of moderate and low latitudes. This feature is associated with the specific nature of the ITW in the Arctic Ocean.

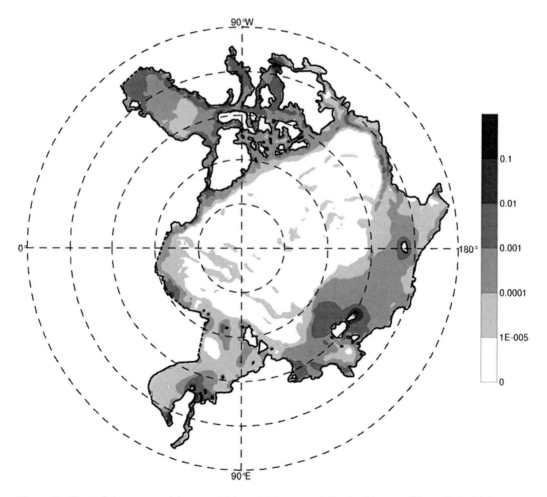

Figure 7. Chart of the averaged (over a tidal cycle) integrated (in depth) rate of baroclinic tidal energy dissipation (W/m^2) for the M2 wave in the Arctic Ocean

We shall discuss the model results for the ITW amplitudes in the Arctic Ocean. Their distribution along the above-mentioned section is displayed in Figure 8a. The first thing, that is note worthy, is a rapid decay of the ITW as the distance moves away from the generation site. So, if in the vicinity of the continental slope, the ITW amplitudes are ~ 4 m, then for the distance of 100 – 400 km from it, they do not exceed ~ 0.5 m. The latter circumstance may be regarded as a manifestation of the ITW trapping by the continental slope and, in general, any large-scale topographic irregularities, being consistent with the estimate of the ITW decay scale.

Figures 8b,c, at which the spatial distributions of the baroclinic tidal energy density and the diapycnal mixing coefficient along the same section are presented, are also interesting because they reinforce the statement that the ITW in the Arctic Ocean are trapped by the continental slope. The diapycnal mixing coefficient is computed using the model results and the standard Osborn' formulae [Osborn, 1980] providing the upper limiting estimate of the sought-for quantity. As can be seen, in the vicinity of the continental slope, the diapycnal mixing coefficient may be larger than the molecular viscosity (~ 0.01 cm^2/s) and the canonic value of the vertical eddy viscosity in the main thermocline (~ 0.1 cm^2/s) and much less than the vertical eddy viscosity in the abyssal over rough topography (~ 10 cm^2/s) (see, e.g., Polzin et al., 1977; Ledwell et al., 1993, 1998, 2000). Clearly, the ITW-induced diapycnal mixing coefficient is great as compared to the canonic value of the vertical eddy viscosity and may make a contribution to the formation of the Arctic Ocean climate.

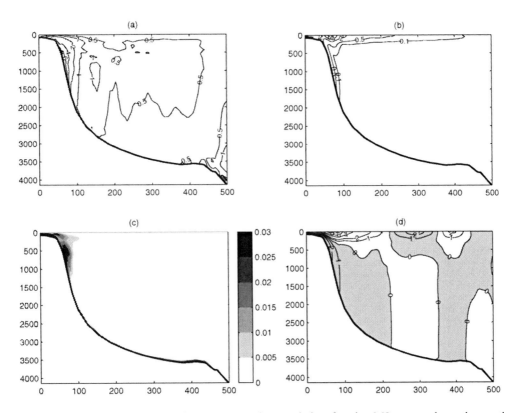

Figure 8. Vertical distributions of some ITW characteristics for the M2 wave along the section indicated in Figure 1. a – ITW amplitudes (m); b – the averaged density of baroclinic kinetic energy (J/m2); c – the diapycnal mixing coefficient (cm^2/s); d – the maximum baroclinic tidal velocity (cm/s). The regions with the southward direction of the baroclinic tidal velocity are shaded. The distance along the section is given in km.

A certain view of the ITW dynamics in the Arctic Ocean may be obtained using the model results for the baroclinic component of tidal velocity, defined as the difference between the predicted velocity and its barotropic value. In Figure 8d, the spatial distribution along the above section of the maximum baroclinic tidal velocity (major semi-axis of the velocity ellipse) is depicted. We notice that the baroclinic tidal velocity in the deep stratified Arctic Ocean is almost uniform: it is $0 - 2$ cm/s both at the continental shelf and in the deep ocean. Another remarkable feature of the baroclinic tidal velocity field is its near-one-mode (corresponding to the first baroclinic mode) vertical structure with opposite (in sign) velocities in the surface and deep layers and almost zero velocities at intermediate horizons. There are no observational evidence, which might support or rule out this feature.

4. Conclusion

A modified version of the 3D finite-element hydrothermodynamic model QUODDY-4 has been applied to simulate the M_2 surface and internal tides in the Arctic Ocean. This version differs from the original one by implementing the rotated coordinate system, which makes it possible to obviate "the pole problem", and considering the effects of the equilibrium tide. It is shown that the qualitative results for the M_2 surface tide are close to those obtained by other authors and that the ITW in the Arctic Ocean are slighter expressed than in other oceans. The ITW in the Arctic Ocean are of nature of trapped waves, localized near the continental slope or large-scale topographic irregularities. The ITW generation site is spaced at a small region of the continental slope to the north-west off the New Siberian islands. The ITW amplitudes are ~ 4 m near the generation region and decrease with distance from it. The field of the averaged (over a tidal cycle) integrated (in depth) baroclinic tidal energy density, like the fields of the maximum baroclinic tidal velocity and the diapycnal mixing coefficient, also attests that the ITW in the Arctic Ocean are of nature of trapped waves. The diapycnal mixing coefficient can significantly exceed the canonic value of the vertical eddy viscosity, implying that the ITW may contribute to the formation of the Arctic Ocean climate.

The vertical profiles of the averaged local baroclinic tidal energy dissipation differ in magnitude at troughs and ridges in the deep Arctic Ocean. Its value increases with approaching the bottom, like at Mid-Atlantic and Hawaiian Ridges . In this respect, the Arctic Ocean is not an exception. The Arctic Ocean is not the exception as applied to the ITW decay scale, too. It turns out that the ITW decay scale for the section going across the ITW generation site is ~ 300 km, what is not beyond the range of its values (100 – 1000 km) for oceans of moderate and low latitudes.

The depth-integrated total (barotropic + baroclinic) tidal energy dissipation rate in the Arctic Ocean as a whole is $\sim 1.4 \times 10^{10}$ W. It is incommensurable with the global tidal energy dissipation rate ($\sim 2.5 \times 10^{12}$ W) following from the data of satellite altimetry [Kagan and Sündermann, 1996]. The contribution of tidal energy dissipation in the Arctic Ocean is negligible, even smaller than the conversion rate of barotropic and baroclinic tidal energy ($\sim 1 \times 10^{12}$ W) in the deep World Ocean [Egbert and Ray, 2000, 2001]. For comparison: according to Kowalik (1981), the rate of tidal energy dissipation in the Arctic Ocean is $\sim 1.3 \times 10^{10}$ W for the horizontal eddy viscosity, varying from 5×10^8 cm^2/s to 5×10^{10} cm^2/s, and $\sim 1.5 \times 10^{10}$ W for the horizontal eddy viscosity, varying from 5×10^{10} cm^2/s to

5×10^{11} cm^2/s. Further, in accordance with Proshutinsky (1993a), in the Arctic Ocean, including the Barents and White Seas, the tidal energy dissipation rate is $\sim 8 \times 10^{10}$ W. The present results testify that our estimate of tidal energy dissipation in the Arctic Ocean, together with the adjacent seas of the Siberian continental shelf, but without Norwegian, Greenland, Barents and White Seas, is near the Kowalik estimate. Clearly, this estimate depends on sizes of the domain under study, topography, a grid resolution etc., so that the question as to what the tidal energy dissipation rate in the Arctic Ocean is, may hardly be regarded as closed.

References

Alford, M.H., Gregg, M.C., Merrifield, M.A. (2006). Structure, propagation and mixing of energetic baroclinic tides in Mamala Bay, Oahu, Hawaii. *Journal of Physical Oceanography*, **36**, 997-1018

Androsov, A.A., Liberman, Yu.M., Nekrasov, A.V., Romanenkov, D.A., Voltzinger, N.E. (1998). Numerical study of the M$_2$ tide on the North Siberian shelf. *Continental Shelf Research*, **18**, 715-730

Baines, P.G. (1973). The generation of internal tides by flat-bump topography. *Deep-Sea Research*, **20**, 179-205

Craig, P.D. (1988). A numerical model study of internal tides on the Australian North West Shelf. *Journal of Marine Research*, **46**, 59-76

Cummings, P.F., Oey, L.Y. (1997). Simulation of barotropic and baroclinic tides off northern British Columbia. *Journal of Physical Oceanography*, **27**, 762-781

D'Assaro, E.A., Marehead, M.D. (1991). Internal waves and velocity fine structure in the Arctic Ocean. *Journal of Geophysical Research*, **96**, 12725-12738

Egbert, G.B., Ray, R.D. (2000). Significant dissipation of tidal energy in the deep ocean inferred from satellite altimeter data. *Nature*, **405**, 775-778

Egbert, G.B., Ray, R.D. (2001). Estimates of M$_2$ tidal energy dissipation from TOPEX/POSEIDON altimeter data. *Journal of Geophysical Research*, **106**, 22475-22502

Gjevik, B., Sraume, T. (1989). Model simulations of the M$_2$ and K$_1$ tides in the Nordic Seas and the Arctic Ocean. *Tellus*, **41A**, 73-96

Holloway, P.E. (1996). A numerical model of internal tides with application to the Australian North West Shelf. *Journal of Physical Oceanography*, **26**, 21-37

Holloway, P.E. (2001). A regional model of the semidiurnal internal tide on the Australian North West Shelf. *Journal of Geophysical Research*, **106**, 19625-19638

Ip, J.T.C., Lynch, D.R. (1995). *User's Manual: Comprehensive coastal circulation simulation using finite elements: Nonlinear prognostic time-stepping* model. Thayler School of Engineering. Dartmouth College, Hanover, New Hampshire. Report Number NML 95-1

Jaine, S.R., St. Laurent, L.C. (2001). Parameterizing tidal dissipation over rough topography. *Geophysical Research Letters*, **28**, 811-814

Joint US-Russian Atlas of the Arctic Ocean, Oceanography Atlas for the winter period (1977). In E. Tanis, L. Timokhov. Environmental Working Group. University of Colorado, Boulder, Media Digital

Kagan, B.A., Romanenkov, D.A., Sofina, E.V. (2008). Tidal ice drift and ice-generated changes in the tidal dynamics/energetics on the Siberian continental shelf. *Continental Shelf Research*, **28**, 351-368

Kagan, B.A., Timofeev A.A. (2005). Dynamics and energetics of the surface and internal semidiurnal tides in the White Sea. Izvestiya, *Atmospheric and Oceanic Physics*, **41**, 550-566 (in Russian)

Kagan, B.A., Sűndermann, J. (1996). Dissipation of tidal energy, paleotides, and evolution of the Earth-Moon system. *Advances in Geophysics,* **38**, 179-266

Katsumata, K. (2006). Two- and three-dimensional numerical tide generation at continental slope. *Ocean Modelling*, **12**, 32-45

Klymak, J.M., Moum, J.N., Nash, J.D., Kunze, E., Girton, J.B., Carter, G.B., Lee, C.M., Sanford, T.B., Gregg, M.C. (2005). An estimate of energy lost to turbulence at the Hawaiian Ridge. *Journal of Physical Oceanography*, **36**, 1148-1164

Konyaev, K.V. (2000). Internal tide at the critical latitude. Izvestiya, *Atmospheric and Oceanic Physics*, **36**, 396-406 (in Russian)

Konyaev, K.V. (2002). Internal tide near the polar front of the Barents Sea from the data of a density survey. In I.V. Lavrenov, E.G. Morozov, *Surface and internal waves in the Arctic Ocean,* 320-321, St. Petersburg: Gidrometeoizdat (in Russian)

Konyaev, K.V., Plüddemann, A., Sabinin, K.D. (2000). Internal tide at the Ermak platea in the Arctic Ocean. Izvestiya, Atmospheric and Oceanic Physics, 36, 542-552 (in Russian)

Kowalik, Z. (1981). A study of the M_2 tide in the ice-covered Arctic Ocean. *Modeling, Identification and Control*, **2**, 201-223

Kowalik, Z., Proshutinsky, A.Yu. (1991). The Arctic Ocean tides. In O.M. Johannessen et al., *Geophysical Monograph Series*, **85**, 137-158, American Geophysical Union. Washington. D.C.

Kozubskaya, G.I., Konyaev, K.V., Plüddemann, A., Sabinin, K.D. (1999). Internal waves at the slope of the Medvezii island from the "Polar front of the Barets Sea (BFPS-92) experimental data". *Okeanologiya*, **39**, 165-175 (in Russian)

Kunze, E., Toole, J.M. (1977). Tidally driven vorticity, diurnal shear, and turbulence atop. Fieberling seamount. *Journal of Physical Oceanography*, **27**, 2663-2693

Kunze, E., Rosenfield, L.K., Carter, G.S., Gregg, M.C. (2002). Internal waves in Monterey Submarine Canyon. *Journal of Physical Oceanography*, **32**, 1890-1913

Ledwell, J.R., Watson, A.J. Law, C.S. (1993). Evidence for slow mixing across the pycnocline from an open-ocean tracer release experiment. *Nature*, **364**, 701-703

Ledwell, J.R., Watson, A.J. Law, C.S. (1998). Mixing of a tracer released in the pycnocline. *Journal of Geophysical Research*, **103**, 21499-21529

Ledwell, J.R., Montgomery, E.T., Polzin, K.L., St. Laurent, L.C., Schmitt, R.W., Toole, J.M. (2000). Mixing over rough topography in the Brazil Basin. *Nature*, **403**, 179-184

Legg, S., Huijts, K.M.H. (2006). Preliminary simulation of internal waves and mixing generated by finite amplitude tidal flow over isolated topography. Deep-Sea Research, II 53, 140-156

Levine, M.D. (1983). Internal waves in the ocean: A review. Review of Geophysics and *Space Physics*, **21**, 1206-1216

Levine, M.D. (1990). Internal waves under the arctic pack ice during the Arctic Internal Wave Experiment: The coherent structure. *Journal of Geophysical Research*, **95**, 7347-7357

Levine, M.D., Paulson, C.A., Morison, J.H. (1985). Internal waves in the Arctic Ocean: Comparison with lower-latitude observations. *Journal of Physical Oceanography*, **15**, 801-809

Levine, M.D., Paulson, C.A., Morison, J.H. (1987). Observations of internal gravity waves under pack ice. *Journal of Geophysical Research*, **92**, 779-782

Lueck, R.G., Mudge, T.D. (1977). Topographically induced mixing around a shallow seamount. *Science*, **276**, 1831-1833

Lyard, F.H. (1997). The tides in the Arctic Ocean from a finite-element model. *Journal of Geophysical Research*, **102**, 15611-15638

Morozov, E.G., Pisarev, S.V. (2002). Internal tide in arctic latitudes (a numerical experiment). *Okeanologiya*, **42**, 165-173 (in Russian)

Morozov, E.G., Pisarev, S.V., Erofeeva, S.Yu., (2002). Internal waves in arctic seas of Russia. In I.V. Lavrenov, E.G. Morozov, *Surface and internal waves in arctic seas*, 217-234, St. Petersburg: Gidrometeoizdat (in Russian)

Moum, J.N., Caldwell, D.R., Nash, J.D., Gunderson, G.D. (2002). Observations of bottom mixing over the continental slope. *Journal of Physical Oceanography*, **32**, 2113-2130

Munk, W.H., Wunsch, C. (1998). Abyssal recipes II: energetics of tidal and wind mixing. *Deep-Sea Research*, **45**, 1977-2010

Muzylev, S.V., Oleinikova, L.N. (2007). To the theory of internal waves under ice cover. *Okeanologiya*, **47**, 191-196 (in Russian)

Niwa, Y., Hibiya, T. (2001). Numerical study of the spatial distribution of the M₂ internal tide in the Pacific Ocean. *Journal of Geophysical Research*, **106**, 22229-22241

Osborn, T.R. (1980). Estimates of the local rate of vertical diffusion from dissipation measurements. *Journal of Physical Oceanography*, **10**, 83-89

Padman, L., Erofeeva, S. (2004). A barotropic inverse tidal model for the Arctic Ocean. *Geophysical Research Letters*, **31**, L02308

Pinkel, R., Munk, W., Worcester, P. (2001). Ocean mixing studied near the Hawaiian Ridge. *EOS Transaction of the American Geophysical Union*, **81**, 545-553

Pisarev, S.V. (1996). Low-frequency internal waves at the Arctic shelf edge. *Okeanologiya*, **36**, 819-826 (in Russian)

Plüddemann, A.J. (1992). Internal wave observations from the Arctic Experimental Buyo. *Journal of Geophysical Research*, **91**, 12619-12638

Polyakov, I.V., Dmitriev, N.E. (1994). The M₂ tide in the Arctic Ocean. 1. Structure of the barotropic tide. *Meteorologiya i Hidrologiya*, no. **1**, 56-68 (in Russian)

Polyakov, I.V., Dmitriev, N.E., Golovin, P.N. (1994). The M₂ tide in the Arctic Ocean. 2. Structure of the stratified ocean. *Meteorologiya i Hidrologiya*, no. **2**, 49-60 (in Russian)

Polzin, K.L., Toole, J.M., Ledwell, J.R., Schmitt, R.W. (1997). Spatial variability of turbulent mixing in the abyssal ocean. *Science*, **276**, 93-97

Proshutinsky, A.Yu. (1993a). Semidiurnal tides in the Arctic Ocean from modeling results. *Proceedings of the Arctic and Antarctic Research Institute*, no. **429**, 29-44 (in Russian)

Proshutinsky, A.Yu. (1993b). *Level oscillations in the Arctic Ocean*. St. Petersburg: Gidrometeoizdat (in Russian)

Rudnik, D.L., Boyd, T.J., Brainard, R.F. (2003). From tides to mixing along the Hawaiian Ridge. *Science*, **301**, 355-357

Sabinin, K.D., Stanovoy, V.V. (2002). Semidiurnal internal waves in arctic seas. In I.V. Lavrenov, E.G. Morozov, *Surface and internal waves in arctic seas*, 265-274, St. Petersburg: Gidrometeoizdat (in Russian)

Savchenko, V.G., Zubkov, L.I. (1976). A numerical model for free internal gravity waves in the Arctic Basin. *Proceedings of the Arctic and Antarctic Research Institute*, no. **332**, 26-50 (in Russian)

Schwiderski, E. W. (1990). On charting global ocean tides. Review of Geophysics and Space Physics, 18, 243–268

Serebryannyi, A.N. (2002). Internal waves in the coastal zone of the Barents Sea. In I.V. Lavrenov, E.G. Morozov, *Surface and internal waves in arctic seas*, 298-309, St. Petersburg: Gidrometeoizdat (in Russian)

Serebryannyi, A.N., Shapiro, G.I. (2001). Observations of internal waves in the Pechora Sea. In: *Experience in System Oceanograhic Investigations in the Arctic*, 140-150, Moscow: Nauchnyi Mir (in Russian)

Simmons, H.L., Jaine, S.R., St. Laurent, L.C., Weaver, A.J. (2004). Tidally driven mixing in a numerical model of the ocean general circulation. *Ocean Modelling*, **6**, 245-263

Smirnov, A.I., Shushlebin, V.G., Korostelev, V.G. (2002). Natural observation results of surface and internal waves in the Arctic Ocean. In I.V. Lavrenov, E.G. Morozov, Surface and internal waves in arctic seas, 280-297, St. Petersburg: Gidrometeoizdat (in Russian)

St. Laurent, L.C., Nash, J.D. (2004). An examination of the radiative and dissipative properties of deep ocean internal tides. *Deep-Sea Research*, *II* **51**, 3029-3042

St. Laurent, L.C., Stringer, S., Garrett, C., Perrault-Joncas, D. (2003). The generation of internal tides at abrupt topography. Deep-Sea Research, *I* **50**, 887-1003

Vlasenko, V.I. (1991). Non-linear model for the generation of baroclinic tidal waves over long irregularities of bottom topography. *Morskoy Gidrofizicheskii Zhurnal*, **6**, 300-308 (in Russian)

Zubov, N.N., (1950). *Basic studies on straits in the World Ocean*. Moscow: Izdatelstvo geograficheskoi literatury (in Russian)

In: Energy Research Developments
Editors: K.F. Johnson and T.R. Veliotti

ISBN: 978-1-60692-680-2
© 2009 Nova Science Publishers, Inc.

Chapter 3

TIDAL POWER- MOVING AHEAD

David Elliott[*]

Energy and Environment Research Unit, The Open University,
Milton Keynes MK 7 6AA, UK

Abstract

The idea of extracting power from the tides is not new, and some tidal barrages have been built across estuaries with large tidal ranges. However, in recent years interest has also emerged in smaller potentially less invasive impoundment options, known as tidal lagoons. In addition, a new concept has emerged, free-standing tidal current turbines, operating on tidal flows.

This paper reviews the state of play with tidal energy developments around the world, looking at barrages and lagoons, but covering tidal current turbine projects in more detail. It draws out some of the key design issues and assesses prospects for the future of tidal energy.

Introduction

The Tides can be exploited for energy collection in two basic ways. Firstly, by using the potential energy in the head of water that can be created behind a damn across an estuary, or in a bounded reservoir in an estuary, to drive a turbine- i.e. tidal barrages or lagoons. Secondly, by harvesting the kinetic energy in the horizontal tidal flows or streams, using tidal current turbines immersed directly in the flow.

Tidal barrages have been the main focus of attention until recently, but tidal lagoons and tidal current turbines have increasingly been portrayed as possibly preferable, in that, it is claimed, they are less environmentally invasive. It is also argued that, being smaller, and in the case of tidal current turbines, modular, they should be easier and quicker to deploy, so that investment costs will be lower.

---

[*] E-mail address: D.A.Elliott @open.ac.uk

There is something of a flurry of innovative activity underway in a number of countries in the tidal current turbine area at present, with forty or more devices of various designs being tested. But there is also still interest in tidal barrages and lagoons.

This paper provides an overview of the state of play with tidal energy developments around the world, looking in particular at progress with tidal current turbine projects. It then explores some of the key design issues and looks at the overall technological and economic prospects for the future use of tidal energy.

Tidal Barrages

Tidal energy is the result of the gravitational pull of the moon acting on the seas, modified by the lesser pull of the more distant sun. The regular rise and fall of the tides has been used for centuries in tidal mills of various designs, but it was only in the last century that electricity production was attempted, using barrages across river estuaries.

The largest tidal barrage so far built is the 240 Megawatt (MW) La Rance barrage in Brittany, France, which was commissioned in 1966. A few smaller projects have been developed e.g. in Canada (20MW) Russia (400 kilowatt) and China (500kW), and there have been many proposals for barrages around the world (Elliott 2004). Most recently, there have been proposals for some large projects in S Korea- a 254MW barrage on Sihwa Lake and an 812MW project at Ganghwa Island.

However the main emphasis has been on the Severn estuary in the UK, which has one of the world's largest tidal ranges. Many detailed studies have indicated that a 11mile long 8.6 Gigawatt (GW) rated barrage could be built between Weston super Mare on the English side and Lavenock Point on the Welsh side. It would generate around 17 Terawatt-hours (TWh) per annum, about 4.6% of UK electricity requirements. The cost however would be significant, around £15 billion ($30 billion). Given that the construction time would also be relatively long (up to 10 years), while it might be feasible as publicly funded project, it seems unlikely to be attractive in the current very competitive UK climate of privatised electricity generation and liberalised energy markets.

It has been estimated that at 2% public sector discount rate, the Barrage would generate at between 2.27 and 2.31pence/kWh (around 4.5 US cents/kWh), depending on how long it took to build, whereas at 10% private sector discount rate, the cost rises to 11.18-12.37p/kWh (around 23 c/kWh) (SDC 2007).

In addition to the high capital costs, there are problems related to the nature of tidal energy. Since the barrage would only operate during the (roughly) twice daily tidal cycles, its 8.6 GW turbine capacity could only offer the same output, averaged out over a year, as around 2 GW of conventional plant. In addition, given the shifting lunar phasing, its output would not always be well matched to the cyclic daily pattern of demand, thus reducing the value of the electricity it produced.

Two-way operation, on the tidal flow (in) as well as the ebb (out) is possible, and would extend the available output pattern to some degree, but having to install two-way turbines adds to costs, and, in the main, barrages are often seen as, in effect, fuel saving hydro damns filled by the tides. It is possible to use off-peak power from other sources to pump water behind the barrage to increase the head of water for later power generation. The barrage could thus be used a pump storage facility, although that would require either separate pumps, or reversible turbines. There have also been proposals for double barrages, with segmented basins allowing for pumping, using the power generated from the other basin, although that would of course add to the cost (Elliott 2004).

Whatever the actual mode of operation, since the barrage would in effect block off an entire estuary to natural tidal flows, it would have significant environmental impacts. For example, in the case of the Severn, the decreased tidal range would reduce the area of mudflats exposed each day, which would have major implications for wading birds. Some of the impacts, however, might be positive: the reduced turbidity would mean that silt would fall out of suspension and the clearer water would be more biologically productive, supporting more, and possibly new, species. That could offset the loss of feeding areas due to the reduced tidal range. However the impacts are complex and need detailed study, and so far most local and national environmental groups have been strongly opposed the idea (FoE 2004).

A study in 2007 by the governments advisory Sustainable Development Commission (SDC) nevertheless backed the Severn barrage concept, although with conditions, including full compliance with European Directives on habitats and birds. It also felt that it should be 'publicly led' as a project and publicly owned as an asset, to ensure the long-term public interest was protected. However it was left open as to who should pay- although a public-private finance arrangement seems to be one possibility (SDC 2007).

Although, given their high capital costs and cyclic outputs, the economics of barrages look problematic at present, they might be more attractive, as a low carbon energy source, if there were significant energy storage facilities. For example, it has been suggested that the electricity produced could be converted to hydrogen gas by electrolysis and this could be stored for later use, either in a gas turbine or a fuel cell, for the generation of electricity when needed.

There have been proposals for smaller barrages on the Severn and also on other estuaries around the UK, including the Mersey. Being smaller, they would be easier to finance and, depending on location, could have less environmental impact. However, while supporting smaller barrages, some environmental groups also favour what they see as a less invasive option than barrages- tidal lagoons

Tidal Lagoons

Whereas a barrage closes off a complete estuary, a bounded lagoon, built entirely out to sea, would have much less impact on tidal flows. The US company Tidal Electric has proposed lagoons at various sites around the world, including the UK. One proposed project is a 60MW lagoon off Swansea Bay, in Wales. The company is also in discussion with the Chinese government, which, it says, has expressed interest in a 300 MW offshore tidal lagoon to be built near the mouth of the Yalu River.

Lagoons would be in relatively shallow water and would be constructed like causeways with rock infill. As well as generating power, they might be seen as an offshore pumped storage facility. Moreover, as with the double barrage idea, segmented lagoons could enable phased operation and pumping between segments.

The disadvantage of lagoons is that the entire containment wall has to be constructed. By contrast, with barrages, the estuary shore provides free containment for the bulk of the water held behind it at high tide. Even so, it is claimed that, given that construction is expected to be easier and quicker than for barrages, costs for power from lagoons can be competitive (Atkins 2004).

In its review of tidal option, the SDC was however concerned that there '*was a lack of available evidence on the costs and environmental impacts, mainly due to the absence of any practical experience'*. But it ' *called on Government to support the development of one or more demonstration project, which would help provide real-life data on their economic and environmental viability.'* (SDC 2007)

Subsequently, the UK government announced that it would allow lagoons, and barrages, of up to 1GW capacity, to be eligible for support under the Renewable Obligation Certificate trading system, with two Renewable Obligation Certificates being offered for each MWh delivered. This is double the level of support compared with more developed renewables like on-land wind farms. The justification for this extra cross-subsidy was that, although barrages and lagoons were capital intensive, they would both continue to operate for decades after the initial costs had been paid back, so providing some extra support was reasonable. As yet no projects have come forward in response, but it seems likely that some will.

A government decision of the much larger Severn Barrage is still awaited. In 2008, the government set up a feasibility study to look into the financial and environmental implications of larger tidal barrages, and also tidal lagoons. It is expected to report in 2010, in conjunction with a public consultation exercise.

Tidal Turbines

While plans for new Tidal Barrages are still being considered, and Tidal Lagoons are still at the conceptual stage, there are many tidal current turbine projects underway or planned in the UK and elsewhere (NATTA 2007). This is not surprising since they are modular and can be installed individually, and are much less invasive then either lagoons or barrages. Most can also easily operate in two-way mode, e.g. by swivelling the rotors around when the tidal direction changes. In which case the system can operate on four tidal movements every 24 hour cycle, instead of just two, as with simple ebb barrages or lagoons.

The UK has led the way in this field. Following work by IT Power on a 15kW pontoon supported prototype in the 1990's, in 2003 its offshoot, Marine Current Turbines Ltd (MCT), installed a 300kW prototype Seaflow machine, essentially a propeller-like device, off the North Devon coast. MCT are installing a larger two rotor 1.2 MW rated commercial version , 'SeaGen', in Strangford Narrows, Northern Ireland. After that the aim is to move up scale to tidal farms with seven machines. The company is looking at possible sites off Wales and N Devon and also possibly Guernsey.

There are many other designs under test. Propeller type designs have dominated so far, with in the case of the MCT devices, the rotors mounted on piles driven into the sea bed. However some designers have tried to get enhancement of the tidal flow using annular ducts around the rotor, as with Lunar energy's device, which sits on the sea bed. There are plans for an eight machine tidal farm off the Welsh coast using this system.

There have also been some novel designs, like the 'Open Centre Turbine' developed by Dublin based Open Hydro. This has blades running in a circular cam ring, which also acts as the generator, thus avoiding gearing. A 250 kW prototype has been tested at the European

Marine Energy Centre on the Orkneys in Scotland, and there are plans for a version to be installed in Alderney in the English Channel.

Although there may be advantages with ducted systems or complex annular rim systems, these configurations have to use two-way turbines if they are to extract power from all four tidal movements, which may increase costs and reduce their efficiency. It may be easier to have rotors which can be turned 180 degrees to face the changed tidal flow direction, as with MCT's Seaflow and Swansea University's 350kW Swanturbine propeller unit. SMD Hydrovisions' twin rotor 500kW 'Tidel' device is held mid stream by tethers between a floating pontoon and an anchor, so it can swivel through 180 degree when the tide changes direction .

However there may also be cost attractions in having fixed units mounted on the sea bed, as with Tidal Hydraulic Generator Ltd's rotor array, Tidal Generation's 500kW sea-bed mounted propeller-type prototype, and Robert Gordon University's 150kW Sea Snail, which has a propeller unit held in place on the sea bed by hydroplanes.

As can be seen, horizontal-axis propeller-type systems dominate. So far, despite the attraction of being able to operate multi-directionally, on both tidal flows and ebbs, free standing vertical-axis designs have not been favoured. However, vertical-axis rotors have been used in the sea bed mounted fixed-duct Proteus system developed by Neptune Renewables and the University of Hull, for use in shallow water.

Rather than using vertical or horizontal axis rotors, another approach is to make use of hydroplanes designed to move up and down in the tidal flow. This idea was developed by The Engineering Business in Northumbria, with a 150kW Stingray prototype tested in the Shetlands Isles in Scotland in 2002. Subsequently, an oscillating hydroplane system called Pulse Tidal has been developed and tested at the University of Hull, and a 100kW prototype is being tested in the Humber estuary.

It should be clear from the sample above that many new ideas are being explored in the UK. Given the it has the largest tidal stream resource in the country, it is not surprising that Scotland has played a major role. Orkney- based Scotrenewables is testing a free-floating rotor, and Tidal Stream have developed a cantilevered multi-turbine 'Semi-Submerged Turbine' system, which allows the rotor blades to be swivelled out of the water for maintenance. It is designed for use in the Pentand Firth. Another Scottish company have developed the StarTider, a novel multi-rotor concept, with arrays of rotors mounted on common axles. In addition, a Scottish consortium has developed the Tidal Delay system, with a turbine which feeds power to a heat store, which is used to raise steam for continuous power generation. That opens up a new approach: a system that can directly compensate for the cyclic availability of tidal energy.

Tidal Current Turbines around the World

There are many tidal current turbine projects underway around the world, in *North America* in particular. For example, six of Verdant Powers three bladed tidal turbines have been on test in New York City's East River near Roosevelt Island, as part of a 10MW $7m project, and the Massachusetts Tidal Energy Company is looking at the possibility of installing up to 150 2MW devices at Vineyard Sound on the New England coast. Meanwhile, Ocean Renewable Power Company is deploying its OCGen tidal turbine in Eastport, Maine.

Moreover, following test with a prototype, the UEK Corporation, based in Annopolis, has plans for a 10MW project at the mouth of the river Delaware, using a series of its two-way 45kW Underwater Electric Kite ducted turbines.

Although the Gulf stream is not the result of lunar gravitational forces, but rather is part of the conveyor system of planetary ocean currents, driven ultimately by solar heat, there is certainly a lot of energy there, and it has been see as an key resource that could be tapped by turbines. Florida Atlantic University has received $5m to establish The Florida Center of Excellence in Ocean Energy, which will focus on South Florida's ocean currents, specifically the Gulf Stream. Several ideas have already emerged in this field. Dr Alexander Gorlov has been developing a novel helical vertical-axis turbine device for use with ocean currents, including the Gulf stream, and Florida Hydro Power and Light, aided by the US Navy, have been developing an 'open center' turbine design.

Further out in the Atlantic, an undersea rotor design has been developed by the Bermuda Electric Light Company and Bermuda Current to Current Ltd, who are planning to install a US built 20MW unit.

On the US West coast, Pacific Gas & Electric Company has signed an agreement with the City and the County of San Francisco, as well as Golden Gate Energy Company, to assess possibilities for harnessing the tides in San Francisco Bay. UK developer Hydro Venturi has also been looking at the possibility of using its innovative venturi turbine device in that location.

Canada has been involved with tidal barrage power for many years, as noted above, and it is now looking to tidal current systems. The pioneering Canadian company Blue Energy has been testing a vertical axis tidal turbine and has plans for very large 'tidal fence' arrays around the world, with turbines mounted in a series of modules forming a causeway. In parallel, a new-comer, Clean Current Power, has a ducted rotor system that sits on the sea bed.

A major focus for many Canadian and overseas projects is the Bay of Fundy, which has a very large tidal range. It is to be the location for test with the Irish Open Hydro device mentioned earlier, in conjunction with Nova Scotia Power (NSP), who already have a 20 MW tidal barrage there. If all goes well with the tests, NSP are hoping to install up to 300 tidal turbines in the area, with a total capacity of 300MW. In addition, the UK company Marine Current Turbines Ltd (MCT) has signed an agreement with Canada's Maritime Tidal Energy Corporation to install one of its 1.2 MW SeaGen devices in the Bay of Fundy. Clearly this location is proving an attractive area for tidal projects: the Atlantic Tidal Energy Consortium says it wants to install around 600MW of tidal systems in the Bay of Fundy over the next 10 years.

However there is also a significant resource on the west coast. In addition to its Bay of Fundy project, MCT have signed a co-peration agreement with Canada's BC Tidal Energy Corporation to deploy at least three of its SeaGen turbines in waters off Vancouver, British Columbia. They should be installed on Vancouver Island's Campbell River by 2009. The agreement is the first step in a plan to develop larger tidal farms off British Columbia's coast, which the company says has a tidal potential of up to 4GW

There are also proposals for tidal schemes in the *Pacific* area. Neptune Power in New Zealand is planning an array of 1 MW floating sub-sea turbines in the tidal currents off the Cook Strait between the North and South Islands, and CREST Energy also have plans for a major $400m 300MW project at Kaipara Harbour near Auckland.

In parallel, Renetec, a Korean renewables development company, have agreed a joint venture with Voith Siemens Hydro for a 600MW 'Seaturtle' tidal current turbine project in the South Korean province of Wando. This has a rotor mounted in a vertical frame on a moveable horizontal bar.

While the UK dominates in *Europe*, Norway has also been very active. A large propeller-type device, fixed to the sea-bed, has been tested at Kvalsund, and Hammerfest Strom and Statoil have plans for a full scale unit. In addition, a large novel floating 'Morild' multi-rotor tidal project is being developed by Hydra Tidal with support from Statkraft.

In France, Hydrohelix Energies have a 10kW demonstration ducted-rotor sea-bed mounted project, and plans for a 1MW Marenegie programme in Brittany, while a German company, Atlantis Strom, has developed a novel sea bed mounted horizontal axis rotor. In the Netherlands, tests have been underway on the two bladed Tocardo turbine.

The EU backed Enemar project in Italy has led to test with a novel vertical-axis Kobald turbine in the straits of Messina, between Sicily and mainland Italy. Interests has been shown in this system by China, which is looking at possible project in the Straits of Jintang in the Zhoushan Archipelaga.

Finally, Russia has also continued its involvement with tidal power, via experiments in the Arkhangelsk region with a 1.5MW turbine developed by Hydro WGC and mounted in a floating pontoon.

The Prospects for Tidal Power

Tidal energy is now becoming recognised as a major potential renewable energy source. It has been estimated that, if all the potential tidal barrages sites around the world were developed, they might supply around 300TWh p.a (Elliott 2004). Location is clearly a crucial issue, in terms of both access to the energy resource and links to power grids. Fortunately, the areas around most large estuaries in the industrial world are relatively highly populated, so there are major potential loads nearby and less need to provide major new gird links over long distances.

In some cases tidal lagoons might be more appropriate as an alternative to barrages, but in others they might be seen as additional options. For example it has been estimated that the UK might expect to obtain up to 8% of its power from lagoons, in addition to around 15% from barrages, although in some locations, e.g. in the Severn estuary, depending on the siting, there could conflicts between them.

As noted above, given the lunar cycles, large barrages and lagoons would at times deliver large bursts of electricity, which, in the absence of major energy storage facilities, may not be well matched to local or national demand patterns. Clearly if there were several barrages and/or lagoons around the coast, all feeding into the grid network, then their aggregate output would be somewhat more continuous, given the different phasing of the tides around the coast. So the system would provide outputs of higher value. However, this benefit could be much more marked if large numbers of tidal current turbines, sited in wider range of locations, were linked to the grid (Sinden 2005).

Suitable locations for tidal current projects may be somewhat less accessible than for barrages and lagoons. Flow rates of around 5m/s are seen as necessary, and these usually only occur in areas where there are natural constrictions in the coastal topography, and these are often relatively remote from centres of population. However, the resource is quite large. The DTI/Carbon Trust has put the world's total practical tidal current turbine resource at around 800TWh p.a., with the EU resource being around 48TWh p.a. (DTI/Carbon Trust 2004).

Tidal barrage technology, and to some extent, tidal lagoon technology, is well developed, with the turbines being similar to those used in hydro projects. There may be some cost cutting civil engineering innovations available in terms of barrage or causeway construction, particularly in the case of lagoons, but otherwise these are relatively mature technologies. By contrast, Tidal current turbine technology is at a relatively early stage of innovation, with many rival designs and concepts emerging- much as in the early phase of wind energy development. As with wind power, significant cost reducing improvements can be expected.

That of course assumes continuing development and continuing financial support- this is a new technology which needs backing to reach maturity. Fortunately, that does seem to be increasing. As indicated, the tidal current projects around the world include some in the USA, where marine power generally now seems increasingly to be seen as a significant option. For example, in 2007, the US House of Representatives Science & Technology Committee backed legislation that would invest approximately $200m in federal funds for RD&D on wave energy over four years. Representative Darlene Hooley (Democrat-Oregon), commented 'Similar to how we helped the wind energy industry get off the ground by providing production tax credits. Congress today made the first significant investment in what is projected to be a promising nonpolluting energy source'. One might hope the tidal energy might also be seen in a similar light. .

The Irish Republic has also recently shown renewed interest in ocean energy, with a Euro 26 allocation for a three years R&D and support programme, although the emphasis seems to be on wave energy, a major potential source for Ireland. .

For the moment, the UK still seems to be well placed in the tidal field. It could take a lead with large barrages and lagoons, if those options are backed by the new government review, and with smaller barrages and lagoons, given the availability of support from the Renewables Obligation system. Moreover, given that tidal current projects can also apply for both capital and revenue support under the UK governments £42m Marine Renewables Deployment Fund, the UK seems likely to continue to dominate in the tidal current field.

The UK's Marine Fund offers eligible tidal current turbine projects grants for up to 25% of the capital cost, up to £5m, and 10p/kWh in revenue support for the first five years. This is on top of the support provided by the UK Renewables Obligation, which is in the process of being adjusted to allow new projects like tidal current systems to receive two Renewable Obligation Certificates (ROCs) for each MWh produced- the same as has also now been agreed for tidal barrages and tidal lagoons under 1GW. .

The value of these ROCs varies, but in effect the adjustment means that projects might get around 8p/kWh extra for the power they produce. In addition, the Scottish government runs a separate grants scheme, as well as its own Marine Supply Obligation support system, for wave and tidal current projects. Scotland has the largest tidal resource in the UK, and is the home of the European Marine Energy Centre, located on the Orkney Islands, which provides testing facilities. It has been estimated that there are sites for Scotland which could be used to install up to l0GW of tidal current turbine capacity, generating around 34TWh p.a..

Conclusions: Issues and Options

The above is far from a complete survey of the many tidal projects underway, but even so, it is clear that a lot of innovation is going on. There are powerful economic motivations. One study suggested that the annual world market for tidal technology might be £155-444bn (ReFocus 2005).

In terms of actual developments, as has been indicted above, while there is interest in tidal barrages and lagoons around the world, with some significant projects being likely to materialise, in recent years tidal current turbines have caught the imagination of many more engineers, and some large developments seem likely to go ahead.

Although the situation now looks quite positive, it has taken some time to build support for tidal current work in the UK. For many years tidal power, like wave power, was treated as a marginal option (Elliott 2007). However, with the issue of climate change being taken more seriously, it is perhaps not surprising that the UK is now pushing ahead quite strongly on tidal current turbine power, as well as wave power. The UK is well placed to do so having extensive marine and offshore engineering expertise to draw on, and a major marine energy resource. A study by ABPmer for npower Juice suggest that ultimately about 36 gigawatts of tidal current device capacity might in theory be installed in the UK. It estimated that devices in under 40m depth might generate around a total of up to 94 TWh per annum- about a quarter of current UK electricity requirements (ABPmer 2007).

However, it may not be possible to exploit all of that, and certainly it could take time. The DTI/Carbon Trusts' Renewables Innovation Review in 2004 put the UK's total practical tidal current resource was put at around 31TWh p.a., about 10% of UK electricity requirements, while a subsequent study by the Carbon Trust, taking economic and other constraints into account, put the tidal figure at only 18TWh p.a. (Carbon Trust 2006).

These estimates may be pessimistic. For example it has been claimed that the UK tidal current resource has been seriously underestimated, by perhaps a factor of 10 or maybe more (MacKay 2007).

There have been some useful estimates of tidal stream resources in the UK, EU and around the world (DTI 2004; Hardisty 2007, 2008), but estimates of what might actually be obtained overall are perhaps inevitably fluid when a technology is at a relatively early stage of development. It is not clear which systems will be successful in the long run, much less where they can best be sited and how much of the resource they can harvest, and at what cost.

Whereas enthusiasts obviously hope for rapid progress, it will take time to get the technology tested and for these issues to be resolved Indeed, as the UK government's Renewables Advisory Board has noted in a review of wave and tidal current project, progress on developing reliable commercial-scaled devices has been slower than was hoped, in part due to the challenging marine environment (RAB 2008). As a result, as yet, no projects had met the eligibility criteria for support under the Marine Renewables Deployment Fund. Clearly there is still some way to go technologically, although as RAB point out, a lot of good experience has been gained.

As indicated above, as in the wind energy field, technologically, the most successful tidal current devices so far seem to be based on standard horizontal-axis propellers, sometimes with ducts. However, some novel vertical-axis rotor designs have also emerged, as well as other novel designs, like the Open Centre turbine and the Pulse Tidal hydroplane system. In addition some designs have emerged which combine tidal current turbines and wind turbines mounted on the same floating platform. Not all areas with good tidal currents will necessarily have good wind speeds, but there could be cost advantages in sharing platforms and marine power cable links to shore. In some locations, combining tidal, offshore wind and wave energy collection may also be an option, as has been proposed by a UK developer (Freeflow 2008)

Since getting access to the devices will be limited by the weather, and using divers will be costly and risky, a key design issue is ensuring ease of access to blades, gearboxes and generators for maintenance. Some devices have gearboxes and/or generators mounted above water, and some of the floating systems can be towed back to harbour, but for those with fixed installations, means have to be provided to lift the blades and generators out of the water. Most devices so far have mechanical systems to do this using their support towers or frames. But low maintenance designs will obviously have major attractions.

Another key issue is environmental impacts. As indicated earlier, this is a major concern for tidal barrages and may also prove to be of some significance for tidal lagoons. By contrast, most studies of tidal current turbines so far have suggested that impacts will be low. Even large arrays will not impede flows significantly and the rotor blades will turn slowly, slower than wind turbines, and much slower than the turbines in barrages and lagoons, and so should not present a hazard to marine life. However all structures put in the sea will have some impact, and this needs to be carefully assessed when considering possible locations.

Perhaps the key issue is cost. Once built, tidal barrages may be able to generate at reasonably competitive generation costs, but their large capital costs makes it hard to see them as viable, except as public sector led projects. By contrast, tidal lagoons might be more viable as private investments.

At present generation costs for tidal current turbines are relatively high, but this is for prototype devices. Some of the device teams have claimed that they can get prices down relatively quickly, if given the right support, with talk of getting to 3-4p/kWh for serially produced devices and then lower once the commercial market expanded.

In its 2006 review, the Carbon Trust used learning curve analysis to identify 'learning rates' for tidal current turbines, based on the fact that costs fall as experience in installing increasing amounts of capacity grows and as economies of volume production are achieved. The slope of the so-called 'learning curve' produced when kWh price reductions are plotted against kW capacity installed, with both put on logarithmic scales, was 5-10% for tidal current turbine technology. On this basis, they suggested that generation costs might reach 2.5p/kWh, competitive with current gas-fired combined cycle turbines, by the time around 2.8 GW of capacity was in place (Carbon Trust 2006).

The large potential of tidal energy has been talked about for many years, but the economics have often been seen as challenging. Now, with increasing concerns about climate change and energy security, it seems that priorities may be changing and more effort is being put into developing new technologies. While interest in barrages and lagoons is increasing, the rapid development of tidal current technology could well mean that it will become the leader in the field.

References

ABPmer. (2007) *Quantification of Exploitable Tidal Energy within the UK waters*. Report for npowerjuice/BritishWind Energy Association, APBmer consultancy, London. July. http://www.abpmer.co.uk/allnews1623.asp

Atkins. (2004) *Feasibility Study for a Tidal Lagoon in Swansea Bay*. Report for Tidal Electric, Atkins Consultants Ltd., Epsom , September

Carbon Trust. (2006) *Future Marine Energy: Results of the Marine Energy Challenge: Cost competitiveness and growth of wave and tidal stream energy. London*. The Carbon Trust.

DTI. (2004) *Atlas of UK Marine Renewable Energy Resources*. Technical Report, London. Department of Trade and Industry.

DTI/Carbon Trust. (2004) *Renewables Innovation Review*. London. Department of Trade and Industry and the Carbon Trust.

Elliott, D. (2004) Tidal Power. In Boyle, G. (ed) *Renewable Energy* Oxford. Oxford University Press/ Open University

Elliott, D. (2007) Sea Power - how we can tap wave and tidal power. In D. Elliott (ed) *Sustainable Energy: Opportunities and limitations.* Bassingstoke. Palgrave.

FoE Cymru (2004) A Severn barrage or tidal lagoons? A comparison Briefing. Cardiff. Friends of the Earth Cymru. January http://www.foe.co.uk/resource/briefings/severn_barrage_lagoons.pdf

Freeflow. (2008) Hybrid tidal, wind and wave concepts: http://www.freeflow69.com/

Hardisty, J. (2007) Assessment of Tidal current resources: case studies of estuarine and coastal sites. *Energy and Environment* Volume 18 No.2 , pp233-249. See also Hardisty, J. (2008) *The Analysis of Tidal Stream Power*. London. Wiley, forthcoming

MacKay. D (2007) Under-estimation of the UK Tidal Resource. Cambridge. Cavendish Laboratory, University of Cambridge. Web paper: http://www.inference.phy.cam.ac.uk/mackay/

NATTA. (2007) *Wave power and Tidal current turbines : a review of progress*. Compilation of reports from Renew on Marine Renewables Vol II. Milton Keynes. Network for Alternative Technology and Technology Assessment.

RAB. (2008) *Marine renewables: current status and implications for R&D funding and the Marine Renewables Deployment Fund*. London. Renewables Advisory Board.

ReFocus. (2005) Refocus Marine Renewable Energy Report: Global Markets, Forecasts and Analysis 2005-2009, Douglas-Westwood Ltd, London. Elsevier.

SDC. (2007) *Turning the Tide*. London. Sustainable Development Commission.

Sinden, G. (2005) *Variability of Wave and Tidal Stream Energy Resources*, Environmental Change Institute report, Oxford University.

In: Energy Research Developments
Editors: K.F. Johnson and T.R. Veliotti

ISBN: 978-1-60692-680-2
© 2009 Nova Science Publishers, Inc.

Chapter 4

New Solid Medium for Electrochemistry and its Application to Dye-Sensitized Solar Cells

Masao Kaneko

Faculty of Science, Ibaraki University; 2-1-1 Bunkyo,
Mito, 310-8512 Japan

Abstract

A tight and elastic polysaccharide solid containing excess water was proposed as a solid medium for electrochemistry. Conventional electrochemical measurements could be performed in this solid the same as in liquid water. Diffusion coefficients (D_{app}) of small molecules and ions in this solid were almost the same as in a liquid. Charge transfer resistance (R_{ct}) as well as double layer capacitance (C_{dl}) at the interface of the electrode/polysaccharide solid was similar as that of electrode/water interface. Ionic conductivity in this solid was the same as in liquid water. It was found that bulk convection does not take place in this solid, which allows discrimination of molecular diffusion from bulk convection. This solid was successfully applied to solidify the redox electrolyte solution of a dye-sensitized solar cell giving almost similar light-to-current conversion efficiency as that of a liquid type cell.

Introduction

It has well been known that hydrophilic polymers form a hydrogel that contains excess water inside [1]. Such hydrogels have been used for water adsorbent, electrophoresis, immobilizing enzymes and medicines, etc. However, diffusion of molecules and ions in these gels has scarcely been studied due to the lack of suitable methodology. The present author has found that a tight and elastic polysaccharide solid containing excess water can be used as a solid medium for electrochemical measurements the same as liquid water, and that diffusion of molecules and ions takes place in this solid the same as in a liquid [2-6]. This allows the solid to be used not only as a medium for electrochemistry, but also as a solid reactor for various chemical reactions. It is of further interest that in this solid bulk convection does not take place [6] meaning that molecular diffusion can be discriminated from bulk convection that should always exist on the earth due to the gravity.

Recently a dye-sensitized solar cell is attracting a great deal of attention to convert solar energy into electricity [7]. Since this cell uses redox electrolyte solution, it is of importance to solidify the liquid in order to stabilize the cell, but the task is not easy to achieve. To overcome this problem solidification of the organic redox electrolyte solution by molten salts and gelator [8] or by polymer film [9,10] has been achieved. We have successfully used the polysaccharide solid to solidify the electrolyte solution [11,12].

In the present review the fundamental properties of polysaccharide solids containing excess water are at first explained (section 2). In the section 3 the characteristics of the polysaccharide solids as media for electrochemistry will then be described in detail. A novel property of the polysaccharide solid will be shown wherein bulk convection does not take place (section 4). The successful application of the polysaccharide solid to solidify the redox electrolyte solution of a dye-sensitized solar cell will be introduced in section 5 followed by future scopes of the solid medium (section 6).

Properties of Polysaccharide Solids Containing Excess Water

The typical polysaccharides reported in the present review are agarose (1) and κ-carrageenan (2). It has well been known that polysaccharides form a tight and elastic solid containing excess water [1,13-15].

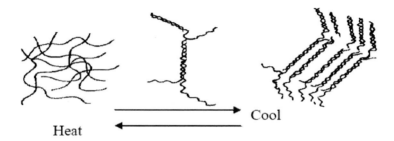

1 (Agarose)   2 (κ-carrageenan)

Gelation mechanism of agarose, for an example, is shown in figure 1 [14].

Figure 1. Gelation mechanism of agarose [14]:On cooling a hot solution, the chains form helical structures which then aggregate to double helix, further aggregating to a bundled structure. The double and/or bundled helical structures act as a bridging point for 3D network structure.

On cooling a hot aqueous solution, the polysaccharide chains form helical structure which then aggregate to double helix. This double helical structure further aggregates to a bundled structure. The double and/or bundled helical structures act as a bridging point for a three-dimensional (3D) network that can contain excess water within the network [1,13-15].

The scanning electron micrograph of agar-agar (major component is agarose) is shown in figure 2 [14]. For the lower molecular weight sample (lower strength sample), the entangled structures are localized and week, but for the higher molecular weight sample (higher strength sample) the bridged network structures are delocalized and strong.

MW=370,000

MW=30,000

Figure 2. Scanning electron micrograph of agar-agar for higher and lower molecular weight (MW) samples [14].

Since κ-carrageenan involves sulfonate anionic groups, the helical structures are bundled by the presence of cations that induce aggregation of helixes by electrostatic interaction.

Very tight polysaccharide solids containing excess water could be obtained by applying microwave very carefully when preparing its aqueous solution [2]. The hardness of polysaccharide solids containing excess water is shown in figure 3.

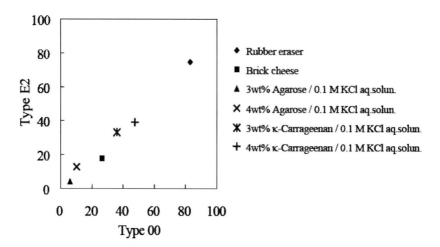

Figure 3. The hardness of the polysaccharide solids measured by the hardness meter (Type 00 and Type E2)[5].

Since the hardness is not an absolute value, it was compared with materials around us. The hardness of a 2 wt% agarose solid was one third of a conventional rubber eraser, and that of a 3 wt% κ-carrageenan involving 0.1 M KCl (M = mol dm^{-3}) half of a rubber eraser.

It is interesting that the solid surface is superhydrophilic: the contact angle with water is almost zero degree. The water inside of the solid is evaporated the same as a liquid water, and after standing under ambient conditions the solid looses all the water to become a very hard and dry solid.

Solid Medium for Electrochemistry

Overview for Electrochemistry in Solid

The electrochemical behavior of redox molecules in polymer films and gels has been investigated [16-25], but such behavior has usually been studied by using a modified electrode coated with a polymer film or gel in the presence of an outer electrolyte solution. For a few examples, entirely solid state voltammetry was also achieved, but by using a microelectrode array [23,24] composed of working, counter and reference electrodes because of a slow ionic or molecular diffusion in the solid matrices. The apparent diffusion coefficient (D_{app}) of a redox substrate in the films or solids coated on an electrode was very small [17-25], usually of the order of less than 10^{-7} cm^2 s^{-1}, and the largest value was of the order of 10^{-6} cm^2 s^{-1} at a carrageenan film (2 mm thick) coated Pt gauze electrode in the presence of an outer aqueous electrolyte solution [16]. Another example of solid state votammetry is a report on the electrochemistry of Prussian blue in silica sol-gel electrolytes [26], but Pt gauze working and counter electrodes for a 1 mm thick silica solid were used; the system was different from a conventional three electrodes electrochemical measurement. Moreover, it is well known that solid electrolytes have been used for various sensors, electrochromic devices, etc. [27,28]. However, in spite of these activities in solid electrolytes, there has been almost no work, to our knowledge, using a solid medium for conventional electrochemical measurement with an ordinary three electrode system. There has been only one report [29] that used agarose gel (1 wt%) for an electrochemical measurement; in this report only a gold wire working electrode was used, and a much lower diffusion coefficient of ferricyanide (1/2.6) in the agarose gel was obtained compared to that in an aqueous solution. If electrochemical measurements using ordinary electrodes could be carried out in a solid medium in the same way as in a pure liquid, a new kind electrochemistry and electrochemical measurements could be initiated.

In the present section 3, tight and elastic polysaccharide solids were used as solid electrolyte media for very conventional three electrode electrochemical measurements. The fundamental electrochemical properties of the polysaccharide solid were investigated by electrochemical impedance spectroscopy including diffusion of redox molecules, charge transfer resistance at the electrode|solid interface, and double layer capacitance on the electrode surface.

Electrochemical Characteristics in Polysaccharide Solid Media

Cyclic voltammograms of 5 mM $Fe(CN)_6^{3-}$ in a 2 wt% agarose solid, in a 2 wt% κ-carrageenan solid, and in an aqueous solution containing 0.5 M KCl are shown in figure 4.

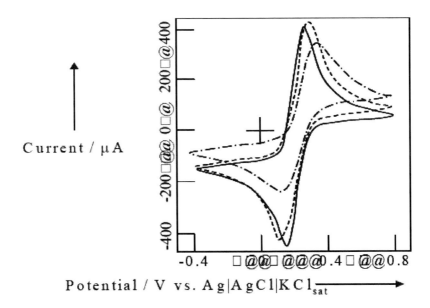

Figure 4. Cyclic voltammograms of 5 mM $Fe(CN)_6^{3-}$ in 2 wt% agarose (dash-dotted line) [3]; 2 wt% κ-carrageenan (dashed line) and aqueous solution (solid line) containing 0.5 M KCl in the range from –0.4 to 0.8 V (vs. Ag|AgCl|KCl$_{sat}$). Working electrode is ITO, and counter electrode Pt. Scan rate, 20 mV s^{-1}.

The voltammograms in the agarose andκ-carrageenan solids show very similar features as in liquid water including the redox potential, and peak currents, but with a slightly larger peak separation for the solid systems. The solid was so tight and stable that CV could be measured without any outer cell or vessel.

Impedance spectra of a 5 mM $Fe(CN)_6^{3-}$ were measured in 2 wt% agarose and κ-carrageenan solids, and in an aqueous solution containing 0.5 M KCl at the rest potentials (figure 5).

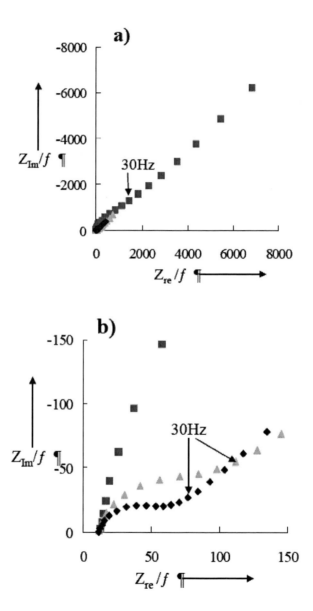

Figure 5. Impedance spectra of 5 mM $Fe(CN)_6^{3-}$ in 2 wt% agarose (squares), 2 wt% κ-carrageenan (triangles), and aqueous solution (diamonds) containing 0.5 M KCl at the rest potentials [3]:. Frequency range is from 20 kHz to 100 mHz (25 data points per spectrum).

All the spectra (figure 5a) show liner relations characteristic for a diffusion-controlled process in the bulk phase at low frequencies (< 30 Hz). However, at high frequencies (figure 5b) semi-circles are clearly seen in theκ-carrageenan solid and in an aqueous solution, while the impedance spectrum in the agarose solid is slightly different showing a larger semi-circle. These semi-circles at high frequencies are determined both by the capacitance (C_{dl}) in the double layer and charge transfer resistance (R_{ct}) at the electrode|solution interface. The estimated equivalent circuit for the impedance spectra of figure 5 is shown in figure 6 based on the Randles circuit. The solution resistance (R_s) was determined from the equivalent circuit, and then D_{app} of the $Fe(CN)_6^{3-}$ was calculated.

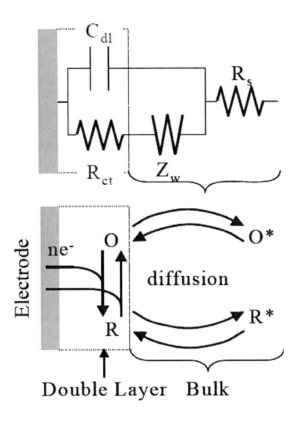

Figure 6. Electrochemical redox system along with its equivalent circuit [3].

The concentration dependence of D_{app} of $Fe(CN)_6^{3-}$ determined by EIS is shown in figure 7 a). D_{app} was estimated also by a PSCA method under a potential step from -0.2 (vs. Ag|AgCl|KCl$_{sat}$) to +0.5 V and is shown in figure 7 (b). It is remarkable that the diffusion of the $Fe(CN)_6^{3-}$ takes place in the solid to the same extent as in an aqueous solution. The D_{app} values from both measurements (figure 7 a) and b)) were almost the same, and were almost independent of the $Fe(CN)_6^{3-}$ concentration. The D_{app} obtained by EIS was slightly lower than that by PSCA; since in the PSCA measurement direct current is used in the potential step, a concentration gradient generated in the double layer would cause slightly higher D_{app} value. The D_{app} of $Ru(bpy)_3^{2+}$ in the 2 wt%κ-carrageenan solid (4.5×10^{-6} cm^2 s^{-1}) was about 70% of the D_{app} in the 2 wt% agarose solid (6.7×10^{-6} cm^2 s^{-1}). This result indicates that the diffusion of $Ru(bpy)_3^{2+}$ in theκ-carrageenan solid is suppressed slightly by the anionic groups of the κ-carrageenan.

The contact angle of water on the present solid surface was almost zero showing that the surface is super-hydrophilic. The concentration dependence of the charge transfer resistance (R_{ct}) at the electrode|solid interface is shown in figure 8 a).

It is remarkable that the R_{ct} in the 2 wt% κ-carrageenan solid was almost the same as that in the aqueous solution, athough the R_{ct} in the 2 wt% agarose solid was much larger than that in the aqueous solution. In spite of the high hydrophilicity of the solid surface, there could be some contact problem between the electrode and the agarose solid. As for the κ-carrageenan, the sulfonate anionic groups would improve the contact between the electrode and the solid.

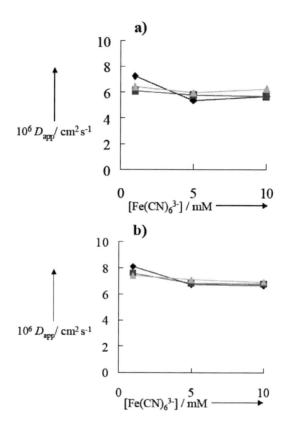

Figure 7. (a) Concentration dependence for the diffusion coefficient (D_{app}) of $Fe(CN)_6^{3-}$ in 2 wt% agarose (squares), 2 wt% κ-carrageenan (triangles), and aqueous solution (diamonds) containing 0.5 M KCl at the rest potential [3]. (b) D_{app} based on Cottrell's equation.

The dependence of the double layer capacitance (C_{dl}) on the $Fe(CN)_3^{6-}$ concentration is shown in figure 8 b). In the aqueous solution, C_{dl} tends to increase with the $Fe(CN)_6^{3-}$ concentration, but for the solid system C_{dl} values are only weakly dependent on the $Fe(CN)_6^{3-}$ concentration. At 5 mM $Fe(CN)_6^{3-}$ concentration, C_{dl} values of the solids are similar to that in the aqueous solution.

The polysaccharide concentration dependence of D_{app} in the agarose and κ-carrageenan solids containing 10 mM $Fe(CN)_6^{3-}$ is shown in figure 9.

The D_{app} in an aqueous solution containing 10 mM $Fe(CN)_6^{3-}$ is also shown (on the y-axis) in figure 9. The D_{app} values in theκ-carrageenan solid are even higher than that in the aqueous solution. It can clearly be seen that the polysaccharide concentration does not greatly affect the molecular diffusion in the solid. Figure 9 shows the plots of charge transfer resistance (R_{ct}) versus polysaccharide concentration for the 10 mM $Fe(CN)_6^{3-}$ in the agarose and κ-carrageenan solids. The plot on the vertical line in figure 9 shows the R_{ct} (32.6Ω) in the aqueous solution containing 10 mM $Fe(CN)_6^{3-}$. The R_{ct} in the κ-carrageenan solid is of the same order of magnitude as in an aqueous solution and even lower in the 4 wt% κ-carrageenan. In figure 8 it was shown thatκ-carrageenan anionic groups improve the charge transfer between the electrode and the redox compound in the solid. A high concentration of the $-SO_3^-$ groups would facilitate charge transfer between the electrode and the redox compound. It is surprising that the R_{ct} decreased with increasing polysaccharide concentration in the solids. The reason is not clear yet, but one reason might be a technical problem; since a higher polysaccharide concentration causes a higher viscosity of the polysaccharide solution when it is warm, the insertion of the electrodes into this solution could give a better and more stable contact between the electrodes and the medium. In the next chapter 4 it will be shown that the ionic conductivity also increases with the polysaccharide concentration (later in figure 11). This fact and the results of figure 9 for the R_{ct} indicate that the increase of the polysaccharide concentration does not increase the resistance for molecular/ionic diffusion probably because the increase of the polysaccharide concentration induces the growth of the chain aggregation rather than increasing the crosslinked structures that can be a resistance for molecular/ionic diffusion.

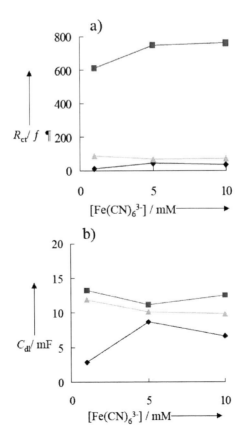

Figure 8. Concentration dependence of (a) the charge transfer resistance (R_{ct}) and (b) the double layer capacitance (C_{dl}) of $Fe(CN)_6^{3-}$ in 2 wt% agarose (squares), 2 wt% κ-carrageenan (triangles) and aqueous solution (diamonds) containing 0.5 M KCl at the rest potential [3].

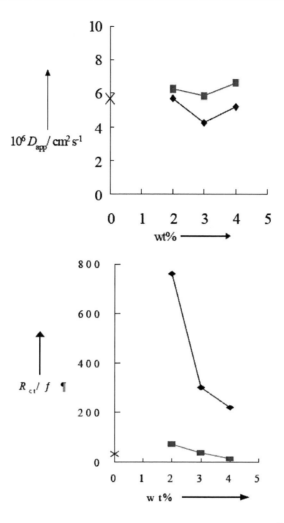

Figure 9. Polysaccharide concentration dependence of R_{ct} for 10 mM Fe(CN)$_6^{3-}$ in 2, 3, 4 wt% agarose (diamonds) and 2, 3, 4 wt% κ-carrageenan (squares) and aqueous solution (crosses) containing 0.5 M KCl at the rest potential [3].

Conclusive Remarks

These results are summarized as follows. Tight and elastic polysaccharide solids could be formed. The cyclic voltammograms in the agarose and the κ-carrageenan solids show almost similar features to those in liquid water including the redox potential and peak currents, but with a slightly larger peak separation for the solid systems. Electrochemical impedance spectra (EIS) were measured successfully in the polysaccharide solid. The D_{app} values of Fe(CN)$_6^{3-}$ in the solids were almost the same as in an aqueous solution. The charge transfer resistance (R_{ct}) of the electrode surface in the κ-carrageenan solid was even smaller than that in an aqueous solution, although it was larger in the agarose solid than that inκ-carrageenan solid. The R_{ct} tended to decrease with the polysaccharide concentration in the solid. The double layer capacitance (C_{dl}) on the electrode surface in the Fe(CN)$_6^{3-}$ aqueous solution tended to increase with the $Fe(CN)_6^{3-}$ concentration, but for the solid system the C_{dl} values were only weakly dependent on the $Fe(CN)_6^{3-}$ concentration, and similar to that in an aqueous solution at 5 mM $Fe(CN)_6^{3-}$ concentration.

Thus, it was elucidated that the polysaccharide solid can work well as a new solid medium in place of liquid water for conventional electrochemistry and electrochemical measurements.

Ionically Conductive Solid of Polysaccharide

Overview for Ionically Conductive Solid

Electrochemical reaction in solid electrolyte can be utilized for various devices such as sensor, energy conversion, etc. Many groups have reported charge transport by redox molecules in polymer thin films modified on an electrode, including our group [16-26]. However, in redox polymer solid phases, the charge transport is usually very slow, showing that other heterogeneous media should be developed where charge transport takes place in the same way as in a solution. It was shown in the last section 3 that molecular diffusion takes place in tight and elastic polysaccharide solids containing excess water in the same way as in a liquid water. This type of solid can offer not only a new solid medium for electrochemical measurement, but also an excellent medium for electrochemical sensors, and devices. For these purposes it is of importance to investigate the fundamental ionic conductivity of such a solid as a medium.

The ionically conductive polymer is applied to the lithium battery, the fuel cell, etc. These polymer electrolytes have high ionic conductivity. However, general electrochemical measurement cannot be preformed in such polymer electrolyte. Electrochemical measurement in such polymer electrolyte has been possible only by special electrode systems described before. The reason is that the diffusion of ions or redox molecules and the rate of electron transfer are slow in these polymer electrolytes.

In this section 4, it is shown that the polysaccharide solids containing excess water can be investigated by a conventional three electrode system because of the high ionic conductivity of the solids. The study was carried out by investigating the ionic conductivity of the polysaccharide solids as a function of pH or hardness of the solid.

Ionic Conductivity of Polysaccharide Solids

Impedance spectra of agarose and κ-carrageenan solids containing excess 0.1 M KNO_3 aqueous solution and of a 0.1 M KNO_3 aqueous solution are shown in figure 10.

All the spectra show almost vertical linear relations characteristic for the circuit composed of serial resistance and capacitance at all the frequencies. Impedance spectra of agarose and κ-carrageenan solids, and an aqueous solution containing other electrolyte are also similar to those in figure 10. The current is a non-Faradaic one because no redox species are present.

The conductivity of polysaccharide solids containing various electrolytes aqueous solution obtained from the impedance spectra are summarized in table 1. The conductivity in the 2 wt% agarose solids and 2 wt% κ-carrageenan solids are almost the same as in an aqueous solution showing evidently that the diffusion of the ions in the solid takes place the same as in a liquid water.

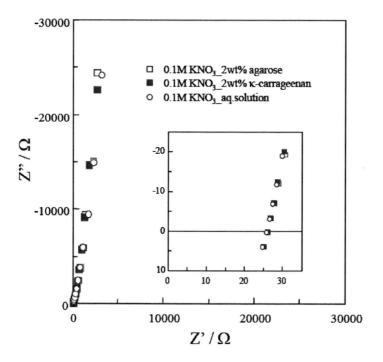

Figure 10. Impedance spectra of 2 wt% agarose, 2 wt% κ-carrageenan solids containing 0.1 M KNO₃ aqueous solution, and 0.1 M KCO₃ aqueous solution at the rest potential [5].

Table 1. Conductivity of 2 wt% agarose and 2 wt% κ-carrageenan solids containing 0.1 M KCl, 0.1 M KNO₃, 0.1 M KI, 0.1 M CsCl, or 0.1 M CaCl₂ [5]

	Conductivity / mS cm^{-1}				
	0.1M KCl	0.1M KNO₃	0.1M KI	0.1M CsCl	0.1M CaCl₂
aq.solution	12.90	12.00	13.05	13.12	20.50
2wt% Agarose	11.35	15.20	13.13	13.27	20.46
2wt% κ-Carrageenan	13.16	13.22	13.17	15.08	21.08

Double-layer capacitance on the electrode surface estimated from impedance spectra are shown in table 2.

Double-layer capacitance values of the solids are similar to that in the aqueous solution for every electrolyte. Generally, a polysaccharide solid is composed of a three-dimensional, fibrous network with pores of 10-100 nm sizes [30]. The results of table 1 show that the polysaccharide structure dose not affect the double layer on the electrode surface. The results of tables 1 and 2 clearly show that the present solids can be used as ionic conductive solid as a medium for electrochemistry the same as an aqueous solution.

Table 2. Double layer capacitance of 2 wt% agarose and 2 wt% κ-carrageenan solids containing 0.1 M KCl, 0.1 M KNO₃, 0.1 M KI, 0.1 M CsCl, or 0.1 M CaCl₂ [5]

	Double Layer Capacitance / mF				
	0.1M KCl	0.1M KNO₃	0.1M KI	0.1M CsCl	0.1M CaCl₂
aq.solution	15.33	18.13	15.33	17.41	18.18
2wt% Agarose	15.56	18.13	15.56	16.91	16.61
2wt% κ-Carrageenan	18.26	18.50	18.26	17.11	16.73

Impedance spectra of 1-5 wt% agarose containing 0.1 M KCl are shown in figure 11. This figure is quite different from figure 10 in that the ionic conductivity changed with changing the concentration of the polysaccharide (inset of figure 11).

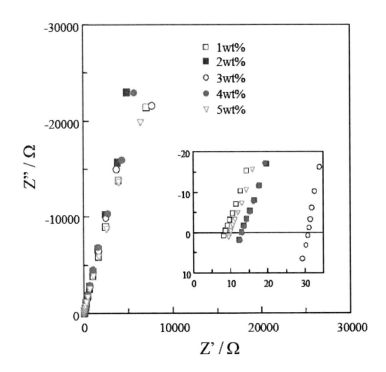

Figure 11. Impedance spectra of 1-5 wt% agarose solids containing 0.1 M KCl aqueous solution at the rest potential [5].

Table 3. Double layer capacitance of 1-5 wt% agarose solid containing 0.1 M KCl [5]

Agarose / wt%	Double Layer Capacitance / mF
1	70.21
2	61.62
3	61.84
4	59.91
5	70.43

Although the ionic conductivity changes much with the polysaccharide concentration (later shown in figure 12), the double-layer capacitance values estimated from impedance spectra (table 3) are almost independent of the polysaccharide concentration.

We examined the ionic conductivity changes of figure 11 in more detail. The agarose concentration dependence of the hardness and the conductivity of the solids containing 0.1 M KCl are obtained from figure 11 and shown in figure 12.

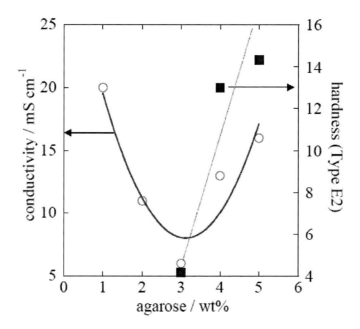

Figure 12. Concentration dependence of the conductivity of agarose solids containing 0.1 M KCl aqueous solution and the hardness of the solids as a function of the agarose concentration [5].

For the 1 and 2 wt% agarose solids, they are so soft that the hardness was not able to be measured with the present hardness meter. For the 3-5 wt% agarose concentrations, they form a tight and elastic solid. The conductivity decreased with the increase in the agarose concentration from 1 to 3 wt%. On the other hand, the conductivity increased with the increase in the agarose concentration from 3 to 5 wt%. That is, the conductivity decreased with the polysaccharide concentration while the solid is very soft, but the conductivity increased with the increase in the hardness when it forms a tight solid. When a polysaccharide becomes a solid including excess water, three dimensional network is formed by single or double helical structure. Now, when the agarose concentration is low (< 3 wt%), it is considered that the three dimensional network structure would not grow up enough. In this range, the network structure is sparse and the chains must be very flexible, so that a steric hindrance by agarose chains in the liquid phase would increase with the increase in the agarose concentration; it is inferred that the diffusion of dissolved ions is prohibited by the agarose chains in this case. However, above the 3 wt% agarose concentrations, the growth of three dimensional networks increase remarkably with the increase in the agarose concentration. In this case, the flexibility of the chains must be lowered by the addition of polysaccharide, so that it is inferred that the further addition of agarose would not block the ionic diffusion in liquid phase, but rather enhances ionic diffusion by forming more rigid network.

The pH dependence of the hardness and the conductivity of the agarose solids containing 0.1 M KCl are shown in figure 13.

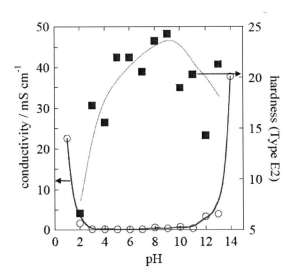

Figure 13. pH dependence of the conductivity and hardness of 4 wt% agarose solids containing 0.1 M KCl aqueous solution [5].

The conductivity has increased both in the strong acidic and the strong basic conditions, which is attributed to the increase in the H^+ or OH^- concentrations. On the other hand, the hardness seems to show a maximum around pH 9, and decreases from neutral region towards acidic conditions although the data show rather high degree of scattering. The agarose solid becomes soft at low pH, because the agarose helical structure are destroyed. Thus, the hardness of the solid is high at pH 4-10, which is ascribable to stabilization of helical structures.

Conclusive Remarks

These results are summarized as follows. Electrochemical impedance spectrum was measured in the polysaccharide solid showing that both the ionic conductivity of the polysaccharide solid and the double-layer capacitance on the electrode surface are the same as those in an aqueous solution of the ions. Impedance spectra of 1-5 wt% agarose containing 0.1 M KCl showed almost linear relations at all the frequencies but the spectra changed with the polysaccharide concentration. From 1 to 3 wt% the conductivity decreased with the increase in the agarose concentration, while from 3 to 5 wt% the conductivity increased with the increase in the hardness of the agarose solid. The conductivity of the agarose solids containing 0.1 M KCl was large both in the strong acidic and strong basic conditions, which is attributed to the increase in the H^+ or OH^- concentration. The conductivity of the solids was discussed in relevant to their hardness. Thus, it was elucidated that the polysaccharide solid containing excess water can be used as new ionically conductive solid for conventional electrochemistry. Other many polysaccharides can be used in principle as a solid medium for electrochemistry.

A Solid Medium wherein Molecular Diffusion Takes Place the Same as in a Liquid but Convection ss Prohibited

Molecular Diffusion and Bulk Convection

In a liquid or gas phase a molecule diffuses both by self-diffusion and convection when there exists no other disturbing factor such as stirring or flow of the medium. Between the self-diffusion and convection, convection contributes much more to the transport of molecules in the whole liquid or gas phase than self-diffusion (*vide infra*). However, it is not recognized in general how convection is important for a molecule to move throughout the liquid or gas phase. It is an important and interesting research subject to investigate the degree of molecular diffusion (movement) by discriminating the self-diffusion and convection. However, it has been impossible to discriminate self-diffusion and convection since convection always take place on the earth due to the presence of gravity except the cosmic space where convection does not occur due to the absence of gravity. Prohibition of convection is also important for crystal growth for which convection prohibits large and good crystal growth by disturbing concentration gradient near the growing crystal surface. On this reason cosmic space has been attracting attention to obtain ideal crystals where convection does not take place due to the absence of gravity.

It has been shown in the last sections 3 and 4 that, in a polysaccharide solid involving excess water, molecular diffusion can take place the same as in a liquid water. Various reactions including photochemical and electrochemical reactions can take place in this solid the same as in an aqueous solution. In this solid molecules can diffuse freely as in liquid water, but we have found that bulk convection does not take place because of the macroscopic solid state of the material [6]. Convection always takes place in a liquid or gas phase on the earth because of the gravity, and so can be prohibited only in the cosmic space. If convection can be prohibited on the earth, it would lead to new basic science, chemical reactions, as well as applications to such as crystal growth. In this section 5 a different aspect of the polysaccharide solid will be mentioned wherein bulk convection does not take place.

Prohibition of Bulk Convection in Polysaccharide Solids

Three -dimensional self-diffusion of thionine dye (**3**) was investigated in an agarose solid involving excess water.

3 (thionine dye) .

A disk shape (diameter 40 mm and height 20 mm) agarose solid (2 wt%) containing excess water was prepared by dissolving agarose in water on a hot plate at about 90°C, and then by cooling the hot solution up to room temperature (25°C) in a glass cell (figure 14).

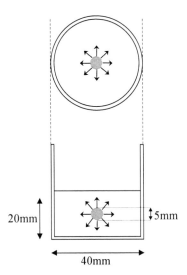

Figure 14. The cell and the size of the solid to investigate diffusion of thionine dye in the solid containing excess water [6].

The 2 wt% agarose solid was tight and elastic, the hardness being almost the same as a brick cheese and one third of a conventional rubber eraser. An air was injected from a syringe to make a round air bubble (diameter 5.0 mm) in the center of the solid, and the bubble was filled with an aqueous solution of 5.0 mmoldm^{-3} thionine (3) by using a syringe. Three-dimensional diffusion of the thionine was observed from the top of the cell, and the diffusion distance was plotted against time at 25°C (figure 15). As a reference, the same thionine solution was injected by a syringe in the middle of the same shape liquid water (as in figure14) in the absence of any stirring and flow, and the diffusion of the thionine was observed.

In the liquid water the injected thionine became homogeneous in the water within 40 min, while in the agarose solid its diffusion was remarkably slow as shown in the figure15.

The diffusion length of a substance (d cm) in a gas or liquid phase in the absence of any other driving forces such as convection and flow of the liquid or gas is represented by the equation (1), where D is the diffusion coefficient (cm^{2}s^{-1}) of the substance and t is time (s).

$$d = (6Dt)^{1/2} \qquad (1)$$

The D values of thionine in a 2 wt% agarose solid and in an aqueous solution were determined by cyclic voltammograms (as for the method see the reference [3]) to be 4.98×10^{-6} cm^{2}s^{-1} (in the 2 wt% agarose solid) and 2.21×10^{-6} cm^{2}s^{-1} (in aqueous solution). The diffusion coefficient in the solid is even larger than that in water showing that self-diffusion of thionine should take place in the agarose solid the same as or even quicker than in a liquid water if only self-diffusion contributes to the diffusion. However, as it was mentioned, the diffusion in liquid water was much quicker than that in the agarose solid. The experimental diffusion distance of thionine in the solid as shown in figure 15 was in good agreement with the calculated value (shown by a solid curve) based on self-diffusion, that is, only self-diffusion contributes to the transport of thionine. These results show evidently that in the agarose solid convection is almost entirely prohibited. It should also be noted that for diffusion of solutes in a liquid, contribution of convection is extremely larger than self-diffusion.

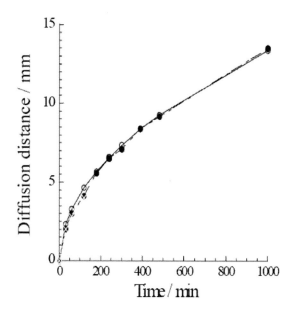

Figure 15. Diffusion distance of thionine against time in agarose solid. (-○-)Theoretical value of diffusion distance of thionine by self-diffusion according to eq.(1) [6]. (●) Observed diffusion distance of thionine in agarose gel. In a pure liquid water thionine became homogeneous within 40 min in a reference experiment under the similar conditions.

In recent years crystal growth in solid phases such as biomaterials, gels, and polymers is attracting a great deal of attention [31,32]. Star-shaped calcite ($CaCO_3$) crystals have been found to be formed in agarose gels (1 wt%) different from the typical rhombohedral calcite crystals [32]. In that work an agarose gel (1 wt%) containing $CaCl_2$ was soaked in an aqueous solution of Na_2CO_3 so that the CO_3^{2-} ions diffuse into the gel to form $CaCO_3$ crystal. The reason for the star-shaped calcite crystal formation was inferred to be due to the slow diffusion of CO_3^{2-} ions from the outer liquid water phase into the gels. In a sense it is true that the diffusion of the carbonate ions from the outer liquid water phase into the gel should be slow. On the other hand convection factor should also be taken into account for crystal growth in matrixes. Convection disturbs the concentration gradient of solutes above a growing crystal surface, which hinders ideal crystal growth. In the absence of convection the solute concentration gradient above a growing crystal is not disturbed, which allows slow solute supply and therefore slow crystal growth to form high quality single crystals, as carried out in the cosmic space where convection does not take place due to the absence of gravity.

As clearly shown in the present paper the polysaccharide solid containing excess water can also provide such conditions where convection does not take place.

Conclusive Remarks

Thus, tight polysaccharide media containing excess water are proposed wherein only self-diffusion of solutes can take place and convection is prohibited. The effect of convection on electrochemistry is unknown but important future research subject; such study is now under way by using the polysaccharide solids. In addition, some crystal growth was tested in a polysaccharide solid giving much larger crystals in the solid than in a conventional liquid water.

Application of Polysaccharide Solids to a Dye-Sensitized Solar Cell

Dye-Sensitized Solar Cell and its Solidification

The photosensitized solar cell composed of nanoparticulated TiO_2 porous film adsorbing dye reported by Graetzel's group (figure 16) [7, 33] has shown a great success in the relevant research area.

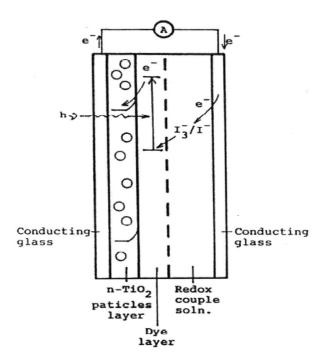

Figure 16. Dye-sensitized solar cell with the configuration, F-doped ITO/TiO_2 nanoparticulated film/dye/(I_3^-/I^-)redox electrolyte solution/Pt-coated ITO.

This so-called Graetzel's cell has reported to give nearly 10% conversion efficiency under AM 1.5 solar irradiation with high cost-performance, and so is attracting a great deal of attention as a future commercial solar cell. One of the problems of this cell to be solved for practical use is that it uses organic liquid. To overcome this problem solidification of the organic redox electrolyte solution by molten salts and gelator [34] or by polymer film [35] has been achieved. Another approach is to use water [36,37], but the efficiency and stability of the cell in an aqueous phase have been low.

As shown in the sections 3 and 4, the present author has reported that polysaccahrides such as agarose (1) and κ-carrageenan (2) can form a tight and elastic solid containing excess water, and that electrochemical and photochemical reactions can take place in the solid the same as in pure water. The hardness of the solid is, for instance, almost the same as a brick cheese and one third of a conventional rubber eraser for a 2 wt%κ-carrageenan solid involving excess water. We have expected that such an interesting solid containing a large excess liquid could offer a solid state medium for the photosensitized cell. We have succeeded to substitute the water in the solid with organic liquid, and found that this solid involving organic solvent and I^-/I_3^- redox electrolyte works well as a medium for the TiO_2 cell with a well-known N3 dye (4) sensitizer [11,12].

(4) N3 dye: *cis*-bis(isothiocyanato)bis(4,4'-dicarboxyl-2,2'-bipyridine)- ruthenium(II), Ru(dcbpy)₂ (NCS)₂.

Experimental and Results for Solid-Type Dye-Sensitized Solar Cell

A colloidal aqueous solution of TiO_2 nanoparticles (P-25) was spin-coated on an ITO electrode (1.0 x 2.0 cm) and heated at 100 °C for 30 min. This procedure was repeated several times and then finally the TiO_2 coated ITO was heated at 450 °C for 30 min to prepare a nanoporous TiO_2 thin film of 10 μm thickness. This ITO/TiO_2 film was soaked in a 1.5×10^{-4} M ethanol solution of *cis*-bis(isothiocyanato)bis(4,4'-dicarboxyl-2,2'-bipyridine)-ruthenium(II), Ru(dcbpy)₂(NCS)₂ (called N3) to adsorb the complex onto the TiO_2. A 2 wt%κ-carrageenan was dissolved in water by applying very carefully a high frequency wave (2.45 GHz). Before solidifying the solution was poured onto the TiO_2/N3 film and solidified by cooling down to room temperature. (C₃H₇)₄NI and I₂ (10:1) were dissolved in a mixture of acetonitrile and 3-methyl-2-oxazolidinone (1:1) so that their concentrations become 0.3 M and 0.03 M, respectively, and the water in the carrageenan solid on the TiO_2/N3 film was substituted by the mixture solution by dipping the TiO_2/N3/carrageenan solid film into the mixture solution. As for the counter electrode, a 5 mM H₂PtCl₆ ethanol solution was spin-coated on an ITO electrode plate, and then the coated ITO was heated at 450 ℃   for 1h to prepare a transparent Pt-coated ITO. The ITO/TiO$_2$/N3/solid (involving organic liquid with $(C_3H_7)_4NI$ and I_2) working and the ITO/Pt counter electrodes were put together to fabricate a dye-sensitized cell. The effective area of the TiO$_2$ was 0.4 x 0.5 cm (0.20 cm^2). This cell was irradiated from the TiO$_2$ side with a 98 mWcm^{-2} visible light from a 500 W xenon lamp using a Toshiba L-42 and IRA-25S cutoff filters.

The I-V characteristics of the cell is shown in figure 17, and the results are summarized in table 4 where the performance of the cell composed of the corresponding liquid medium is also given with the same conditions as the solid type cell.

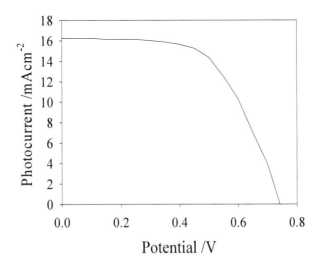

Figure 17. I-V curve of the solid-state dye-sensitized solar cell, TiO$_2$/N3/Carrageenan solid involving aceonitrile/3-methyl-2-oxazolidinone (1/1) and $(C_3H_7)_4NI/I_2(0.3$ M/0.03 M)/Pt [11]. The data correspond to Run 4 of table 4.

Table 4. Solid-state dye-sensitized solar cell, TiO$_2$/N3/Carrageenan solid involving aceonitrile/3-methyl-2-oxazolidinone (1/1) and $(C_3H_7)_4NI/$ $I_2(0.3$ M/0.03 M)/ Transparent Pt [11]. Under 98 mW cm^{-2} irradiation from a 500 W xenon lamp

Run	Electroyte medium	J_{sc} /mAcm^{-2}	V_{oc} /V	ff	η /%
1	Liquid	13.75	0.68	0.46	4.39
2	Solid	13.50	0.68	0.51	4.82
3*	Solid	17.50	0.71	0.41	5.22
4**	Solid	16.25	0.72	0.61	7.23
5**	Liquid	13.25	0.76	0.56	5.42

In the table 4 short circuit photocurrent (J_{sc}), open circuit photovoltage (V_{oc}), Fill Factor (ff) and light-to-electricity conversion efficiency (η) are given.   In the Run 3 the TiO$_2$/dye electrode was treated with a 2 mol% t-butylpyridine/acetonitrile solution before cell fabrication in order to suppress back electron transfer from the injected electron in the TiO$_2$ layer to the oxidized dye. In the Runs 4 and 5 a 0.5 M t-butylpyridine solution was added in the cell medium. Although optimum conditions were not investigated, the solid-state cell showed even better performance than the corresponding conventional (reference) liquid type cell. In the recent investigation it was elucidated that the difference between the solid-type and liquid-type cell is ascribable to the delicate difference of the liquid phase thickness for both cells. A careful comparison between the two type cells showed that both the type cells exhibit very similar performance [12].

The diffusion   coefficient of holes via the I^-/I_3^- redox couple in the solid involving 2.5 wt% carrageenan, 0.3 M KI and 0.03 M I_2 and excess water was investigated by an impedance spectroscopy: it was ca. 1.7×10^{-5} cm^2s^{-1}, almost the same as in a liquid water showing that the hole transport in the solid is not problem in comparison with the liquid medium system. A similar results have been obtained also for a solid involving redox electrolyte and organic liquid [38]. In the present solid transport of small ions and molecules takes place the same as in a liquid showing that the liquid phase in this solid behaves as in a pure liquid.

Conclusive Remarks

A solid-type dye-sensitized solar cell was thus easily fabricated by using an inexpensive and commercially available polysaccharide solid involving organic medium and I^-/I_2 redox electrolyte. Polymer gels are usually soft and fragile. However, the present solid is tight and elastic. Although a long-term performance was not tested yet, the photocurrent did not decrease in a 3 h irradiation without any sealing of the solid-type cell showing that it is stable against evaporation of the organic liquid.

Conclusion and Future Scopes

It has been shown that in a tight and elastic polysaccharide solid involving excess water molecular and ionic diffusion takes place the same as in liquid water. This allows its use as a solid electrolyte in place of liquid water for conventional electrochemical measurements and other electrochemical devices. Application of the solid to solidify a dye-sensitized solar cell was successfully achieved. Another interesting behavior in this polysaccharide solid with excess water is the entire prohibition of bulk convection in contrast to the free diffusion of molecules, which would allow its application to fundamental investigations to discriminate diffusion and convection in chemical processes that has been impossible on the earth because of the gravity. The concept of chemical reactions including electrochemistry, photochemistry, chemical dynamics and other applications to devices and sensors could be changed in the near future by using this interesting solid.

References

[1]   Y.Osada and K.Kajiwara eds, *Gel Handbook*, NTS, Tokyo, 1997.
[2]   M.Kaneko, N.Mochizuki, K.Suzuki, H.Shiroishi, K.Ishikawa, *Chem.Lett.*, 2002, 530 (2002).

[3] H.Ueno and M.Kaneko, *J.Electroanal.Chem.*, **568**, 87 (2004).

[4] N.Mochizuki, H.Ueno, and M.Kaneko, *Electrochim.Acta*, in press (2004).

[5] H.Ueno, Y.Endo, Y.Kaburagi, and M.Kaneko, *J.Electroanal. Chem.*, in press (2004).

[6] M.Kaneko, N.Gokan, and K.Takato, *Chem.Lett.*, **33**, 686 (2004).

[7] B.O'Regan and M.Graetzel, *Nature*, **353**, 737 (1991).

[8] W. Kubo, T. Kitamura, K, Hanabusa, Y. Wada, and S. Yanagida, *Chem. Commun.*, 2002, 374.

[9] S. Mikoshiba, H. Sumino, M. Yonetsu, and S. Hayase, Abst. of *13th International Conference on Photochemical Conversion and Storage of Solar Energy* (W6-70), Snowmass (2000).

[10] T.N.Murakami, Y.Kijitori, N.Kawashinma, and T.Miyasaka, *J.Photochem.Photobiol.*, **164**, 187 (2004).

[11] M.Kaneko, T.Hoshi, *Chem.Lett.*, **32**, 872 (2003)

[12] M.Kaneko, T.Hoshi, Y.Kaburagi, and H.Ueno, *J.Eloectroanal.Chem.*, in press (2004)

[13] S.Arnott, A.Fulmer, and W.E.Scott, *J.Mol.Biol.*, **90**, 269 (1974).

[14] Y.Uzuhashi and K.Nishinari, *FFI J.*, **208**, 791 (2003).

[15] S.Ikeda, *FFI J.*, 208, 801 (2003).

[16] A.L. Crumbliss, S.C. Perine, A.K. Edwards, and D.P. Rillema, *J. Phys. Chem.* **96**, 1388 (1992).

[17] R.W.Murray (Ed.), *Technology of Chemistry*, John Wiley, New York, 1992.

[18] N. Oyama and F.C. Anson, *J. Am. Chem. Soc.* **101**, 3450 (1979).

[19] D.N. Blauch, and J.-M. Savéant, *J. Am. Chem. Soc.* **114**, 3323 (1992).

[20] C.M. Martin, I. Rubinstein, and A.J. Bard, *J. Am. Chem. Soc.* **104** ,4817 (1982).

[21] J.A.R. Sende, C.R. Arana L. Hernández, K.F. Potts, M. Keshevarz-K, and H.D. Abruna, *Inorg.Chem*, **34**,3339 (1995).

[22] J. Zhang, F. Zhao, and M. Kaneko, *J. Porphyrins and Phthalocyanines* **4**, 65 (2000).

[23] M. Watanabe, H. Nagasaka, and N. Ogata, *J. Phys. Chem.* **99**, 12294 (1995).

[24] M.E. Williams and R.W. Murray, *J. Phys. Chem.* **103**, 10221 (1999).

[25] M. Kaneko, *Photoelectric Conversion by Polymeric and Organic Materials*, in H.S. Nalwa (Ed.), Handbook of Organic Conductive Molecules and Polymers, John Wiley & Sons, Chichester, 4 (1997) 661.

[26] S. Zamponi, A.M. Kijak, A.J. Sommer, R. Marassi, P.J. Kulesza, and J.A. Cox, *J. Solid State Electrochem.* **6**, 528 (2002).

[27] O. Lev, Z. Wu, S. Bharathi, V. Glezer, A. Modestov, J. Gun, L. Rabinovich, and S. Sampath, *Chem. Mater*, **9**, 2354 (1997).

[28] K.S. Alber, J.A. Cox, and P.J. Kulesza, *Electroanalysis*, **9**, 97 (1997).

[29] M.-H. Lee and Y.-T. Kim, *Electrochemical & Solid-State Lett*, **2**, 72 (1999).

[30] A. Medin, *Studies on Structure and Properties of Agarose, Fyris Tryck*, Uppsala, (1995)

[31] S.Mann, *Angew.Chem.Int.Ed.*, **39**, 3392. (2000).

[32] D.Yang, l.Qi and J.Ma, *Chem.Commun.*, 2003, 1180.

[33] M.Kaneko and I.Okura eds., *Photocatalysis –Science and Technology-,* Kodansha-Springer, Tokyo (2002),

[34] W. Kubo, T. Kitamura, K, Hanabusa, Y. Wada, and S. Yanagida, *Chem. Commun.*, 2002, 374.

[35] S. Mikoshiba, H. Sumino, M. Yonetsu, and S. Hayase, Abst. of *13th International Conference on Photochemical Conversion and Storage of Solar Energy* (W6-70), Snowmass (2000).

[36] Q. Dai and J. Rabani, *Chem. Commun.*, 2001, 2142.

[37] M. Kaneko, T. Nomura, and C. Sasaki, *Macromol. Chem. Rapid Commun.*, **24**, 444 (2003).

[38] M.Kaneko, T.Hoshi, Y.Kaburagi, and H.Ueno, *J. Eloectroanal. Chem.*, in press (2004)

In: Energy Research Developments
Editors: K.F. Johnson and T.R. Veliotti

ISBN: 978-1-60692-680-2
© 2009 Nova Science Publishers, Inc.

Chapter 5

PERFORMANCE CHARACTERIZATION OF A MULTI-STAGE SOLAR STILL

S. Ben Jabrallah[1,2]*, B. Dhifaoui[2], A. Belghith[2,3] and J.P. Corriou[4]

[1] Faculté des Sciences de Bizerte, Zarzouna, Bizerte, Tunisie
[2] Laboratoire d'Energétique et des Transferts Thermique et Massique de Tunis
[3] Faculté des Sciences de Tunis. Campus universitaire. Tunis, Tunisie
[4] Laboratoire des Sciences de Génie Chimique, CNRS-ENSIC, Nancy Cedex, France

Abstract

The present paper concerns the experimental study of heat and mass transfer in a distillation cell. The cell is a parallelepiped cavity, of high form factor and with vertical active walls. Pure water is evaporated from a film of salted water which falls along a heated wall while the opposite wall is maintained at a relatively low temperature and is used as a condensation wall. The experimental results show that that the mass transfer inside the distillation cell is dominated by the latent heat transfer associated to evaporation. These results allowed us to perform a parametric study of the operating parameters of the distillation cell. A convenient choice of these parameters is necessary to optimize the distillation yield. The performance of a three-stage distiller derived from the characteristics of a single-stage distiller has been studied.

Keywords: Heat and mass transfer, evaporation, condensation, thin film, energy, desalination, multi-stage distiller.

Nomenclature

b distance between the two active walls (m)
C_p heat capacity (J/kg/K)
H form factor of the cavity

---

* E-mail address: sadok.jabrallah@fsb.rnu.tn

h depth of the cavity (m)
l length of the water film (m)
L_v latent heat of vaporization (J/kg)
m_{feed} flow rate feed (l/h)
m_v flow rate evaporated water (l/h)
m_d flow rate distilled water (l/h)
m_{out} flow rate of the rejected brine (l/h).
q_f heat flux density provided to the water film (W/m²)
R yield
T temperature of the fluid (°C).
T_c temperature of reference, equal to that of the cooled plate (°C).
T_{feed} temperature of the feed water (°C).
x, y Cartesian coordinates (m).

Greek Letters

δ thickness of the water film (m).

Subscripts

c reference (relative to the cold plate)
v water vapor
$feed$ feed

Introduction

In distillation, two antagonist phenomena with respect to energy intervene: evaporation which consumes energy and condensation which returns it. In some situations, in order to conceive robust and reasonably low-cost processes, these phenomena are gathered in a single cell: this is the case of the capillary film solar distiller [1, 2]. To be efficient, an optimization study is then necessary. Several works dealt with the modeling of these exchanges so as to understand the mechanisms of heat and mass transfer that accompany evaporation. However, because of the physical complexity, simplifying hypotheses are adopted which prevent from taking into account all the phenomena which occur in the process. To make the study more reliable, additional experiments are necessary to quantify the influence of the operating parameters on the yield of distillation. Evaporation comes from a thin film which falls along a wall heated with a constant heat flux density. The vapor thus formed condenses on the opposite wall maintained at a uniform and constant temperature. Both walls are vertical and form the active walls of a rectangular cavity of large form factor, equal to 10. The evaporation of a film falling along a heated wall presents great practical interests: it involves large heat exchange coefficients for low temperature gradients. According to the expected application, the interest concerns the mass transfer (drying a film), or the heat transfer (cooling the wall which supports the film). Several authors have studied the heat and mass transfer which accompany the evaporation of a liquid film. Such a film can flow on a vertical wall [3, 4], or on an inclined wall [5, 6, 7]. The wall itself can present a porous surface [8, 9, 10]. Most works deal with evaporation in a free medium [15, 16] or a semi-confined medium like a canal [11, 12]. However the works dealing the evaporation of a thin film in a closed cavity are rare. In the latter case, the yield of distillation depends both on evaporation and condensation. To improve the yield of this process, we need a better understanding of competition between the transfer phenomena that accompany distillation in a cavity and of the manner according to which the energy is distributed during the operation of the cell.

The complexity of the studied process and of the occurring phenomena needs the use of a combination between experience and modeling. In a previous work [14], a model describing the behavior of the distillation cell where the parameters influencing its yield are incorporated has been produced. In the present work, an analysis of the experimental results from the energetic point of view has been performed. Particular care is set on the distribution of the energy to better explain the yield of the process.

Position of the Problem

The cavity is a parallelepiped of small width b, formed by an adiabatic of height h, of length l, closed on the two main sides by two vertical plates, distant of b. The plates are square, identical and of dimensions: 0.5m x 0.5m. On the internal face of the plate heated from the outside by a constant heat flux density q_f, a very thin film of water flows, of thickness δ (y) varying with the height y, entering with a mass flow rate m_{feed} and exiting with a mass flow rate m_{out}. These mass flow rates refer to the surface area of the heated wall. The opposite plate is maintained at a constant temperature T_c allowing the condensation on its internal face.

Experimental Setup and Measurement Device

An experimental setup including a distillation cell, separate devices for heating, cooling for the condensation plate, and water feeding has been performed. The distillation cell is a parallelepiped cavity of form factor $H = h/b = 10$, of dimensions: $l = 50$ cm, $h = 50$ cm and $b = 5$ cm. The active walls of the cell, i.e. the vertical walls between which the heat and mass exchange take place, are two square stainless steel plates of thickness 1 cm, of side 50 cm. These walls are distant of 5 cm and are separated by a thick perspex frame to reduce the side heat losses. The wall supporting the film is provided with the heating system. A resistance cut in a carbon sheet allows us to distribute the flux imposed at the wall. A water cooling system has been machined in the opposite wall to control its temperature and ensure its cooling. To realize a thin falling film, a very thin fabric with meshes of very low dimensions has been spread on the heated wall. When wetted, this fabric adheres to the wall by capillarity. Furthermore, this fabric presents the advantage to reduce the thermal resistance at the contact of the wall surface. The water feeding is provided by a tube perforated along a generating line and linked to a constant level tank. The feeding tube placed at the top of the cell is surrounded by the fabric in order to form the liquid film.

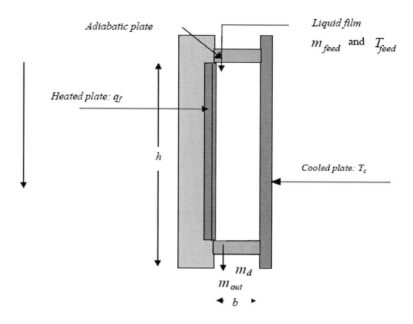

Figure 1. Description of the distillation cell.

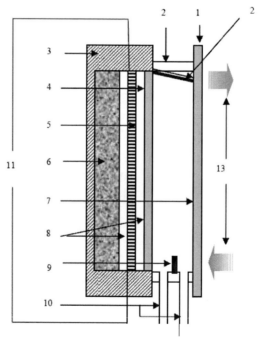

1. Condensation plate
2. Perspex frame
3. Wood frame
4. Heated plate
5. Resistance made of carbon sheets
6. Thermal insulator (cork)
7. Thermocouples (K type)

8. Samicanite plates (electrical insulator)
9. Brine-distillate separation plate
10. Recovery of distillate and brine
11. Continuous voltage generator
12. Inclined plate
13. Water cooling circuit

Figure 2. Schematic cross-section of the cavity and its measurement device.

This experimental setup is completed by a measurement device: about ten thermocouples have been placed along the height on each of the internal faces of the active walls. To measure the flow rates, a weighing machine with a sensitivity of 0.1g and a chronometer have been used. A typical experience consists in imposing a heat flux density while simultaneously fixing the flow rate and the temperature of the feed water as well as the temperature of the condensation wall. When the steady state regime is reached, the temperature of the liquid film along the heated wall and the distillated flow rate are recorded. The temperature acquisition is automated by means of an interface card and a micro-computer. It must be noticed that the thermocouples have been installed at the contact of the wetted face of the heated wall and that the recorded measurement consequently indicates the temperature of the solid-liquid interface.

Results

In a previous work [14], it was shown that the behavior of the cell depends on parameters:

- the temperature T_c of the condensation plate.
- the heat flux density q_f,
- the salted water mass flow rate m_{feed},
- the temperature of the feed water T_{feed},

In practice, the temperature of the condensation wall is imposed by the initial climatic conditions where the distiller operates and varies in a relatively low domain. This, in the present work, the influence of the last three parameters on the yield of the distillation cell has been studied through the use of heat balances obtained from the experimental results.

The heat balance at the level of the liquid film can be formulated as

$$q_f = q_l + q_s + q_{cv} + q_p \tag{1}$$

with q_f: heat flux density provided to the liquid film

q_l: heat flux density consumed by the phase change
q_{out}: heat flux density evacuated by the salted water
q_{cv}: heat flux density exchanged by convection between the liquid and the gas
q_{loss}: heat flux density lost
The heat flux density transformed into latent heat is expressed as

$$q_l = m_v L_v \tag{2}$$

where m_v is the evaporated mass flow rate.
The heat flux density evacuated by the brine can be evaluated by using

$$q_{out} = m_{out} \, Cp \, \Delta T = m_{out} \, Cp \, (T_{out} - T_{feed}) \qquad (3)$$

where T_{out} is the temperature of water at the outlet.

The second member of equation (1) makes use of the different heat flux densities which occur in the behavior of the cell. Two of these heat flux densities do not vary much during the operation of the distillation cell:

- q_{loss} is a heat flux density lost through different parts of the cell which can be in a major part controlled by an efficient thermal insulation.
- q_{cv} is the heat flux density exchanged by pure convection. In reference [14], it has been shown that this heat flux density remains lows compared to the other concerned heat flux densities.

The two other heat flux densities are directly related to the quantity of produced pure water and can be varied by means of the operating parameters of the cell. Special care concerns these heat flux densities.

Influence of the Heat Flux Density

Equation (1) describes the energy balance at the level of the liquid film through evaporation: the first member q_f representing the heat flux density applied to the wall which supports the film, appears as a parameter which acts a deciding role in the energy balance. It represents the source of energy occurring in the system.

In figure 3 a, the variations of the densities for the latent heat flux density q_l and sensitive q_{out} heat flux density with respect to the heat flux density provided to the heated wall q_f.

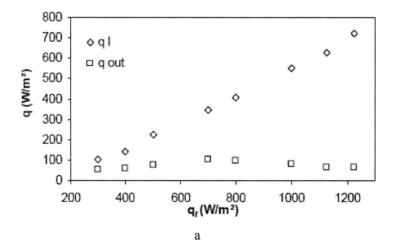

a

Figure 3. Continued on next page.

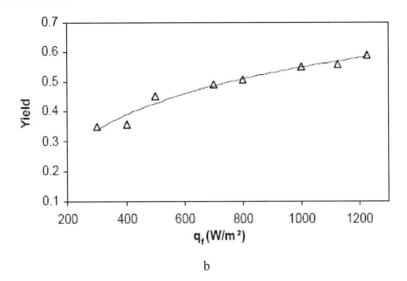

b

Figure 3. Influence of the heating flux on the exchanged fluxes (a) and on the yield (b).(m_{feed} = 1.46l/h, T_c = 26°C, T_{feed} =23.87 °C).

The latent heat flux density, associated to the evaporation increases with the heating flux density, hence the evaporated mass flow rate increases. On the same figure, the sensitive heat flux density q_{out} evacuated by the brine has been represented. q_{out} increases with q_f but remains always very low compared to q_l. It means that during evaporation the heat exchanges are controlled by the latent heat flux associated to the phase change, in agreement with the conclusion of [6].

To describe the behavior of the cell, the thermal yield has been defined as the ratio q_l/q_f of the latent heat flux density to the heat flux density provided to the system. Figure 3-b presents the variation of this ratio with respect to the heat flux density. The yield of the cell increases with the provide heat flux density, thus also with the latent heat flux density, but always remains limited by a maximum value around 0.6.

Influence of the Temperature of Feed Water

According to Fick's law, the mass transfer between the film and the gas mixture inside the cell is proportional to the concentration gradient at the liquid-gas interface. Thus, before being able to evaporate, the water that runs on the heated wall first heat itself and reaches a temperature so that its saturated concentration is larger than that of the surrounding gas mixture. It results that if the temperature of the liquid which feeds the film, the phase change and the yield of the cell are favored. This result is demonstrated on figure 4.

According to equation (1), when the heat flux density q_f is constant, a decrease of the sensitive flux q_{out} must be compensated by an increase of the other terms of the second member, in particular of the latent heat flux density. This result is confirmed in figure 4-a where the values of q_{out} are decreasing with respect to T_{feed} and can even be for values of T_{feed} larger than about 40°C. Negative values of q_{out} mean that this term represents an inflow of energy to the system that accompanies the inlet of preheated water in a medium of lower temperature. The improvement of the yield that results from the increase of the feed temperature can also be explained from an energy point of view by this income of energy. For that reason, the heat flux density q_{feed} which comes from the preheating of water has been defined and evaluated with respect to the reference temperature, taken equal to the temperature of the condensation plate T_c. Thus, the heat flux density brought by the feed can be expressed as

$$q_a = m_a.C_p.(T_a - T_0) \qquad (4)$$

Figure (3 a) presents the variation of the exchanged heat flux densities during distillation. It appears that q_{feed} increases with temperature T_{feed}, hence the increase of q_l and thus of the quantity of desalted water. The preheating of the feed water contributes to the improvement of the yield of distillation.

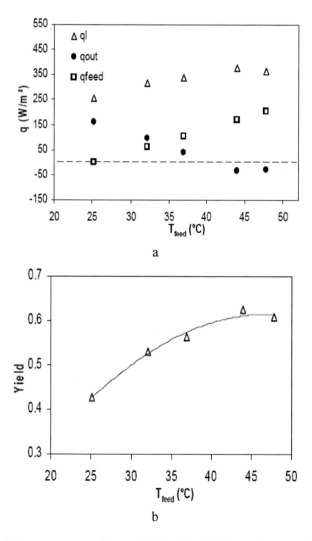

Figure 4. Influence of the temperature of water feed on the distribution between the exchanged fluxes (a) and on the yield (b). ($q_f = 600$ W/m², $m_{feed} = 1.95$ l/h, $T_c = 24.14$°C).

Influence of the Water Feed Flow Rate

Just after its inlet in the cavity, the water is successively heated then evaporated. Both operations depend on the mean velocity of the falling film along the heated wall which depends on the water feed flow rate. This parameter thus influences the behavior of the distillation cell. Figure (3-b) shows that the yield decreases when the water feed flow rate increases. The energy aspect appears on figure (3-a) where it can be observed, that when the water feed flow rate increases:

- The sensitive heat flux density evacuated by the brine increases, as the outlet flow rate m_{out} increases influencing q_{out} according to Equation (3).
- The latent heat flux density decreases q_l as a consequence of the increase of q_{out}.

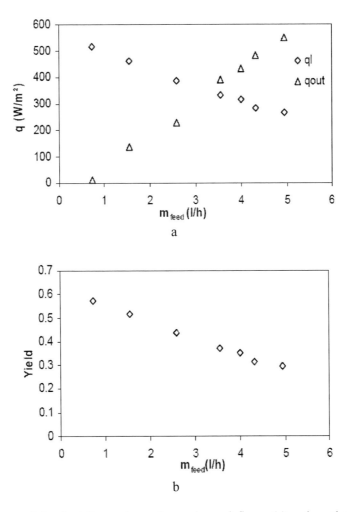

Figure 5. Influence of the feed flow rate on the exchanged fluxes (a) and on the yield (b). *(q_f = 900W/m², T_c = 23.7 °C, T_{feed}= 25°C).*

The best yield is obtained at low values of the feed flow rate. However, in real operating conditions, imposing very low feed flow rates presents the risk of provoking a drying along the heated wall which modifies the behavior of the cell. Some authors like [25] recommend to work at a feed flow rate larger than twice the distillated flow rate to avoid the drying phenomenon and the subsequent salt deposits.

Determination of the Production of a Multi-Stage Solar Distiller

To remedy the drawbacks of the classical solar distiller which limit its yield, some authors [25] proposed a multi-stage configuration. Each stage is constituted by a distillation cell analogous to the one previously described and studied. The latent heat released by the condensation on the wall of a cell is used to evaporate the liquid in the next cell. This regeneration of energy allows us to increase the production of the process. The present work is performed in the frame of a research work the objective of which is the optimization of the behavior of this type of solar distiller and the improvement of its yield. To estimate the production of pure water of the distiller, the results previously obtained in the case of a single cell have been used.

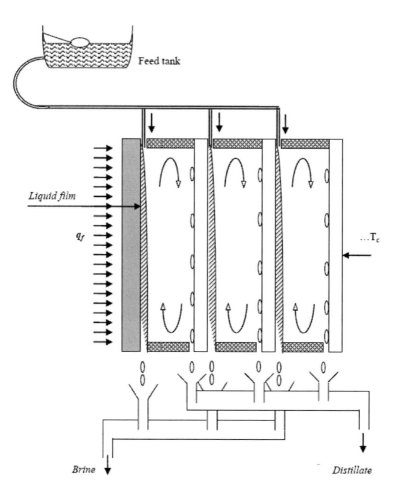

Figure 6. Scheme of a multi-stage solar distiller.

The number of cells intervening in such an apparatus is not fixed. Some prototypes having been used in North Africa (Senegal, Algeria, Libya, Tunisia, ...) include up to six stages. The number of cells depends on the operating conditions. Some works have shown that the production of the two last cells of a distiller that includes six cells is very low. In a previous work [28], we have shown that in the Tunisian climatic conditions, it is not necessary to implement more than three stages. In the present work, it will be considered that the solar distiller includes three stages.

The three cells run under the same conditions. It is assumed that they have the same thermal yield

$$R = \frac{q_{l,i}}{q_{f,i}}, i = 1,2,3.$$ (5)

where $q_{f,i}$. is the heat flux density provided at the level of the the cell. Thus $q_{f,i}$. is the heat flux density imposed to the face of the first cell, thus q_f for $i=1$. For the other values of i, $q_{f,i}$ represents the density of the latent heat flux released by the condensation at the level of the previous cell. Knowing the value of the yield R previously determined, the different values of the latent heat flux density $q_{l,i}$, thus of the distillation mass flow rate $m_{d,i}$, can be deduced as

$$m_{d,i} = \frac{q_{l,i}}{Lv}$$ (6)

The total production of the apparatus is obtained by summing

$$m_{d,t} = m_{d,1} + m_{d,2} + m_{d,3}$$ (7)

In this calculation, it is assumed that the latent heat flux released by condensation at the level of a cell is totally transmitted to the liquid film of the following cell.

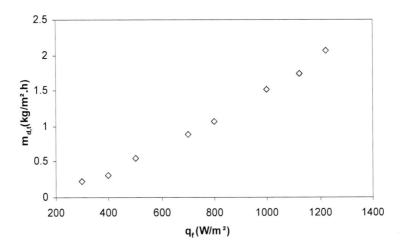

Figure 7. Distillate flow rate of a three-stage solar distiller: Influence of the heat flux density.

This method has allowed us to evaluate the distillate flow rate produced by a three-cell distiller. This flow rate increases with the heat flux density (figure-7). In practice, the solar heat flux density does not go beyond 1000 W/m^2, which corresponds to a production of 1.5 kg/m².h, thus to a total daily production limited to 12 kg/m² if it is supposed that in summer, the apparatus works during eight hours in mean. This distillate flow rate is larger than that of a classical one-stage distiller which is around de *4* kg/m² per day in similar conditions.

Note that the results thus obtained for a three-stage distiller are in agreement with those of Bouchekima [27].

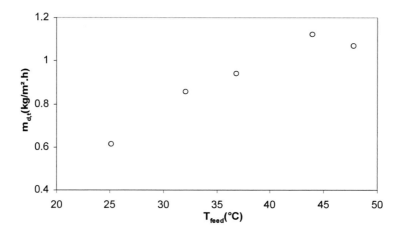

Figure 8. Distillate flow rate for a three-stage distiller: Influence of the feed temperature.

The production of the distiller increases with the temperature of the feed water (figure 8), from 0.6 to 1.1 kg/m^2.h when the feed temperature increases from de 25 to 45°C.

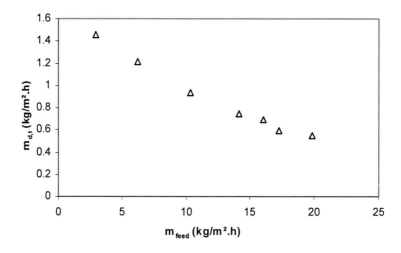

Figure 9. Distillate flow rate of a three-stage distiller: Influence of the feed flow rate.

The influence of the feed flow rate on the production of the distiller is shown on figure 9, where the total distillation flow rate decreases when the feed flow rate increases: respectively from *1.4* to *0.5 kg/m².h* for the production and from 3 to 20 kg/m².h for the feed flow rate.

The variation of the distillate of a multi-stage distiller with respect to the different operating parameters is in agreement with the variation of the production for a single cell. The obtained results show that the production of a multi-stage distiller can reach 1.5 kg/m².h, corresponding to a daily production of about 12 kg/m². This result is well above that a classical distiller the daily production of which varies between 4 and 6 kg/m² in the same conditions.

Conclusion

Several phenomena playing opposite roles in energy balances intervene during distillation of salted water in a cavity. Their complexity and their competition make the use of experimental studies necessary in order to better approach the couplings and antagonisms and their influence on the distillation yield. The present study has been performed in this way and examines experimentally the energy aspect and the analysis of the contribution of the different types of heat flux densities occurring in the process of salted water distillation in a cavity.

It has been shown that when the feed flow rate is low, the latent heat flux density associated to the change of phase overcomes the other occurring heat flux densities, in agreement with the conclusion of several authors [3, 6, 11, 21, 22, 23, 24]. The distillation yield can be improved by increasing the temperature of the feed water. It is explained by the income of energy that results as a preheating of the cell.

The study of the feed flow rate has shown a great sensitivity of the yield of the cell to this parameter. The production of pure water drops by a factor equal to two when the feed flow rate increases from 1 à 5 l/h. This decrease is explained by the increase of the sensitive heat flux density evacuated by the brine.

From the results obtained in the case of a single cell, the production of a three-stage solar distiller could be determined. Thus the influence of the different operating parameters on the distillate flow rate was studied and shown to be in agreement with the results from a single cell. The final results conform to those of Bouchekima [27] who followed the behavior of a three-stage distiller on a real site, in South Algeria.

The production of a three-stage distiller can reach in some conditions 12 kg/m² per day, which represents a noticeable improvement with respect to a classical distiller the daily production of which remains comprised between 4 and 6 kg/m².

Finally, this experimental study has shown that the thermal yield of the distillation cell remains limited and always below 0.6. The increase of the yield of solar distillation can be performed by using systems where a separation between the compartment of evaporation and the compartment of condensation is realized.

References

[1]   Storis A. Kalogirou, Design of a new spray–type seawater evaporator. *Desalination* **139**: 345-352, 2001.

[2]   K. Bourouni, J. C. Deronzier, L. Tadrist. Experimentation and modeling of an innovative geothermal desalination unit. *Desalination*, **125**: 147-153, 1999.

[3]   Tsay Y. L., Lin T.F. Evaporation of a heated falling liquid film into a laminar gas stream. *Experimental Thermal and Fluid Science,* **11**: . 61-71, 1995.

[4]   A. Ali Cherif, A. Daïf . Etude numérique du transfert de chaleur et de masse entre deux plaques planes verticales en présence d'un film liquide binaire ruisselant sur l'une des plaques chauffée. *Int. J. Heat Mass Transfer,* **42**: 2399-2418,1999.

[5]   El H. Mezaache, M. Daguenet . Etude numérique de l'évaporation dans un courant d'air humide laminaire d'un film d'eau ruisselant sur une plaque inclinée. *Int. J. Ther. Sci,***39**: 117-129,2000.

[6]   A. Agunaoun, A. Il Idrissi, A. Daïf, R. Barriol . Etude de l'évaporation en convection mixte d'un film liquide d'un mélange binaire s'écoulant sur un plan incliné soumis à un flux de chaleur constant. *Int. J. Heat Mass Transfer*, **41**: 2197-2210, 1998.

[7]   B. Song, H. Inaba, A. Horibe, K. Ozaki Heat, mass and momentum transfer of a water film flowing down a tilted plate exposed to solar irradiation . *Int. J. Ther. Sci,* **38**: 384-397, 1999.

[8]   K.T. Lee, H.L. Tsay .Mixed convection heat and mass transfer in vertical rectangular ducts. *Int. J. Heat and Mass Transfer,* **40**: 1621-1631,1997.

[9]   J. Wang, I. Catton . Enhanced evaporation heat transfer in triangular grooves covered with a thin fine porous layer. *Applied Thermal Engineering,* **21**: 1721-1737, 2001.

[10]   A. Brautsch, P.A. Kew. Examination and visualization of heat transfer processes during evaporation in capillary porous structures. *Applied Thermal Engineering,* **22**: 815-824, 2002.

[11]   A.G.Fedorov, R. Viskanta, A.A.Mohamed . Turbulent heat and mass transfer in asymmetrically heated, vertical parallel plate channe. *Int. J. Heat and Fluid Flow,* **18**: 307-315,1997.

[12]   A.Belhadj Mohamed . Etude du ruissellement et de l'évaporation en convection mixte d'un film liquide dans un canal vertical chauffé. *Journées Tunisiennes d'Ecoulement et de Transferts*, 2002.

[13]   Z.Ait Hammou. Etude comparative du transfert de chaleur et de masse dans un canal vertical en convection mixte: Effet des forces d'Archimède. *Journées Tunisiennes d'Ecoulement et de Transferts,* 2002.

[14]   S.Ben Jabrallah, A.Belghith et J. P. Corriou. Etude des transferts couplés de matières et de chaleur dans une cavité rectangulaire: Application à une cellule de distillation. *Int. J. Heat Mass Transfer,* **45**: 891-904, 2002.

[15]   V. Srzic, H.M. Soliman, S.J. Ormiston . Analysis of laminar mixed-convection condensation on isothermal plates using the full boundary-layer equation: mixtures of a vapor and a lighter gas. *Int. J. Heat Mass Transfer,* **42**: 685-695, 1999.

[16]   DE-YI Shang, B.X.Wang . An extended study of steady state laminar film condensation of superheated vapor on an isothermal vertical plate. *Int. J. Heat Mass Transfer* PII S0017 9310, (96) 00101-9, 1996

[17] S. He, P. An, J. Li, J.D.Jackson. Combined heat and mass transfer in a uniformly heated vertical tube with water film cooling . *Int. J. Heat and Fluid Flow,* 401-417, 1998

[18] S. He, P. An, J. Li, J.D. Jackson. Study of the cooling of a uniformly heated vertical tube by an ascending flow of air and a falling water film . *Int. J. Heat and Fluid Flow,* **20**, 268-279, 1999

[19] A.Bejan, Mass transfer principles In *Heat transfer,* John Wiley and Sons, Inc, 581-582, United States of America, 1993.

[20] J.F. Sacadura, Transmission de la chaleur à l'évaporation, In *Initiation aux transferts thermiques,* édition TEC and DOC, 352-353. Paris, 2000.

[21] Y. C. Tsay, T. F. Lin and M. Wyan, Cooling liquid film through interfacial heat and mass transfer. *Int. J. Multiphase Flow,* **16**: 853-865, 1990

[22] W. M. Yan and T. F. Lin. Combined heat and mass transfer in natural convection between vertical parallel plates with film evaporation. *Int. J. Heat Mass Transfer,* **33**: 529-541, 1990.

[23] W. M. Yan and T. F. Lin. Evaporative cooling of liquid film with through interfacial heat and mass transfer in a vertical channel-II: Numerical study. *Int. J. Heat Mass Transfer,* **34**: 1113-1124, 1991.

[24] W. M. Yan and C. Y. Song. Convective heat and mass transfer along an inclined heated plate with evaporation. *Int. J. Heat Mass Transfer,* **38**: 1261-1269, 1995.

[25] P. Le Goff, J. Le Goff, C. Ouhaes et R. Ouhaes, Modélisation et expérimentation du distillateur solaire multi-étage à film capillaire. *Journées Internationales de Thermique,* 832-841, 1989.

[26] S. Ben Jabrallah, *Transferts couplés de chaleur et de masse dans une cavité: application à une cellule de distillation.* Thèse de Doctorat. Université de Tunis II, 1998.

[27] B. Bouchekima, A solar desalination plant for domestic water needs in arid areas of south Algeria, *Desalination,* **135**: 65-69, 2002.

[28] F. Fahem, S. Ben Jabrallah, A. Belghith and R. Benelmir, Technico-economical optimization of the number of cells for a multi-stage solar still. International Journal of Energy, *Environment and Economics,* **12**: 33-44, 2004.

In: Energy Research Developments
Editors: K.F. Johnson and T.R. Veliotti

ISBN: 978-1-60692-680-2
© 2009 Nova Science Publishers, Inc.

Chapter 6

DESIGN AND SIZING OF A DIGESTER COUPLED TO AN AIR SOLAR COLLECTOR

S. Igoud[1], N. Said[1] and R. Benelmir[2]

[1] Centre de Développement des Energies Renouvelables, Bouzaréah, Alger
[2] LEMTA (umr 7563 CNRS-UHP-INPL), ESSTIN 2 Rue Jean Lamour,
Vandoeuvre, France

Abstract

The bio-mechanization of organic waste offers energetic, economic, environmental and social advantages which harmoniously integrate it in the concept of continuous development. Among the principal recovered products, we are interested in agricultural biogas whose production was carried out successfully through the mechanization of cows dungs in an experimental digester of 800 liters. Following these first experiments, we present a feasibility study for the realization of a family digester of 15 m³ which can produce between 68.4 and 185.4 m³ of biogas according to the digestion temperature. In order to improve the efficiency of biogas production, we heat the digestion substrate up to a temperature of 40°C by means of an air low temperature heating system using 20 m² of solar collectors. The proposed digester will produce monthly the energy equivalent of three standard cans of butane gas (13 kg/can) and avoid the use of an equivalent of 927 kg of heating wood.

Nomenclature

I_a :   solar heat flux (W/m^2)

$(\tau\alpha)$:   transmission-adsorption coefficient

U_l :   heat losses coefficient (W/m2 K)

T_a :   ambient air temperature (K)

F_R, F' :   correction coefficients

T :   fluid temperature (K)

T_p :   stone temperature (K)

h_v :   volumic exchange coefficient (W/m³/K)

U :   loss coefficient of the storage (W/m²/K)

D: diameter (m)

f :   porosity

ρ_a :   air density (kg/m³)

C_p :   air specific heat (J/kg/K)

A: storage cross section (m²)

G :   surface mass flow rate (kg/m²/s)

ρ_p :   stone density (kg/m³)

C_{pg} :   stone specific heat capacity (J/kg/K)

k_p :   conduction coefficient (W/m²/K)

Introduction

Continuous development rests on the conciliation between economic and social development, the protection of environment and the conservation of the natural resources. Hence, the contribution of renewable energies is of major interest. It is the case of the biogas. The objective of this project, which also fits with Algerian national concerns, aims at the energy autonomy of the populations installed in the rural, agricultural and wedged zones, characterized by their far distance from distribution networks of conventional energies. The biomass used must be abundant on the site of the digester. For this purpose, the use of cows' dungs was retained. Moreover, a complement of substrate must be locally available through other biomass contributions as livestock dejections, agricultural and kitchen wastes and domestic worn water. Besides energy and financial advantages and the life improvement of rural population, environmental impact is not the least. Indeed, in Algeria, the principal preserved resource would be the forest biomass which constitutes, in rural zone, a privileged energy source for cooking and heating.

In accordance with first experiments of the bio-methanisation of cows' dungs [1] our study presents a second phase of our objective, namely:

- the design and sizing of a digester with a floating bell
- the sizing of a low temperature air solar collector for the heating of the digester

Family Digestor

Operating Basis

This family Digester is inspired by an Indian model [2]. It is made up of an underground waterproof digestion enclosure which offers anaerobic conditions and the retaining of moisture. It thus allows the bio-methanisation of a fluid biodegradable biomass (animal dejections, agricultural wastes and waste water). The feed from the storage tank is ensured by gravity. The partial draining is carried out through the pond thanks to the phenomenon of the communicating vases. The daily supply of the digester in biomass ensures a continuity of the bio-methanisation process which allows, beyond a digestion temperature of 15° C, a renewable production of biogas.

Description

Digestion enclosure: of cylindrical form, the digestion enclosure is characterized by a lower section in V shape for the deposit as well as the evacuation of the large particles and a higher section in dome for the recovery of the produced biogas.

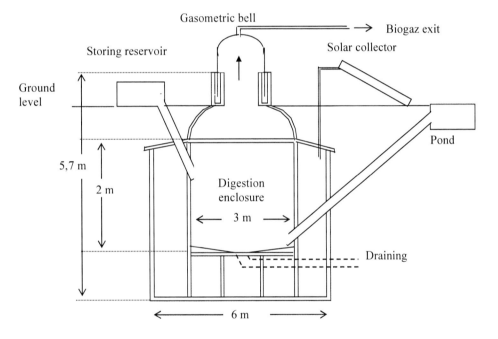

Figure 1. Family Digestor.

Storage reservoir: of rectangular or cubic form, it allows the deposit of animal dejections and their dilution for the preparation of a fluid and homogeneous digestion substrate.

Pond: of identical form to that of the storage reservoir, it allows the recovery of the digested substrate.

Gasometrical Bell: it caps the dome of the digestion tank and allows the storage of biogas by keeping it at constant pressure. Once withdrawn, it is also used as an inspection pit at of the digester.

Piping: Materialized by tubes, they ensure the arrival of the biomass towards the digester. After digestion, they allow the discharge of the digested biomass in the pond. A drain at the bottom of the digestion enclosure is used in the complete draining of the digester. A smaller diameter piping (21/27 mm) is used for the routing of the produced biogas towards the family kitchen.

The Sizing

Digestion enclosure: diameter = 3 m height = 2 m capacity = 11.77 m^3
Dome: central height = 1 m.
Evacuation and storing reservoirs: 2 m^3
Piping diameter: 30 cm
The digester enclosure maximum load is of 8240 kg substrate of digestion. For a residence time fixed to 90 days, the daily provisioning of the digester is about 130 liters/day.

Table 1. Sizing of the digester

Digester units				Diameter (m)	Length (m)		Volume (m^3)
Enclosure	Top	Dome	3	1	-	-	3.63
	Center	Cylindrical	3	1.5	-	-	1.77
	Bottom	V shape	3	0.5	-	-	
Storage tank and ponds		Rectangular	-	1	2	1	2

Low Temperature Solar System

Operating Principle

The heating up to a temperature reaching 40 °C improves the production of biogas. In order to keep a positive energy balance, we use renewable energy for the heating of the family digester: an air static solar collector at low temperature (figure 2). A depression causes a circulation of hot air through a stones bed surrounding the digester. Due to heat transfer from air to the stones bed, a thermal seasonal storage is carried out. This system ensures the restitution of the heat stored throughout the cold periods for the heating of the digester.

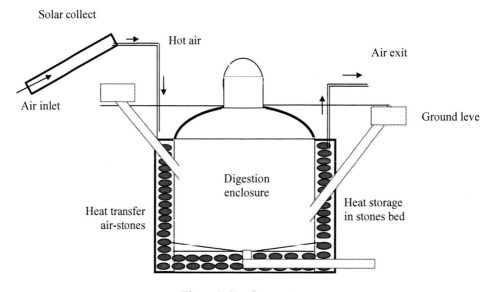

Figure 2. Heating system.

Description

The heating system is made up of:

- a field of air solar collectors
- a stone bed seasonal storage
- a thermal regulation

Solar Collector

The single pass air solar collector was retained due to the simplicity of its realization and to its low cost. It also allows us to reach average seasonal temperatures largely higher than that required by the substrate. The selected thermal model is that of Hottel-Whiller-Bliss [3] which was used successfully in the majority of the simulation cases of solar systems. According to this model, the instantaneous efficiency η of the solar collector can be written according to the inlet temperature of the fluid T_i:

$$\eta = F'[(\tau\alpha) - U_l(T_i - T_o)/I_a] \tag{1}$$

The exit temperature of the fluid is calculated by carrying out a total energy balance between the entry and the exit of the solar collector:

$$T_s = T_i + q_c/(m_c C_p) \tag{2}$$

with:

$$q_c = I_a\eta \tag{3}$$

where:

m_c : surface mass flow rate of the liquid

C_p : specific heat of the liquid

The coefficients F_R and F' are dependent with respect to the following relation:

$$F_R = (m_c C_p)/U_l[1 - \exp\{U_l F'/(m_c C_p)\}] \tag{4}$$

Heat Storage in the Packed Bed

The storage unit is an isolated space containing a stone bed crossed by air coming from the solar collector. The collected heat results leads an increase of stones temperature. It is a sensible heat storage. Hence, temperatures are only function of time and distance covered according to the direction of the air flow.

The heat balance [3]:

on air:

$$h_v A(T - T_p) dxdt + C_p GA \frac{\partial T}{\partial x} dxdt + UD\pi(T - T_a) dxdt = -\rho C_p Af \frac{\partial T}{\partial t} dxdt \qquad (5)$$

on stones:

$$k_{gs} A \frac{\partial^2 T_p}{\partial x^2} dxdt + h_v A(T - T_p) dxdt = \rho_p C_{pg} A(1 - f) \frac{\partial T_p}{\partial t} dxdt \qquad (6)$$

Sizing

The collector surface A_c and the storage volume V_p are essential sizing parameters which cannot be considered independently. We must solve simultaneously the balance equations with the particular conditions of this installation which are in our case, the maintaining of the substrate temperature at 40°C for all the year with a solar supply rate of 100 %.

We chose a sizing method in two stages:

- a pre-sizing by holding account of the characteristic parameters of the installation,
- an adjustment of the sizing according to the results obtained.

The adjustment is carried out through several simulations. The idea is to find a compromise by gradually reducing the acceptable ranges for significant sizing parameters. Hence, we define the parameters which have to be validated through the calculation of the performances. The final result must lead to the evaluation of the collector surface and the storage volume knowing the temperature and mass of the substrate.

The annual supply E of a commercial air collector in average operating conditions is from 450 à 500 kWh/m²/year. With a daily provisioning of the digester of 130 l/jour, the annual energy need B is about 10 MWh/year. The collector surface $A_c = B\ C_s/E = 20\ m^2$.

Estimation of Biogaz Production

According to the sizing and to the digestion temperature we estimate the biogas production and the energy equivalences (table 2). It is possible to produce 68.4 m³ of biogas monthly without consequent contribution of heat [4]. In Algeria, the ground temperature ranges from 15 to 20 °C for the mayor part of the year, except in winter so the contribution of solar and storage heat could raise the underground temperature to 40°C. For our sizing (digester enclosure: 11,77 m3 charge: 8240 kg of digestion substrate residence time: 90 days daily supply: 130 litters/day) the production of biogas [1] is estimated to 185,4 m³/month and could be continuous during all year.

**Table 2. Estimation of biogas production with respect
to digestion temperature - energy equivalent [1, 4, 5]**

Digestion temperature (C°)	Biogas monthly Production (m^3)	Equivalence in wood for heating (kg)	Equivalence in butane gas standard cans (13 kg)
15 – 20	68,4	342	1
20 – 25	126,9	634,5	2
25 – 30	145,5	727,5	2
35 – 40	185,4	927	3

Conclusion

The proposed family digester is inspired by the an Indian model. It is primarily made up of an digestion enclosure, an evacuation and feeding system of the digestion substrate and a biogas storage bell. Its sizing allows a daily contribution of 130 litters of wet biomass made up mainly of cows' dung. The digestion temperature is ensured thanks to the use of a low temperature heating system which uses air solar collectors and of a storage system made up of a stones bed. The sizing of the system is carried out by taking into account a digestion temperature of 40°C. At the end of a residence time fixed to 90 days, the monthly production of biogas is estimated to 68,4 m^3 for a digestion temperature ranging between 15 and 20°C. This production can reach 185.4 m^3 of biogas for a digestion temperature ranging between 35 and 40°C, which is the energy equivalent of approximately 3 butane cans (13 kg) or 927 kg of wood which could be preserved. Hence, the use of biogas in the rural zones would constitute a very good substitute for heating wood and an important support to the national programs against deforestation and desertification.

References

[1]   S. Igoud, I. Tou, Kehal.S, N. Mansouri et A. Touzi. Première approche de la caractérisation du biogaz produit à partir des déjections bovines. *Rev. Energ. Ren.* Vol.4. 123-128 (2002).

[2]   S. Igoud. *Réalisation d'un digesteur familial de 15 m³.* Fiche technique. Document interne, Centre de Développement des Energies Renouvelables (2004).

[3]   P. Lunde. *Solar Thermal Engineering.* Edition John Wiley and sons, 612 p. (1980).

[4]   N. Baek. *Analyse Technico-économique d'un Système Solaire de Production d'Air Chaud pour le Séchage du Tabac « Virginie » et le Chauffage d'un Bâtiment.* Thèse de Doctorat, Univ. de Perpignan, 195 p. (1984).

[5]   Ludwig Sasse. *L'installation de biogaz,* Ed. GATE-GTZ, 89 p. (1986).

[6]   B. Lagrange. *Biométhane. T1: Une alternative crédible.* Ed. EDISUD, 204 p. (1979).

In: Energy Research Developments
Editors: K.F. Johnson and T.R. Veliotti

ISBN: 978-1-60692-680-2
© 2009 Nova Science Publishers, Inc.

Chapter 7

FLUID INCLUSION MICROTHERMOMETRY AND GAS CHEMISTRY IN GEOTHERMAL SYSTEM, JAPAN AND ITS APPLICATION FOR THE STUDY OF FLUID EVOLUTION

Yoichi Muramatsu

Department of Liberal Arts, Faculty of Science and Technology,
Tokyo University of Science, 2641 Yamazaki,
Noda, Chiba 278-8510, Japan.

Abstract

Fluid inclusions in transparent minerals, which trapped different hydrothermal fluids at various stages after precipitation of the minerals till now, offer us information about the physicochemical properties of the fluid such as temperature, salinity and chemical composition. From the point of view, fluid inclusion studies have been widely used to estimate reservoir temperature during drilling of a well and interpret the physicochemical evolution of reservoir fluid at geothermal field.

The minimum Th value of fluid inclusion in quartz, calcite and anhydrite from a geothermal well can be used to estimate the reservoir temperature with real time at a development site in most liquid-dominated and vapor- dominated geothermal fields, Japan. In numerous methods developed for fluid inclusion analysis, semi-quantitative convenient gas analysis of liquid-rich inclusion using a crushing experiment and microthermometry is available to estimate CO_2 content in inclusion fluid lower than about 2.0 mol%. To the contrary, a quadrupole mass spectrometer (QMS) is one of most useful quantitative analytical methods to clarify gas chemistry of tiny fluid inclusion. Based on the analytical results using the QMS on fluid inclusions in anhydrite and quartz from wells in the Japanese geothermal fields, the main component is H_2O with small amounts of CO_2, N_2 and CH_4 and Ar. The CO_2 and CH_4 contents in the inclusion fluids are nearly equal to or slightly higher than those in the present reservoir fluids. In contrast, the N_2 contents in the former are generally about one to three orders of magnitude higher than those in the latter. The differences in the CO_2 content and the CO_2/CH_4 and CO_2/N_2 ratios between the inclusion fluid and present reservoir fluid could be explained by degassing and/or dilution. In the degassing case, early stage fluid was trapped in the fluid inclusions before considerable boiling and vapor-loss. It appears that the

CO$_2$ content in the reservoir fluid decreased more or less during progressing of the exploitation of geothermal field.

1. Introduction

Fluid inclusions in mineral are undoubtedly most important geological material to study physicochemical change of the reservoir fluid in geothermal field, because they contain evolutional fluids at various stages till now in geothermal reservoir. From the point of view, fluid inclusion studies have been widely used to interpret the thermal evolution of many of the liquid-dominated geothermal fields (e.g. Shimazu and Yajima, 1973; Taguchi and Hayashi, 1983; Takenouchi and Shoji, 1984; Muramatsu, 1984; Takenouchi, 1986; Muramatsu and Komatsu, 1996; Browne et al., 1974; Roedder, 1984) and the vapor-dominated geothermal fields (e.g. Sternfeld et al., 1983; Belkin et al., 1985; Moore and Gunderson, 1995; Gianelli et al., 1997; Moore et al., 2000). On the other hand, although gas chemistry of fluid inclusion is very important to interpret the fluid chemical evolution of geothermal field, it has been scarcely reported because of difficulty of measurement for a little gas content in fluid inclusions in mineral. In numerous methods developed for fluid inclusion analysis (Roedder, 1984; Shepherd et al., 1985), quadrupole mass spectrometry is one of most useful methods to measure gas chemistry of tiny fluid inclusion.

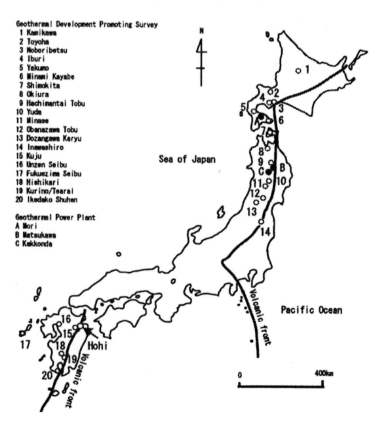

Figure 1. Location of Japanese geothermal fields in this chapter.

This chapter describes application of homogenization temperature to estimate current reservoir fluid temperature in liquid-dominated and vapor- dominated geothermal fields, Japan. This chapter presents moreover two gas analytical methods (individual and bulk gas analyses) of fluid inclusions using a quadrupole mass spectrometer (QMS), and its application on fluid inclusions in anhydrite and quartz collected during drilling of production wells in Japanese geothermal fields. The location of this study fields in this chapter is shown in figure 1.

2. Microthermometry

2.1. Estimation of Reservoir Temperature

It is significantly important to know exactly temperature of reservoir around a production well on evaluating the potential of a geothermal field. The reservoir temperature is usually estimated by the temperature build-up test. To be concrete, a developer measures usually a static borehole temperature (Tb) several times after the end of drilling of a production well because of cooling down by circulation of mud on drilling, afterward estimates the recovery temperature (RT) using Horner plots (e.g. Hanano and Matsuo, 1990). The validity of the method is discussed below.

Since 1980, the New Energy and Industrial Technology Development Organization (NEDO) had taken the initiative in surveying the promising Japanese geothermal energy fields where development survey has not been conducted because of the risk of exploration. This survey aims to verify the presence of geothermal reservoir in a regional field of 50 to 70 km^2 for the promotion of the development of geothermal power generation by private enterprises. Figure 2a illustrates the relation of Tb values between after a short standing time of 120 hours and a long standing time of 30 days to two years for many wells drilled on all over Japan from 1980 to 1988 by the "Geothermal Development Promotion Survey (GDPS)" project. The latter Tb value reflects a true pre-drilling reservoir temperature because of the data obtained after a long period of thermal recovery. The Tb value is 30 to 80 °C lower after a standing time of 120 hours than after a long standing time of 30 days to two years in reservoir temperature over 150 °C. The former Tb thus shows a value on the recovering way of the reservoir temperature. Figure 2b illustrates the relation between the Tb value after a long standing time of 30 days to two years and the RT value. The RT values estimated from data of the temperature logging sets (Standing times of 8, 16, 32, 48, 96 and 120 hours) using Horner plots are in good agreement with the static borehole temperatures, suggesting that the RT value is nearly equal to the reservoir temperature.

The estimation method of the reservoir temperature using the temperature build-up test is unfortunately defective in high cost as consequence of long standing times and so on. By contrast, microthermometry of fluid inclusion is useful method to resolve the problem. From the reason, fluid inclusion has been widely used to determine reservoir temperature during drilling of a well with real time at a development site. When we measure homogenization temperature (Th) of many fluid inclusions in mineral at a depth of a well using a heating stage, we can usually obtain Th values with various ranges (figures 3, 4). To solve how Th data reflects the reservoir temperature, it is necessary to clarify the relation between Th and RT in a geothermal field. The relationship between the Th and the RT in each well of geothermal field is divided into four patterns as follows:

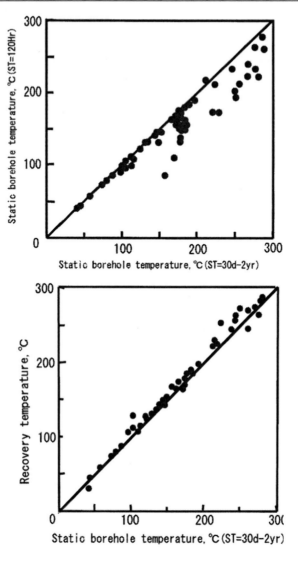

Figure 2. (a) Relationship of static borehole temperatures after standing times between 120 hours and 30 days to two years (Muramatsu,1993) (b) Relationship between static borehole temperature after standing times between 30 days to two years and estimated recovery temperature (Muramatsu,1993).

a   Minimum Th is largely higher than RT (figure 5a). Reservoir temperature has largely decreased after precipitation of the measured mineral in the reservoir, and the fluid activity had ceased at the past days.

b   Minimum Th is coincident with RT (figure 5b). Temperature of the reservoir fluid has decreased to the present days.

c   Mean Th is nearly equal to RT (figure 5c). Temperature of the reservoir fluid has not largely deceased after precipitation of the measured mineral to the present in the reservoir.

d   Maximum Th is nearly equal to RT (figure 5d). Temperature of the reservoir fluid has increased to the present.

**Table 1. Underground temperature profile type of each well
in the Geothermal Development Promotion Survey fields
and of the geothermal fields established the power plant in Japan**

| Pattern | Promoting Geothermal Development | | | Geothermal Power Plant |
	Conduction type well name	Convection type well name	Numbers of wells*	Field name
Pattern (a) (Minimmum Th RT)	KK-2, TH-7, IB-7, NB-1, NB-2, NB-4, YK-2, YK-7, MK-1, YD-1, MS-1, MS-2, MS-7, FS-1, FS-2, FS-3, FS-4, HK-3, HK-4	NB-3, MS-5, UZ-2, UZ-5, HK-2, HK-5, HK-7, ID-1	27 (36)	
Pattern (b) (Minimmum Th RT)	KK-1, NB-6, YK-5, MK-4, MK-5, MK-7, SK-5, DZ-6, KJ-5, KJ-7	TH-1, TH-3, TH-5, IB-6, OU-8, YD-2, YD-3, OB-1, UZ-1, UZ-3, UZ-4, UZ-7, KT-1, KT-8, HK-6, ID-3, ID-5, ID-6	28 (37)	Mori, Kakkonda, Uenotai, Takigami, Otake, Hatchobaru, Kirishima
Pattern (c) (Mean Th RT)	YK-4, YK-6, MK-3, YD-5, OB-2, OB-3, DZ-4, MS-6, IN-1	TH-2, TH4, MK-2, MK-6, SK-6, HT-2, MS-3, KT-5	17 (22)	Sumikawa, Okuaizu
Pattern (d) (Maximum Th=RT)	DZ-3, DZ-5	TH-6, KT-7	4 (5)	Sumikawa, Okuaizu Uenotai
Total	40	36	76 (100)	

* Number in parenthesis shows percentage. For example of well name, well KK-2 shows a well drilled for the second times at the Kamikawa field. Abbreviation: KK, Kamikawa; TH, Toyoha; NB, Noboribetsu; IB, Iburi; YK, Yakumo; MK ,Minami Kayabe; SK, Shimokita; OU, Okiura; HT, Hachimantai Tobu; YD, Yuda; MS, Minase; OB, Obanazawa Tobu; DZ, Dozangawa; IN, Inawashiro; KJ, Kuju; UZ, Unzen Seibu; FS, Fukuejima Seibu; HK, Hishikari; KT ,Kurino-Tearai; ID, Ikedako Shuhen.

Through a long history of exploration of geothermal systems in Japan, there are now seventeen geothermal power plants and total geothermal generated capacity exceeds 500 MWe. As pattern (b) is recognized in most geothermal fields where the power plant was constructed in Japan (table 1), such as Mori (Muramatsu et al., 1997), Kakkonda (Muramatsu, 1984), Takigami (Takenaka and Furuya, 1991), Otake and Hachobaru (Fujino and Yamazaki, 1985: Taguchi and Nakamura, 1991), Kirishima (Taguchi et al., 1984 Kodama and Nakajima, 1988), the minimum Th is useful to know the present reservoir temperature in almost Japanese geothermal fields. As an example, figure 3 shows distributions of Th values of liquid-rich inclusions in the analyzed samples with Tb profile for well ND-6 at the Mori geothermal field (Muramatsu et al., 1997). The Tb value obtained after a long period (13 days) of thermal recovery reflects a true pre-drilling underground temperature. The minimum Th value is in general agreement with the Tb value, suggesting that the inclusion which has a minimum Th value was trapped at the recent stage of geothermal activity. The similar feature is recognized in the shallow reservoir ($\leqq$1400 m depth; Muramatsu,1984; Muramatsu et al., 1996) and the deep reservoir (>1400 m depth) at the Kakkonda geothermal field (Komatsu and Muramatsu, 1994; discussed in section 3.3.3). Taking the results into consideration, the minimum Th value has been used to estimate temperature of the reservoir fluid when it encountered a perfect or large lost circulation during drilling of a production well, afterwards judge immediately whether the fluid is suitable to use as a power generation in the Mori and Kakkonda geothermal fields.

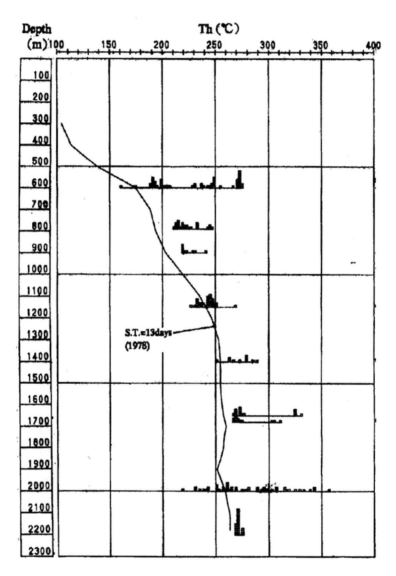

Figure 3. Distributions of Th values of liquid-rich inclusions in the analyzed samples with static borehole temperature profile for well ND-6 in Mori geothermal field (Muramatsu et al., 1997) Open and solid squares show quartz and anhydrite, respectively. S.T. is a standing time.

Patterns (c) or (d) are however exceptionally recognized in the Okuaizu and Sumikawa geothermal fields (Nitta et al., 1987; Takenouchi, 1990; NEDO, 1989). An example of well N58-OA-6 in Okuaizu is shown in figure 4 (Nitta et al., 1987). The Tb value was obtained after a standing time of 96 hours at the deeper depth and 120 hours at the shallower depth.

The RT value seems to be a few ten degrees higher than the Tb value after a standing time of 120 hours at 600 to 870 m depths. This mean the well must belong to pattern (d).

In conclusion, geothermal activity had a large cooling after precipitation of the measured mineral in the reservoir of most geothermal fields in Japan. The minimum Th value of fluid inclusions in transparent mineral from the production well have been used to judge whether reservoir temperature is suitable to use as a production well with real time at a drilling site, in addition to evaluate broadly the possibility of development of the reservoir in most Japanese geothermal fields. The transparent mineral suitable to use for microthermometry is not only hydrothermal minerals such as quartz, calcite and anhydrite, but also primary quartz in igneous rock and quartz fragment in pyroclastic rock because the present reservoir fluid were trapped as a secondary inclusion in these quartz formed by different mechanism.

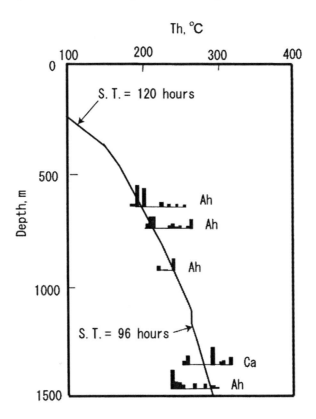

Figure 4. Distributions of Th values of liquid-rich inclusions in the analyzed samples with static borehole temperature (Tb) after standing times of 96 and 120 hours for well N58-OA-6 in the Okuaizu geothermal field (Nitta et al., 1987).

To arrange the thermal evolution of Japanese geothermal fields, table 1 shows predominant Th pattern for each well drilled by the GDPS project, and for most geothermal fields where the power plant was constructed. Patterns (a) and (b) occupy 73 per cent of all wells in the GDPS fields. Single pattern is recognized at the Kuju and Fukuejima Seibu fields, but plural patterns are recognized at the other fields except for Okiura, Hachimantai Tobu and Inawashiro. It thus reveals that the reservoir temperature has decreased to the present days in almost geothermal fields surveyed by the GDPS project, and especially the fluid activity had ceased at the past days in half of the fields.

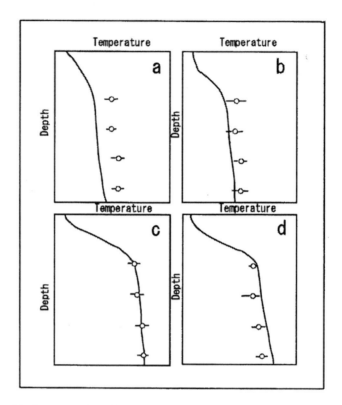

Figure 5. Relationship between Th and recovery temperature in a well. Pattern (a) Minimum Th is largely higher than recovery temperature. Pattern (b) Minimum Th is coincident with recovery temperature. Pattern (c) Mean Th is nearly equal to recovery temperature. Pattern (d) Maximum Th is nearly equal to recovery temperature.

2.2. Evaluation of Geothermal Potential

The activity index (AI) is useful to evaluate numerically the potential of a geothermal field as well as a geothermal well (Hayashi et al., 1981). Based on the AI value, a geothermal system or a geothermal well can be classified seven categories with decreasing reservoir temperature as follows: AA above 100, A 100-80, B 80-60, C 60-40, D 40-20, E 20-0 and F below 0. Very active field belongs to Type A, whereas less active one to Types B or C. According to Hayashi (1981), almost geothermal fields where the power plant was established in Japan, such as Matsukawa, Otake, Onikobe and Kakkonda belong to Type A (AI of 84 to 89), but exceptionally Onuma field belongs to Type B (AI of 68).

Divided the temperature gradient with depth into the conduction and convection types, the AI value of each exploration well drilled in GDPS fields is shown in figure 6. A symbol shows a maximum recovery temperature within the depths where Th values were measured in each well. The AI value of geothermal field is relatively lower for the conduction type than for the convection type. Patterns (a) and (b) have activity indices lower than 40 (except for wells MS-5 and NB-3) and from 25 to 85, respectively, while patterns (c) and (d) have activity indices from 61 to 98 and from 77 to 90, respectively. Patterns (b), (c) and (d) of convection type belong mainly to Types A or B. The proposed field of geothermal power plant construction, where is necessary to have a reservoir temperature above 180 °C, shows patterns does not belong to pattern (a).

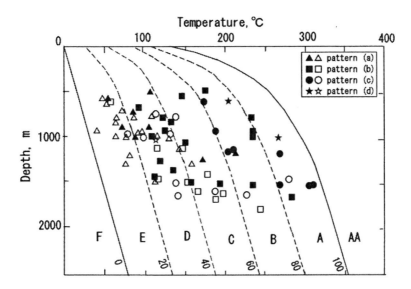

Figure 6. Activity index of each exploration well drilled in the "Geothermal Development Promotion Survey" fields by NEDO. A symbol shows a maximum recovery temperature within the depths where Th values were measured in a well.

3. GAS Chemistry

3.1. Crushing Experiments

The semi-quantitative gas analytical method using a crushing stage was devised by Sasada et al. (1986). When a transparent mineral was crushed using the stage under the microscope (Roedder,1970), behavior of vapor bubble in individual liquid-rich inclusion related to the internal pressure of the non-condensable gases gives semi-quantitative information on gas content in the inclusion. As CO_2 is a major component of the non-condensable gases in most geothermal fields (Ellis and Mahon, 1977), Sasada et al. (1986) assumed a simplified model in which all the non-condensable gas is CO_2 in a liquid-rich inclusion. The convenient and cost-effective method can be used to low saline liquid-rich inclusion frequently found in geothermal environment. The crushing experimental results were previously performed on transparent minerals from the four geothermal fields in Japan, containing Hohi (Sasada et al., 1986), Kakkonda (Muramatsu et al., 1996), Kirishima (Sawaki et al., 1997) and Matsukawa (Muramatsu et al., 2006).

For example, the crushing experimental results for the Matsukawa and Kakkonda geothermal fields are shown in table 2. The experiment was conducted on hydrothermal anhydrite collected from 916 m depth of well M-11 in Matsukawa. The collapse of vapor bubble was observed in six liquid-rich inclusions longer than 30μm. Based on the gas composition of individual liquid-rich inclusion obtained by QMS as mentioned in section 3.2.2, the dominant non-condensable gas in the inclusion is CO_2. Upon crushing, the vapor bubble gradually expanded until they filled the entire volume of two primary (T_h of 229 °C) and three secondary inclusions (T_h of 223 to 236 °C). This behavior suggests that the CO_2 content in these inclusions exceeded 0.2 mol%. Whereas, the bubble fills 80 % of the inclusion volume in one other primary inclusion (T_h of 227 °C), suggesting CO_2 content ranges from 0.1 to 0.2 mol%. According to Hedenquist and Henley (1985), dissolved CO_2 will contribute to the depression of the freezing-point in an inclusion. If the ice-melting temperatures (T_m) of the inclusions depend solely on their CO_2 contents, their maximum possible CO_2 contents can be calculated from their T_m values. In the case of the secondary inclusions in anhydrite from a depth of 916 m, their T_m values of - 0.8 and - 0.9 °C correspond to CO_2 content of 0.9 mol% (Sasada et al., 1992). The CO_2 contents of the secondary inclusions, therefore, range in 0.2 to 0.9 mol%. They are in good agreement with CO_2 content of 0.4 mol% (table 3) in the inclusions in the anhydrite analyzed by the QMS.

Table 2. Bubble behavior on crushing and estimated CO_2 contents of liquid-rich inclusions from the Matsukawa and Kakkonda geothermal fields

Well	Depth (m)	Mineral	Type	Bubble behavior	Th ( )		Tm ( )		CO_2 (mol%)	
Matsukawa										
M11	916	AH	Pr	fill the inclusion (2)*	229		3.0 to	4.5	0.2≦	
			Pr	expand, but not fill the inclusion (1)	227		3.8		0.1	0.2
			Sc	fill the inclusion (3)	223	236	0.8 to	0.9	0.2	0.9
Kakkonda										
Well-12	650	QZ	Pr	fill the inclusion (1)	254		0.9		0.2	0.9
			Sc	ditto (1)	229	231	0.4		0.2	0.4
			Sc	expand, but not fill the inclusion(1)	229	231	0.4		0.1	0.2
Well-73	50	CA	Sc	fill the inclusion (2)	150	157	0.1		0.1	
Well-42	600	CA	Pr	fill the inclusion (2)	275	302	0.4 to	0.7	0.2	0.8

* Number in parenthesis shows percentage.

The similar method by the crushing run was performed on hydrothermal quartz and calcite from three production wells in the Kakkonda geothermal field. The CO_2 content is 0.1mol% for two secondary inclusions in calcite from Well-73 and range in 0.2 to 0.8 mol% for two primary inclusions in calcite from Well-42. On the other hand, the CO_2 contents of the primary and secondary inclusions in quartz from 650 m of Well-12 range in 0.2 to 0.9 and 0.1 to 0.4 mol%, respectively. The CO_2 content of 0.11 mol% (table 3) in the inclusions analyzed by the QMS for quartz from 500m depth of Well-13 is the lowest of these ranges. According to Sasada et al. (1992) and Sawaki et al. (1997), the CO_2 contents of fluid inclusions estimated for quartz and anhydrite by microthermometry and crushing technique are consistent with those obtained by the QMS analysis at the Sengan, Yuzawa, Kurikoma, Hohi, Hishikari, Broadlands and Kirishima geothermal fields.

Table 3. Bulk gas compositions of fluid inclusions from geothermal fields

Field	Well	Depth (m)	Mineral	Gas composition (mol %)					Reference
				H_2O	CO_2	N	CH	Ar	
Mori	NT-306	360	Qz	99.1	0.95	0.085	0.00080	tr	Muramatsu et al.(2000)
	NB-1	395	Qz	98.7	1.06	0.075	0.0059	tr	
	NC-1	1135	Ah	98.7	1.13	0.10	0.0052	0.00006	
	ND-1	1200	Ah	98.9	0.87	0.088	0.0055	tr	
	ND-1	2260	Qz	92.4	2.0	5.1	0.058	0.0007	
	ND-1	2265	Qz	92.6	2.0	5.0	0.060	0.0004	
	ND-1	2265	Qz	91.5	2.4	5.8	0.063	0.0007	
	ND-6	600	Qz	99.2	0.81	0.034	0.0075	0.001	
	ND-6	600	Ah	98.7	1.23	0.11	0.0025	0.0006	
	ND-6	790	Ah	97.9	1.85	0.17	0.019	0.0004	
	ND-6	1680	Qz	98.4	1.56	0.055	0.0054	0.0001	
	ND-6	2000	Qz	98.1	1.21	0.24	0.032	tr	
	ND-6	2000	Ah	98.0	1.92	0.095	0.012	0.0002	
	ND-7	1585	Qz	99.3	0.56	0.045	0.010	tr	
	NF-3	1000	Ah	98.7	1.35	0.046	0.0019	tr	
	NF-6	2155	Ah	99.0	0.97	0.029	0.0014	0.00002	
	NF-9	1970	Qz	99.0	0.73	0.13	0.0078	tr	
	N2-KX-3	1288	Ah	99.1	0.88	0.073	0.00043	0.001	
Matsukawa	M11	850	Ah	99.4	0.60	0.031	0.0024	0.00006	Muramatsu et al.(2006)
	M11	916	Ah	99.5	0.40	0.034	0.0049	0.0017	
	M12	1097	Ah	96.5	2.6	0.48	0.018	0.0006	
Kakkonda	Surface	0	Qz	100.0	0.011	0.0055	0.00026	0.00001	Sawaki et al.(1999
	Surface	0	Qz	99.9	0.057	0.030	0.0013	0.00006	
	Surface	0	Qz	99.9	0.071	0.021	0.0019	0.00002	
	Surface	0	Qz	99.9	0.061	0.030	0.0021	0.00003	
	Well-13	500	Qz	99.8	0.11	0.046	0.0020	tr	
	Well-86	140	Qz	100.0	0.031	0.0083	0.0011	0.00009	
	Well-86	170	Qz	100.0	0.01	tr	tr	tr	

Table 3. Continued

Field	Well	Depth (m)	Mineral	Gas composition (mol %)					Reference
				H₂O	CO₂	N	CH	Ar	
	Well-18	982	Ah	99.3	0.67	0.19	0.0059	0.0004	Muramatsu et al.(2000)
	Well-18	1535	Ah	99.8	0.14	0.038	0.0050	0.00004	
	Well-18	1830	Ah	97.6	2.0	0.13	0.020	0.00008	
	Well-18	1911	Ah	99.6	0.35	0.041	0.0053	0.0006	
Sengan	SN-3	322	Qz	99.5	0.21	0.088	0.017	0.0001	Sasada et al. (1992)
Yuzawa	YO-3	1099	Qz	99.8	0.17	0.014	0.0010	0.00006	
Kurikoma	KR-1	1351	Qz	98.8	1.3	0.034	0.0024	0.0001	
Okuaizu	87N-18t	1365	Ah	99.4	0.64	tr	0.0001	nd	
Hohi	DW-5	1283	Qz	99.6	0.41	0.074	0.012	0.0002	
Kirishima	KT-5	1095	Qz	99.7	0.20	0.034	0.0050	0.0001	
Hishikari	HOSEN3-1	+25	Qz	99.9	0.14	0.00052	nd	nd	
Broadlands	BR-45	490	Qz	99.1	1.0	0.016	0.012	0.0002	
Kirishima	KT-1	969	Qz	99.6	0.30	0.022	0.0064	tr	Sawaki et al. (1997)
	KT-1	1269	Qz	99.2	0.68	0.13	0.012	0.00007	
	KT-4	382	Qz	99.7	0.21	0.030	0.0059	nd	
	KT-4	1157	Qz	99.6	0.20	0.047	0.0045	tr	
	KT-5	771	Qz	99.7	0.21	0.030	0.010	tr	
	KT-5	1095	Qz	99.7	0.22	0.034	0.0056	0.00007	
	KT-7	930	Qz	99.7	0.23	0.024	nd	nd	
	KT-7	1730	Qz	97.5	1.56	0.035	0.36	tr	
	KT-8	1179	Ah	99.6	0.21	0.052	tr	nd	
	KE1-3	1139	Ah	99.7	0.18	0.033	0.0056	nd	
	KE1-3	1139	Ah	99.8	0.097	0.031	0.0041	nd	
	KE1-4	1125	Ah	99.4	0.51	0.11	0.024	0.0004	
	KE1-5	1414	Qz	99.4	0.38	0.020	0.0019	0.00003	
	KE1-11	1100	Qz	99.5	0.34	0.011	0.0012	0.0001	
	KE1-19S	1300	Qz	99.4	0.32	0.089	0.013	0.0002	

In conclusion, the crushing experimental method is available for a convenient and semi-quantitative gas analysis of liquid-rich inclusion with CO_2 content lower than about 2.0 mol% in geothermal fields.

3.2. Quadrupole Mass Spectrometry

Gas chemistry of fluid inclusions in minerals provides us useful information of the evolution of ore-forming fluid and geothermal fluid, and furthermore the exploration for mineral deposit and geothermal field. For these purposes, various gas analytical methods such as gas chromatography, mass spectrometry, Raman spectrometry, FT-IR have been researched by several investigators (Shepherd et al., 1985). Although it is desirable to obtain the gas composition of individual inclusion, the volume of gases released from individual inclusion is generally small to obtain the quantitative value. But, the QMS can be obtained gas composition semi-quantitatively and quantitatively. The analytical system and method using ANELVA AGA-360 were reported in detail by Sasada et al. (1992). The gas chemistry of fluid inclusion performed by the equipment at the Institute of the Geological Survey of Japan (AIST) has been reported for several Japanese geothermal fields by Sasada et al.(1992), Sawaki et al. (1997, 1999), Sawaki and Muraoka (2002), Muramatsu et al. (1997, 2000, 2006).

3.2.1. Analytical Method

Quartz and anhydrite samples were treated with concentrated HCl, organic solvent and H_2O_2 to eliminate contamination by organic matter and other minerals such as calcite which produce excess gases by their decompositions. The samples were also checked microscopically to verify that they did not include other minerals. As calcite and wairakite decompose during heating for gas analysis and produce excess gases and water, the fluid inclusions have been previously studied only in quartz and anhydrite. Equipment for fluid inclusion gas analysis is shown in figure 7 (Sasada et al., 1992). The spectrometer was connected by stainless steel tubing to a capacitance manometer (MKS315), a cold trap, an infrared furnace and a vacuum system. Individual and bulk analyses of the inclusions were performed following the methods.

Individual Gas Analytical Method

Because a mineral has commonly trapped the various fluids progressed from its precipitation stage to the present in a inclusion, the gas analysis of individual inclusion is ultimate purpose to clarify fluid evolution in detail in a geothermal reservoir. Individual gas analysis determines semi-quantitatively the major gaseous species in the mass range from 2 to 70 for individual inclusion. About 10 mg of each sample was placed in a quartz glass tube and heated to 500 °C at a slow rate of 10 °C/min. Water vapor and other gases, released when an individual inclusion decrepitated during the heating run, were analyzed by the QMS run in a rapid-scanning mode. The analyzed fluids were mainly derived usually from the liquid-rich inclusions because the vapor-rich inclusions did not decrepitate at <500 °C.

Bulk Gas Analytical Method

The bulk gas analysis determines total compositions of fluid released from whole liquid-rich inclusions in each sample of about 0.3 g decrepitated during heating to 500 °C within a few minute. If one inclusion in large size is contained in the whole inclusions decrepitated, the bulk gas composition not always indicates mean value of whole inclusions because of significantly affected large inclusions. The gases released by decrepitation were collected in the stainless steel tubing. The gas pressure was measured by the capacitance manometer after separating the water with a cold trap to obtain the gas/water ratio. Ion counts were measured by QMS for each gas species except H_2O and converted to pressure. Calibrations were conducted using standards for CO_2, CH_4, N_2 and Ar. Quantification of reactive gases such as H_2S and SO_2 was not successful owing to the adsorption and reaction of these gases with the stainless steel tubing. Experimental precision is estimated to be better than 10 % for each of the gas species.

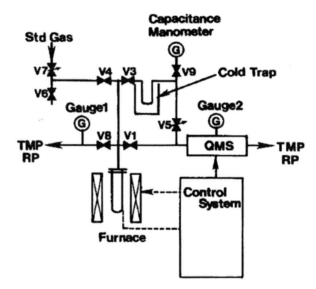

Figure 7. Equipment for fluid inclusion gas analysis (Sasada et al., 1992). Gauges 1 and 2, ionization gauges; V1-V4, V6, V8 and V9, bellows valve; V5 and V7, leak valve; TMP, turbomolecular pump; RP, rotary pump; QMS, quadrupole mass spectrometer.

3.2.2. Interpretation of Gas Analytical Data

Individual Gas Analytical Data

The gas analytical chart of individual liquid-rich inclusion trapped a liquid homogeneously in anhydrite from the Matsukawa geothermal field is shown in figure 8A.

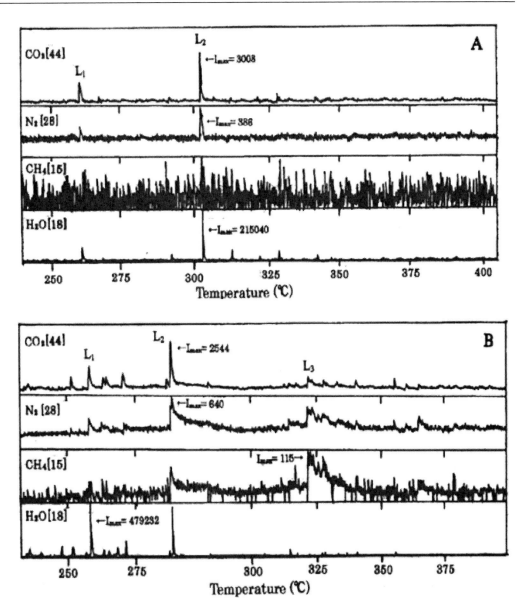

Figure 8. Gas analytical chart of individual liquid-rich inclusions in anhydrite from well M11 of the Matsukawa geothermal field (Muramatsu et al., 2006). (A) Anhydrite from 850 m depth; (B) Anhydrite from 916 m. Each peak corresponds to the decrepitation of a different fluid inclusion as the sample is heated. The charts show the mass spectrometer response at a mass/electron charge ratio of 44 (CO_2), 28 (N_2), 15 (CH_4) and 18 (H_2O). The mean heating rate was 10 °C/min. The Imax is the maximum ion count. The composition of liquid-rich inclusions L1, L2 and L3 are discussed in the text.

Decrepitation temperature ranges from 260 to 310 °C. Four charts show the mass spectrometer response at mass/electron charge ratio (m/z) = 44 for CO_2, 28 for N_2, 15 for CH_4 and 18 for H_2O. The main component is H_2O with small amounts of CO_2, N_2 and CH_4. The similar peak ratios of CO_2/H_2O (I_{CO2}/I_{H2O}) and N_2/H_2O (I_{N2}/I_{H2O}) were measured at two temperatures (e.g. L1 and L2 in figure 8a), suggesting that the anhydrite trapped a homogeneous liquid with nearly constant CO_2 and N_2 contents. The gas species such as

CO_2, N_2, CH_4 and Ar were usually detected in fluid inclusions from eleven geothermal fields, but H_2S was detected in fluid inclusions from the Kirishima geothermal field (Sasada et al., 1992).

The gas analytical chart of individual liquid-rich inclusion trapped a liquid heterogeneously in anhydrite from the Matsukawa field is shown in figure 8b. The peak ratios of CO_2/H_2O (e.g. L1, L2 and L3 in figure 8B) decrease with decrepitation temperature. These anhydrites may have trapped liquids that have undergone different degrees of boiling and vapor-loss. In this case, later trapped liquids are likely to have lower gas contents because of CO_2-rich vapor-loss during boiling of CO_2-rich fluid.

Bulk Gas Analytical Data

Degassing trend on a plot of log X_{CO2} versus temperature of liquid is illustrated in figure 9. The decrease in CO_2 content by vapor-loss was calculated for a single step separation process with adiabatic cooling by assuming the hypothetical initial liquid composition of 1.5 mol % CO_2 at 350 °C (open star). Equations for the vapor-liquid fractionation are taken from Giggenbach (1980). As boiling proceeds, CO_2 content decreases in the residual liquid phase because CO_2 preferably fractionate to vapor phase. If a fluid had boiled and all inclusions

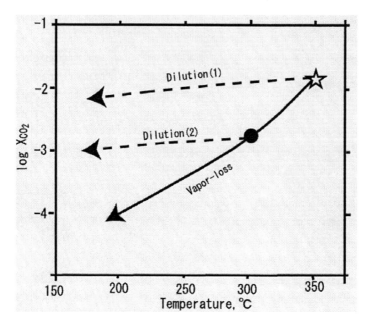

Figure 9. Schematic model to show general trends produced as a result of two processes operating on an initial reservoir liquid in terms of temperature and CO_2 content The decrease in CO_2 content by vapor-loss was calculated for a single-step separation process with adiabatic cooling by assuming the hypothetical initial reservoir liquid composition of 1.5 mol% CO_2 at 350 °C (open star). Equations for the vapor-liquid fractionation are taken from Giggenbach (1980). The schematic dilution curve (1) and curve (2) are the mixing lines between the air-saturated groundwater at 20 °C and the hypothetical initial reservoir liquid, and between the groundwater and the reservoir liquid differentiated at 300 °C by vapor-loss, respectively.

were trapped simultaneously in a hydrothermal mineral from the reservoir fluid, we can use the minimum Th values of all Th data because of the same as trapping temperature (Bodnar et al., 1985). If a fluid had no boiled, we may use Th value with high frequency. In this case, we must check whether inclusions are much the same in size, because the bulk composition is influenced by fluid composition of a large inclusion. Degassing trend on a plot of log (X_{CO2}/X_{N2}) versus log (X_{CO2}/X_{CH4}) of a liquid is illustrated in figure 10. The schematic vapor-loss curve was calculated for a single-step separation with adiabatic cooling by assuming the liquid contained 2.6 mol % CO_2, 0.5 mol % N_2 and 0.03 mol % CH_4 at 240 °C (open star). As boiling proceeds, less soluble gases preferably fractionate to vapor phase in the order of N_2, Ar, CH_4 and CO_2 (Giggenbach,1980; Potter and Clynne,1978). As a result, both the CO_2/CH_4 and CO_2/N_2 ratios increase in the residual liquid phase with progressive degassing.

Dilution trends on a plot of log X_{CO2} versus temperature of liquid are also illustrated in figure 9. The schematic dilution curve (1) and curve (2) are the mixing lines between air-saturated groundwater at 20 °C and the hypothetical initial liquid of 1.5 mol %CO_2 at 350 °C (open star), and between the groundwater and the liquid differentiated at 300 °C by vapor-loss, respectively.

3.2.3. Bulk Gas Composition of Fluid Inclusion from Geothermal Fields in Japan

Table 3 lists the bulk gas composition of liquid-rich inclusions in quartz and anhydrite from twelve geothermal fields in Japan and New Zealand, including Mori (Muramatsu et al., 1997), Matsukawa (Muramatsu et al., 2000), Kakkonda (Sawaki et al., 1999; Muramatsu et al., 2000), Sengan, Yuzawa, Kurikoma, Okuaizu, Hohi, Hishikari, Broadlands (Sasada et al., 1992) and Kirishima (Sasada et al., 1992; Sawaki et al., 1997). The water contents range from 91.5 to 100.0 mol%. Carbon dioxide is the major non-condensable gas in all samples, containing less than 2.6 mol%. Nitrogen and CH_4 contents are less than 5.8 mol% and less than 0.063 mol%, respectively, and generally about one and two orders of magnitude lower than CO_2 content, respectively. Argon is present at less than 0.002 mol%. Gas contents in quartz from 2260 to 2265 m depths in well ND-1 in the Mori field are the highest of all measured samples.

The gas compositions of the present reservoir fluids from the geothermal fields are shown in table 4. They were calculated using the chemical compositions of steams discharged from the geothermal wells and the quartz or the Na-K-Ca geothermometers. The CO_2 and CH_4 contents in the inclusion fluids are slightly to one order of magnitude higher than those in the present reservoir fluids. In contrast, the N_2 contents are generally about one to three orders of magnitude higher in the former than in the latter.

Table 4. Gas compositions of present geothermal fluid from geothermal fields

Field	Well	Tr (°C)	Gas composition (mol %)						Reference
			CO_2	N	CH	Ar	H_2S	H_2	
Mori	NT-306	224[*1]	0.472	0.00078	0.00057	-	-	-	Muramatsu et al.(2000)
	NC-1	219[*1]	0.937	0.00301	0.00170	-	-	-	
	ND-1	249[*1]	0.724	0.00398	0.00399	-	-	-	
	ND-6	259[*1]	0.515	0.00359	0.00286	-	-	-	
	NF-9	245[*1]	0.619	0.00429	0.00388	-	-	-	
Matsukawa	M5		0.129	0.00110	0.00025	-	0.058	0.00245	Muramatsu et al.(2006)
	M6		0.255	0.00268	0.00075	-	0.058	0.00361	
	M7		0.189	0.00207	0.00058	-	0.055	0.00385	
	M11		0.188	0.00432	0.00143	0.00006	0.042	0.00499	
Kakkonda	Well-12	233[*1]	0.008	-	-	-	-	-	Muramatsu and Komatsu (1996)
	Well-62	236[*1]	0.007	-	-	-	-	-	
	Well-73	215[*1]	0.011	-	-	-	-	-	
Sengan (Sumikawa)	S-2	240[*2]	0.0044	0.010	0.00011	0.00022	-	-	Sasada et al. (1992)
	S-4	300[*2]	0.0172	0.0015	0.000069	0.000023	-	-	
Yuzawa	YO-3	290[*2]	0.0047	nd	nd	nd	-	-	

Table 4. Continued

Field	Well	Tr (°C)	Gas composition (mol %)						Reference
Okuaizu	84N-6T	283*2	0.45						
	87N-14T	262*2	0.57						
	87N-15T	299*2	0.16						
Hohi	DY-1	186*2	0.0032	0.00073	0.000007	0.000013	-	-	
Broadlands	BR-11	284*2	0.178	0.00152	0.00333	nd	-	-	
	BR-25	300*2	0.524	0.00653	0.0145	nd	-	-	
Kirishima	KT-5	317*2	0.2080	0.0057	0.012294	-	-	-	Sawaki et al. (1997)
	KT-8	281*2	0.0004	0.0002	0.000184	-	-	-	
	KE1-3	224*2	0.0026	0.0003	0.000161	-	-	-	
	KE1-4	229*2	0.0018	0.0008	0.000121	-	-	-	
	KE1-5	236*2	0.0053	0.0035	0.001068	-	-	-	
	KE1-11	265*2	0.0004	-	0.000006	-	-	-	
	KE1-19S	234*2	0.0015	0.0000	0.000008	-	-	-	

Tr, reservoir temperature; nd, not detected. *1Temperature was calculated using the quartz geothermometer with adiabatic cooling (Truesdell, 1976), *2Temperatures estimated from the Na-K-Ca geothermometer (Fournier and Truesdell, 1973). Gas composition was calculated using the chemical composition of steam discharged from geothermal well and Tr.

According to Sasada et al. (1992) and Sawaki et al. (1997), the CO_2 contents in the inclusion fluids of hydrothermal quartz and anhydrite are generally about one order of magnitude higher than those in the present discharged fluids of the Sengan, Yuzawa, Kurikoma, Okuaizu, Hohi, Kirishima, Hishikari and Broadlands geothermal fields. It suggests that early stage fluids were trapped in the fluid inclusions before considerable boiling and vapor-loss. Based on the differences in CO_2/CH_4, CO_2/N_2 and N_2/Ar ratios between the inclusion fluids and the present reservoir fluids, the fluid evolution in the Kirishima and Kurikoma fields could explained by degassing and dilution, respectively (Sasada et al.,1992; Sawaki et al.,1997). The fluid must be evolved by both degassing and dilution in Sengan (Sasada et al., 1992). On the contrary, the similarity in CO_2/CH_4 and CO_2/N_2 ratios between the inclusion fluids and the present reservoir fluids in Broadlands might show a small time interval between fluid trapping in the hydrothermal quartz and present geothermal activity (Sasada et al., 1992).

The gas analytical results of fluid inclusion from the Mori, Matsukawa and Kakkonda geothermal fields are discussed in section 3.3.

3.3. Case Studies of Fluid Inclusion from Geothermal Fields, Northeastern Japan

The northeastern Japan is one of the most important regions for Japanese geothermal exploitation, and produces about 60 per cent of the nation's total installed geothermal power capacity. The results of microthermometric measurement and gas analysis on fluid inclusions from the Mori and Kakkonda liquid-dominated and the Matsukawa vapor-dominated geothermal fields (figure 1), northeastern Japan are discussed in this section. The Mori geothermal field is located in the Nigorikawa Basin, about 20 km northwest of Mt. Komagatake, an active volcano, southwestern Hokkaido. The Hachimantai volcanic area located in northeastern Honshu is one of the important areas for Japanese geothermal exploitation, contains the Matsukawa and Kakkonda geothermal fields. The Th and ice melting temperature (Tm (ice)) of fluid inclusion were measured on a USGS-type gas flow heating/freezing stage, and gas composition of liquid-rich inclusion was analyzed by the QMS at the Institute of the Geological Survey of Japan (AIST).

3.3.1. Mori Geothermal Field

The Mori geothermal power plant (50MWe) has been in operation since 1982. Microthermometry and gas analysis of fluid inclusion from the Mori geothermal field were published by Muramatsu et al. (1997, 1999). The genesis of the inclusion fluid and the gas evolution of the reservoir fluid based on the fluid inclusion studies are discussed below.

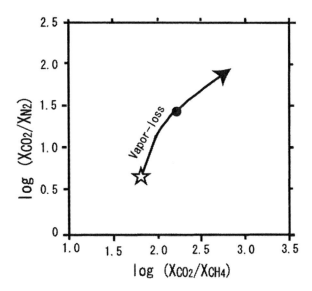

Figure 10. Schematic model to show general trend as a result of degassing process operating on an initial reservoir liquid in terms of log (X_{CO2}/X_{N2}) versus log (X_{CO2}/X_{CH4}) of an initial reservoir liquid. The schematic vapor-loss curve was drawn through calculation for a single-step separation with adiabatic cooling by assuming the hypothetical initial reservoir liquid composition of 3.2 mol% CO_2, 0.6 mol% N_2 and 0.05 mol% CH_4 at 265 °C (open star). Equations for the vapor-liquid fractionation are taken from Giggenbach (1980). The schematic dilution curve was drawn by assuming the mixing of the air-saturated groundwater at 20 °C with the differential reservoir liquid at 260 °C formed by boiling of the hypothetical initial reservoir liquid.

Formation of Ca-Rich Hypersaline Brine and CO_2-Rich Fluid

Analyzed samples were hydrothermal quartz and anhydrite (ϕ0.5mm thick) selected from cuttings and a boring core from geothermal wells in the field. Fluid inclusions in the minerals are divided into four types based on phase relations at room temperature and cooling microthermometry. Type A is a liquid-rich inclusion with low- to moderate-salinity homogenized to liquid phase. Type B is a vapor-rich inclusion homogenized to vapor phase. Type C is a vapor- and CO_2-rich inclusion with low-salinity less than 3.8 wt% NaCleq., homogenized to vapor phase. Type D is a Ca-rich polyphase inclusion homogenized to liquid phase. In particular, Type D inclusion is associated with Types B and C inclusions.

Because the minimum Th of Type A inclusions is in general agreement with the static borehole temperatures after a long standing time of 13 days as mentioned in section 2.1 (figure 3), the reservoir fluid formed by the recent geothermal activity must be trapped in parts of the inclusions. Relations among Tm (ice), Tm (halite) and Th of type D inclusions in anhydrite from 2000 m in depth of well ND-6 are shown in table 5. Tm (halite) and Th range from 132 to 160 °C and from 250 to 265 °C, respectively. Tm (ice) of -40.1 to -43.6 °C in the presence of hydrohalite are lower than the eutectic temperature of the system NaCl-H_2O (-21.2 °C; Hall et al., 1988). It indicates that the brine is not simply NaCl-H_2O binary fluid, but contains significant amounts of additional aqueous components. Considering the eutectic temperatures of the ternary system NaCl-$FeCl_2$-H_2O (-37.0 °C; Borisenko,1977), NaCl-KCl-H_2O ternary (-22.9 °C; Sterner et al.,1988), NaCl-$CaCl_2$-H_2O ternary (-55.0 °C;

Borisenko,1977), the ice melting data strongly suggests that aqueous Ca comprises a significant component, even though K or Fe may be contained as a minor component in these liquid-rich and polyphase inclusions. The NaCl-CaCl$_2$-H$_2$O ternary model composition was therefore determined by a graphical method (Vanko et al., 1988) based on Tm (ice) and Tm (NaCl) for the type D inclusions. As a result, CaCl$_2$/NaCl weight ratio ranges from 2.2 to 2.9, and the ternary fluid compositions are 9 to 11 wt% NaCl and 24 to 26 wt% CaCl$_2$, or a total salinity of 33 to 35 wt% (table 4).

Table 5. Microthermometric data and ternary fluid compositions of type D inclusions in anhydrite from the Mori and Kakkonda fields (Data from Muramatsu et al., 1999, 2000)

Inclusion number	Tm(ice) (℃)	Tm(halite) (℃)	Th (℃)	Fluid composition (wt%)		
				H$_2$O	NaCl	CaCl$_2$
2000m depth of well ND-6 in the Mori field						
1	41.2	160	250	65	11	24
2	43.6	152	263	65	9	26
3	42.2	155	259	65	10	25
4	40.1	132	265	67	9	24
5	43.3	147	258	65	9	26
1875m depth of Well-19 in the Kakkonda field						
6	40.3	163	300	65	11	24

All Tm (ice) values were measured in the presence of hydrohalite.

According to Muramatsu et al. (1999), four types of the inclusion fluids have been generated by the following process. Phase separation due to boiling of initial homogeneous hot seawater at the deeper parts of the reservoir might have produced Ca-rich hypersaline brines, and vapor- and CO$_2$-rich low saline fluids at the earliest stage of fluid evolution in the Mori geothermal system. Afterwards, the low saline aqueous fluids derived from the mixture of seawater and meteoric water have been dominant in the reservoir since the recent natural state stage before operation of the Mori geothermal power plant in 1982.

Gas Evolution in the Reservoir Fluid

Figure 11 illustrates variation of CO$_2$ content with temperature for the inclusion fluids (table 3) and the present reservoir fluids at the initial stage of exploitation (table 4). The CO$_2$ contents in the inclusion fluids in quartz are roughly plotted along the schematic vapor-loss curve by a single step separation, suggesting that the CO$_2$ contents in the inclusion fluids in quartz are considerably dominated by CO$_2$ degassing. Figure 12 illustrates the correlation between CO$_2$/N$_2$ and CO$_2$/CH$_4$ ratios for the inclusion fluids and the present reservoir fluids. quartz are considerably controlled by degassing.

The CO$_2$/N$_2$ ratios generally increase with increasing the CO$_2$/CH$_4$ ratios for the inclusion fluids in quartz. This trend is roughly plotted along the schematic vapor-loss curve by a single step separation, suggesting that the CO$_2$/N$_2$ and CO$_2$/CH$_4$ ratios of the inclusion fluids in quartz are considerably controlled by degassing.

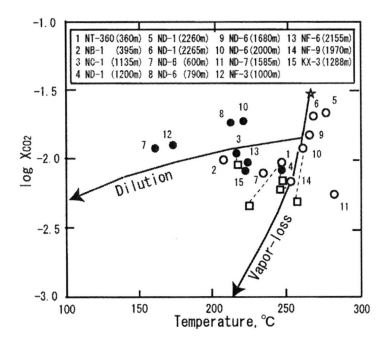

Figure 11. Plot of log XCO2 versus temperature diagram for the inclusion fluids and the present reservoir fluids at the initial stage of exploitation of the Mori geothermal field (Muramatsu et al., 1997). Open and solid circles are quartz and anhydrite, respectively. Open rectangular shows the present reservoir fluid at the initial stage of exploitation. The schematic vapor-loss curve was drawn through calculation for a single step separation with adiabatic cooling by assuming the hypothetical initial reservoir liquid composition of 3.2 mol% XCO2 at 265 °C (open star). The schematic dilution curve is the mixing line between the air-saturated groundwater at 20 °C and the differential reservoir liquid at 260 °C formed by boiling of the hypothetical initial reservoir liquid. Dotted line joins the present reservoir fluid from the main inflow depth of five wells and the inclusion fluid in the analyzed samples from the depth near the point.

On the contrary, $\log CO_2$ value of the inclusion fluids in anhydrite has relatively narrow ranges between -1.7 and -2.1, although fluid temperatures are widely distributed toward the low temperature side (figure 11). It thus reveals that the inclusion fluids in anhydrite are affected by dilution in addition to degassing. The difference of CO_2 contents in anhydrite between 600 m depth and the two other deeper depths (790 and 2000m) in well ND-6 can be explained by the degree of dilution. Additionally, the inclusion fluids in quartz from the shallow depth of 395 m in well NB-1 might contain the most abundant groundwater of all measured quartz samples (figure 11). As shown in figure 12, the inclusion fluids in anhydrite are plotted on the higher CO_2/N_2 ratio side of the vapor-loss curve. It might indicate that anhydrite have trapped the reservoir fluids which are affected by degassing, in addition dilution with air-saturated groundwater which has relatively low CO_2/N_2 ratio and relatively high CO_2/CH_4 ratio. According to Arai (1994), anhydrite had started to form from hydrothermal solution posterior to quartz, which has formed till after intrusion of hornblende andesite. From these results and the mineral paragenesis, compared to the earlier quartz formation stage, the reservoir fluid seems to be still affected by dilution with groundwater at the later anhydrite formation stage. Moreover, the reservoir fluid seems to be affected by dilution more at the shallower depths than at the deeper depths.

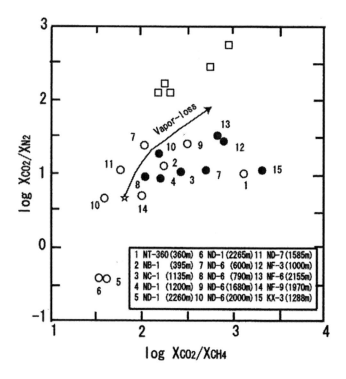

Figure 12. Plot of log (X_{CO2}/X_{N2}) versus log (X_{CO2}/X_{CH4}) diagram for the inclusion fluids and the present reservoir fluids at the initial stage of exploitation of the Mori geothermal field (Muramatsu et al., 1997) Symbols are the same as for figure 11. The schematic vapor-loss curve is the same as for figure 10.

Dotted lines in figure 11 join the present reservoir fluid from the main feed point of five wells (table 4) and the inclusion fluid in the analyzed samples from the depth near the point. These lines, especially of wells NT-306 and ND-6, are nearly parallel to the schematic degassing curve, probably suggesting that the CO_2 degassing has proceeded from the quartz-anhydrite mineralization stage to the present.

3.3.2. Matsukawa Geothermal Field

The Matsukawa geothermal field is the only vapor-dominated geothermal resource so far developed in Japan. The country's first geothermal power plant and the world's fourth, a 9.5 MWe facility, was brought on line at Matsukawa in 1966, and upgraded to 23.5 MWe in 1993. The conceptual model of the Matsukawa geothermal system has been discussed by several authors (e.g. Nakamura, 1967; Ide, 1985; Akazawa and Muramatsu, 1988; Hanano and Matsuo, 1990; Ozeki et al., 2001). Hydrothermal alteration in the Matsukawa geothermal field and the surrounding area has been described by Sumi (1966, 1968) and Kimbara (1983), who defined two alteration periods: an early alkaline stage and a later acidic stage. The widespread occurrence of pyrophyllite and diaspore within the reservoir rocks in the western upflow zone of the field are indicative of extensive fluid-rock interactions with acidic fluids (Hemley et al., 1980). According to Muramatsu et al. (2006), the textural relationships of the hydrothermal minerals in cores from well M11 drilled in this zone suggest that the acidic hydrothermal fluids were involved in the hydraulic fracturing of the diorite porphyry, and caused the subsequent formation of the early-stage pyrophyllite and diaspore, and the late-stage deposition of anhydrite and quartz in open fractures. Microthermometric measurements and gas analyses of fluid inclusions were performed on anhydrite and quartz, which were formed from the late acidic reservoir fluid, collected from the production wells drilled in the western upflow zone (Muramatsu et al., 2006). Fluid evolution of the acidic reservoir fluid in the western upflow zone as specific topics of the inclusion studies of the field are discussed below.

The T_h values of liquid-rich inclusions in anhydrite and quartz from wells M5, M7 and M11 suggest that reservoir temperature had decreased from the pyrophyllite-diaspore to the anhydrite-quartz stage. The reservoir temperature was up to several tens of degrees lower at the anhydrite-quartz stage than at the present reservoir temperature. It suggests that subsequent reheating of the reservoir has occurred in response to renewed magmatic activity. Absolute ages of the altered rocks and stratigraphic relationships lead us to infer that this reheating may have occurred within the last 100,000 - 5000 years.

The majority of the inclusions mostly less than 10 μm in length contain only liquid and water vapor at room temperature. Liquid- and vapor-rich inclusions are intimately associated in many samples, providing widely unequivocal evidence of boiling at the anhydrite-quartz stage. The precipitation of anhydrite from a boiling fluid is also supported by the inverse correlation between their T_m and T_h values of individual liquid-rich inclusions in anhydrites from a depth of 916 m in well M11, and the semi-quantitative gas analyses mentioned in section 3.2.2 (figure 8B). The reservoir fluids trapped in anhydrite had temperatures of up to 257 °C, CO_2 concentrations in the 0.4 to 2.6 mol% ranges, and salinities of 1.9 to 11.3 wt% NaCleq.. This compositional variation may be ascribed to the vapor-loss occurred by boiling.

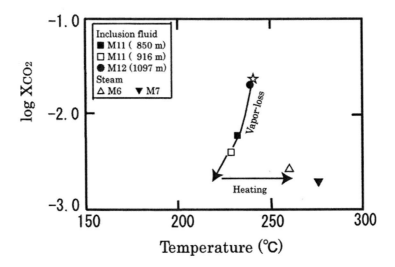

Figure 13. Plot of log X_{CO2} versus temperature of liquid-rich inclusions in anhydrites from wells M11 and M12 of the Matsukawa geothermal field (Muramatsu et al., 2006). The decrease in CO_2 content during vapor-loss was calculated for a single-step separation process with adiabatic cooling by assuming that the initial reservoir liquid contained 2.6 mol % CO_2 at 240 °C (open star).

Figure 13 illustrates the relationship between temperature and CO_2 content of liquid-rich inclusions in anhydrite from wells M11 and M12 and of steam from wells M6 and M7 drilled near well M11. The CO_2 contents decrease just along a vapor-loss curve with temperature. As the inclusion liquids from depths of 850 and 916 m in well M11 have undergone more extensive CO_2 degassing than the liquid trapped at 1097 m depth in well M12, the reservoir fluid seems to have flowed upward with vapor-loss during the formation of anhydrite. The relationships in figure 13 moreover reveal that the present-day steam could have formed as a result of the recent heating of degassed fluids similar to those from well M11.

The CO_2 and CH_4 contents of the inclusion fluids in anhydrite from well M11 are slightly higher than those in the steam discharged from the well (table 1; Yoshida and Ishizaki, 1988). In contrast, the N_2 contents of the inclusion fluids are one order of magnitude higher than that in the steam. Figure 14 illustrates the correlation between CO_2/CH_4 and CO_2/N_2 ratios of the inclusion fluids in anhydrite from wells M11 and M12. As the CO_2/N_2 ratios in the inclusion fluids increase roughly along a vapor-loss curve with CO_2/CH_4 ratios, vapor-loss due to boiling provides a reasonable explanation for the variations in the gas ratios.

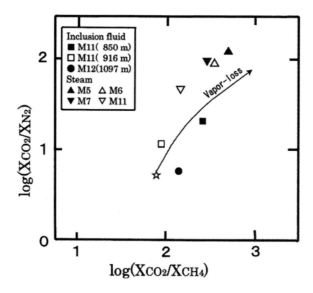

Figure 14. Plot of log (X_{CO2}/X_{N2}) versus log (X_{CO2}/X_{CH4}) of liquid-rich inclusions in anhydrites from wells M11 and M12 of the Matsukawa geothermal field (Muramatsu et al., 2006). The schematic vapor-loss curve was calculated for a single-step separation with adiabatic cooling by assuming the reservoir liquid contained 2.6 mol % CO_2, 0.5 mol % N_2 and 0.03 mol % CH_4 at 240 °C (open star).

3.3.3. Kakkonda Geothermal Field

The Kakkonda I geothermal power plant (50 MWe) has been in operation since 1978, afterward the Kakkonda II geothermal power plant (30 MWe) began to generate in 1996. In the Kakkonda geothermal field, more than 70 geothermal wells ranging from 500 to 3000 m in depth have been drilled, and furthermore the deep geothermal exploration well WD-1a was drilled from 1994 to 1995 to a depth of 3729 m, deepest of all geothermal wells in Japan as a principal component of the Japanese national project for the "Deep-Seated Geothermal Resources Survey" by NEDO. Fluid inclusion from these wells has been researched by several investigators (Muramatsu, 1984, 1987; Komatsu and Muramatsu, 1994; Sasaki et al., 1995; Muramatsu and Komatsu, 1996; Komatsu et al., 1998; Sawaki et al., 1999; Muramatsu t al., 2000). The conceptual model of the Kakkonda geothermal system, and the gas evolution and the origin of the reservoir fluid are discussed below as specific topics of the inclusion studies of the field.

Characterization of the Upflow Zone

Based on the fluid inclusion study, the conceptual model of the Kakkonda geothermal system is shown in figure 15 (Komatsu et al., 1998). The deep seated Kakkonda granite intrusion (K-Ar ages of hornblende of 0.24 to 0.11 Ma) is a heat source of the Kakkonda geothermal system (Doi et., 1998). The system is strongly controlled by fractures (Doi et al., 1988), and the present reservoir fluid is flowing upward through the open space of the fracture where hydrothermal minerals such as quartz were precipitated in the shallow reservoir (Muramatsu and Komatsu, 1996).

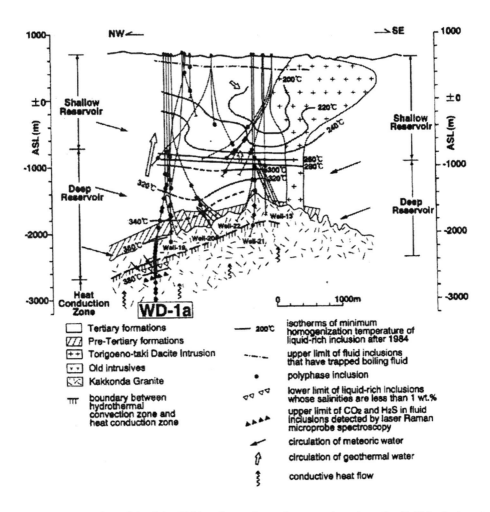

Figure 15. Conceptual model of the Kakkonda geothermal system based on the fluid inclusion study (Komatsu et al., 1998).

Fluid inclusions in transparent minerals are classified into two-phase, vapor-rich and liquid-rich inclusions, and polyphase inclusions comprising liquid, vapor and solids. As mentioned in section 2.1, the minimum Th of the inclusions in hydrothermal quartz, anhydrite and calcite, and igneous quartz from production wells is in good agreement with the measured borehole temperature after a long standing time in the shallow (1400 m depth) and deep reservoirs (1400 m depth). Three dimensional distribution of the minimum Th of the liquid-rich inclusions from many production and reinjection wells suggests that high temperature reservoir fluid has been ascending from the northwestern depth during the present magmatic activity related to Kakkonda granite. The upflow zone is presented not only by the three dimensional distribution of Th value at the surface, but also by three dimensional distribution of polyphase inclusion. Polyphase inclusion observed in quartz and anhydrite is widespread in the deep reservoir, whereas is restricted at the northwestern area in the shallow reservoir. The upper limit of three dimensional distribution of anhydrite corresponds with the up flow zone, being indicative the upflow zone of the field.

Based on the results and the geological, geochemical and geophysical data in the field, the exploration well WD-1a was drilled to the direction of the deep reservoir in the upflow zone at the contact of the Kakkonda granite at a depth near 3000 m. According to Komatsu et al.(1998), the minimum Th of liquid-rich inclusions in both igneous and hydrothermal quartz samples from depths of 3700 and 3728 m of the well has a maximum value ranging from 483 to 507 °C of all wells in the field. The similar temperature was obtained using melting tablets after standing times of 129 and 159 hours at 3700 m (500 to 510 °C), and moreover estimated using Horner plots at 3500 m (501 °C; Ikeuchi et al.,1998).

Gas Evolution in the Reservoir Fluid

Gas chemistry of fluid inclusion was performed on hydrothermal quartz and anhydrite in order to discuss gas evolution on the reservoir fluid at the Kakkonda geothermal field (Sawaki et al., 1999; Muramatsu et al., 2000). Under the microscope, liquid-rich inclusions are mostly observed on these minerals from the shallow reservoir. But both liquid-rich and vapor-rich inclusions are observed on anhydrite from the deep reservoir, suggesting that fluid inclusions were widely trapped under boiling condition. As shown in figure 16, the CO_2 contents in the liquid-rich inclusions of these minerals decrease with decreasing temperatures. Figure 16 also reveals that the reservoir liquid was made by differentiation of the initial reservoir fluid due to vapor-loss at deeper depths, and was made by dilution of the residual liquid with the meteoric water at shallower depths. The dilution degree increases with decreasing the depth. On the other hand, the present reservoir fluids in the shallow reservoir are plotted on the higher temperature side of the vapor-loss curve calculated by a single-step vapor separation, suggesting that they may be formed through multi-step vapor separations.

It is significant problem when the CO_2 content in the reservoir fluid decreased. Figure 17 shows variation of temperature and CO_2 content of the reservoir fluid from three production wells drilled in the shallow reservoir with time during about two years after the end of drilling. Although the steam and hot water discharged from these wells were collected after stopping of affection of a drilling mud to fluid composition, the CO_2 content of 0.232mol% in the reservoir fluid at the end of drilling of well-62 is nearly equal to that in the inclusion fluid (about 0.1 to 0.7 mol%; table 3). As shown in figure 17, after the CO_2 content in the reservoir fluid from well-62 abruptly decreases during discharge testing of the well, it is constant of low values (0.006 to 0.007 mol%). The CO_2 contents in the reservoir fluid from well-12 and well-31 are also constant at the low values since three month intervals after drilling of the wells. It thus appears that the CO_2 content in the reservoir fluid decreases two orders of magnitude during progressing of the exploitation of the field.

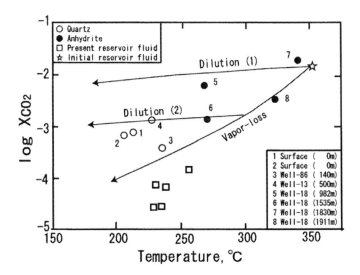

Figure 16. Plot of log X_{CO_2} versus temperature of liquid-rich inclusions in anhydrites from Well-18 of the Kakkonda geothermal field (modified from Muramatsu et al., 2000). The data for quartz is after Sawaki et al. (1999). The schematic vapor-loss curve and schematic dilution curves (1) and are the same as for figure 9.

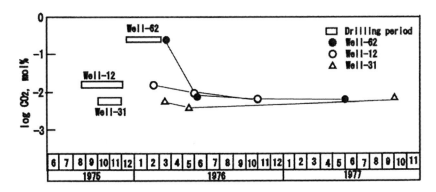

Figure 17. Variation of CO_2 content in the reservoir fluid from three shallow production wells with time during about two years after the end of drilling of the wells in the Kakkonda geothermal field (Muramatsu and Komatsu, 1996).

Origin of the Reservoir Fluid

Muramatsu et al. (2000) performed microthermometry of fluid inclusion in anhydrite to discuss origin of the reservoir fluid at the Kakkonda geothermal field. The Tm(ice) values of liquid-rich inclusions in anhydrites from the shallow reservoir are in high value and narrow range above -0.8 °C, whereas those from the deep reservoir range from -0.2 to -40.6 °C (table

3). The Tm (ice) values of -40.3 to -40.6 °C of the liquid-rich and the polyphase inclusions in anhydrite from the deep reservoir strongly suggest that aqueous Ca comprises a significant component, even though K or Fe may be contained as a minor component in these liquid-rich and polyphase inclusions. The NaCl-CaCl$_2$-H$_2$O ternary model composition was therefore determined by a graphical method (Vanko et al., 1988) using Tm (ice) and Tm (NaCl) for the polyphase inclusion. The estimated composition is 11 wt.%NaCl and 24 wt.%CaCl$_2$ with total salinity of 35 wt.% (table 6). Total salinities of the liquid-rich inclusions were estimated to range from 22 to 29 wt.%NaCl+CaCl$_2$eq. on the basis of the ice melting isotherms and the liquid+ice+liquid+hydrohalite boundary (Oakes et al., 1990).

According to microthermometric and SEM-EDX studies of secondary polyphase inclusions in igneous quartz in the Kakkonda granite (Sasaki et al., 1995), the inclusions contain hypersaline fluids of 35 to 75 wt%NaCl eq., and daughter minerals consist of halite, Fe-bearing chloride (containing minor amounts of Mn, K, Ca), sylvite (containing minor amounts of Mn, Fe, Na), iron oxide (magnetite or hematite). Based on chemical analysis of polyphase inclusion from a quartz vein in the Kakkonda granite using micro LA-ICP-MS, major components are Na, K, Ca, Mn, Fe and Cl with minor amounts of B, Cu, Zn, Pb and Ba, and trace elements of Li, Mg, Al, Rb and Sr (Sasaki et al., 1998). These hypersaline fluid in polyphase inclusions was produced either by direct exsolution of the hypersaline fluid from a magma (Roedder and Coombs, 1967; De Vivo and Frezzotti, 1994) formed the Kakkonda granite or boiling of an aqueous solution (Takenouchi and Kennedy, 1965; Bowers and Helgeson, 1983). Above polyphase inclusion data suggests that the hypersaline fluid was released from the crystallizing magma itself.

The metal bearing hypersaline fluid was found from the deep exploration well WD-1a in the Kakkonda granite (Kasai et al., 1998). Since the well did not encounter productive fracture zone, fluid sample could not be directly obtained from the well. It was therefore collected from 3708 to 3589 m in depths by reverse circulation using the drillpipe like a straw to avoid contamination of the deep reservoir fluid with the shallow reservoir fluid within the well. The hypersaline fluid consists of NaCl (15.0 wt%), FeCl$_2$ (9.7 wt%), KCl (7.0 wt%), CaCl$_2$ (4.5 wt%) and MnCl$_2$ (2.5 wt%), ZnCl$_2$ (0.64 wt%) and PbCl$_2$ (0.14 wt%) with a total salinity of 39.5 wt%. According to the estimation of the original hypersaline liquid prior to dilution by circulation water using a tritium-salinity mixing model by Kasai et al.(1998), it has a salinity of about 55 wt%NaCl eq., consisting of Na-Fe-K-Ca-Mn chloride, rich in Zn and Pb but poor in Cu, Au and Ag. It thus has become clear that the hypersaline fluid in polyphase inclusions is derived mainly from a residual magmatic fluid formed the Kakkonda granite.

Figure 18 illustrates variation of total salinities with homogenization temperatures (Th) of fluid inclusions (Muramatsu et al., 2000). The salinities of the liquid-rich inclusions with lower saline fluids widely vary from 0.9 to 29 wt%NaCl+CaCl$_2$ eq. at Th values of 320 to 360 °C in the deep reservoir. The lower saline fluids might have been produced by dilution of the hypersaline fluid and boiling of the dilute fluids. The dilution should have been caused by mixing with low saline fluids with temperatures of 320 to 360 °C. As the liquid-rich inclusions from the shallow reservoir contain low saline fluids ($\leqq$1.2 wt%NaCl eq.), the low saline fluids are likely to be derived from a meteoric water conductively heated by the Kakkonda granite intrusion.

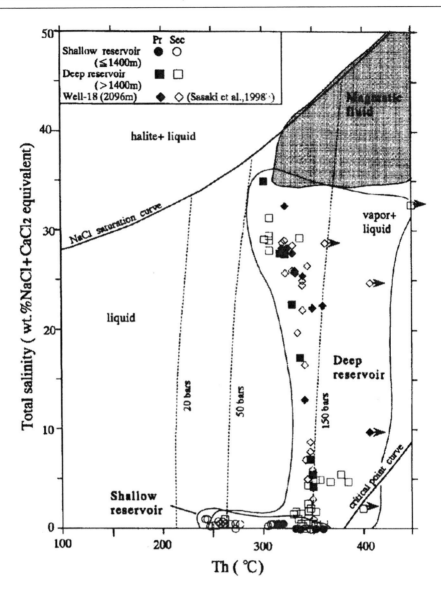

Figure 18. Variation of total salinities with homogenization temperatures (Th) of fluid inclusions in anhydrites from the Kakkonda geothermal field (Muramatsu et al., 2000) The Pr and Sec indicate primary and secondary inclusions, respectively. Short arrows indicate that the Th values were obtained as minimum estimates. The NaCl saturation curve and critical point curve were drawn for the system NaCl-H$_2$O after Sterner et al. (1988) and Bischoff and Pitzer (1989), respectively. The isobaric curves at pressures of 20 and 50 bars are after Haas (1976), and 150 bars after Sourirajan and Kennedy (1962).

4. Conclusion

Fluid inclusion trapped ancient to present reservoir fluid in transparent mineral offers us information about the change of the physicochemical properties such as temperature, salinity and chemical composition of the fluid in geothermal fields. The minimum Th value can be used to estimate the reservoir temperature with real time at a development site in most geothermal fields, Japan. Semi-quantitative convenient gas analysis of liquid-rich inclusion using a crushing experiment and microthermometry is available to estimate the CO_2 content in inclusion fluid lower than approximately 2.0 mol%. To the contrary, the gas analytical method using a quadrupole mass spectrometer is very useful to obtain not only semi-quantitatively non-condensable gas content in individual inclusion, but also bulk gas composition in transparent mineral. This method offers furthermore us significant information about the fluid evolutions such as degassing and dilution by groundwater.

Based on the gas analysis of liquid-rich inclusions in hydrothermal quartz and anhydrite from most Japanese geothermal fields, the main component is H_2O with small amounts of CO_2, N_2 and CH_4 and Ar. The CO_2 content in the inclusion fluid is generally higher than those in the present reservoir fluid, mainly suggesting that early stage fluid was trapped in the fluid before considerable boiling and vapor-loss. The decrease of CO_2 content in the reservoir fluid occurs more or less during progressing of the exploitation of the geothermal field. The CO_2/CH_4 and CO_2/N_2 ratios could be explained by degassing and/or dilution by groundwater.

References

Akazawa, T. & Muramatsu, Y. (1988). Distribution of underground fractures at the Matsukawa geothermal field, northeast Japan. *J. Geotherm. Res. Soc. Japan* **10**, 359-371 (in Japanese with English abstract).

Arai,F.(1994). Significance of the distribution of skarn minerals in the Mori geothermal field, Southwest Hokkaido. *1994 Annual Meeting Geotherm.Res.Soc. Japan, Abstr. with Progr.,* **B12** (in Japanese).

Belkin, H., De Vivo, B., Gianelli, G. & Lattanzi, P. (1985). Fluid inclusions in minerals from the geothermal fields of Tuscany, Italy. *Geothermics* **14**, 59-72.

Bischoff,J.L. Pitzer,K.S.(1989). Liquid-vapor relations for the system NaCl-H_2O: Summary of the P-T-x surface from 300 to 500℃. *Amer. Jour. Sci.* **289**,217-248.

Bodnar,R.J., Reynolds,T.J. & Kuehn,C.A.(1985). Fluid-inclusion systematics in epithermal systems. *Rev.Econ. Geo.,* **2**,73-97.

Borisenko,A.S.(1977) Study of the salt composition of solutions in gas- liquid inclusions in minerals by the cryometric method. *Soviet Geol. Geophys.* **18**,11-19.

Bowers,T.S. & Helgeson,H.C.(1983). Calculation of the thermodynamic and geochemical consequences of nonideal mixing in the system H_2O-CO_2-NaCl on phase relations in geologic systems: Equation of state for H_2O-CO_2-NaCl fluids at high pressures and temperatures. *Geochim. Cosmochim. Acta,* **47**, 1247-1275.

Browne, P.R.L., Roedder, E. & Wodzicki, A. (1974). Comparison of past and present geothermal waters from a study of fluid inclusions, Broadlands field, New Zealand. *Proc. Int. Symp. Water-Rock Interaction,* **43-47**.

De Vivo, B. & Frezzotti,M.L. (1994). Evidence for magmatic immiscibility in Italian subvolcanic systems. Fluid inclusions in Minerals; Methods and Applications. *Inclusions in Minerals, Short Course of the Working Group (IMA),*345-362.

Doi,N., Muramatsu,Y., Chiba,Y. & Tateno,M.(1988) Geological analysis of the Kakkonda geothermal reservoir. *Proc. Int. Symp. Geotherm. Energy, 1988, Kumamoto and Beppu, Japan,* 522-525.

Doi,N., Kato,O., Ikeuchi,K., Komatsu,R., Miyazaki,S.-I., Akaku,K. & Uchida, T.(1998). Genesis of the plutonic-hydrothermal system around Quaternary granite in the Kakkonda geothermal system, Japan. *Geothermics* **27**,663-690.

Ellis,A.J. & Mahn,W.A.J.(1977). Chemistry and geothermal systems. Academic Press, New York, N.Y.,392pp.

Fournier,R.O. & Truesdell,A.H.(1973). An empirical Na-K-Ca geothermometer for natural water. *Geochim. Cosmochim. Acta,* **37**, 1255-1275.

Fujino,T. & Yamazaki,T. (1985). The use of fluid inclusion geothermometry as an indicator of reservoir temperature and thermal histry in the Hatchobaru geothermal field, Japan. *Geothermal Resources Council transactions,***9**, Part 1,429.

Gianelli, G.., Ruggieri, G. & Mussi, M. (1997). Isotopic and fluid inclusion study of hydrothermal and metamorphic carbonates in the Larderello geothermal field and surrounding areas, Italy. *Geothermics* **26**, 393-417.

Giggenbach, W.F. (1980). Geothermal gas equilibria. *Geochim. Cosmochim.Acta* **44**, 2021-2032.

Haas, J.L., Jr.(1976). Physical properties of coexisting phases and thermochemical properties of the H_2O component in boiling NaCl solutions. *U. S. Geological Survey Bulletin* **1421-A**,73 p.

Hall, D.L., Sterner, S.M. & Bodnar, R.J. (1988). Freezing point depression of $NaCl-KCl-H_2O$ solutions. *Econ. Geol.* **83**,197-202.

Hanano, M. & Matsuo, G. (1990). Initial state of the Matsukawa geothermal reservoir: Reconstruction of a reservoir pressure profile and its implications. *Geothermics* **19** 541-560.

Hayashi, M., Taguchi, S. & Yamazaki, T. (1981). Activity index and thermal history of geothermal system. *Geothermal Resources Council transactions,***5**, 177-180.

Hedenquist, J. W. & Henley, R.W. (1985). The importance of CO_2 on freezing point implications for epithermal ore deposition. *Econ. Geol.* **80**,1379-1406.

Hemley, J.J., Montoya, J.W., Marinenko, J.W. & Luce, R.W. (1980). Equilibria in the system $Al_2O_3-SiO_2-H_2O$ and some general implications for alteration/mineralization process. *Econ. Geol.* **75**,210-228.

Ide, T. (1985). Geothermal models of the Matsukawa and Kakkonda areas. *J. Geotherm. Res. Soc. Japan* **7**, 201- 213 (in Japanese with English abstract).

Ikeuchi,K., Doi,N., Sakagawa,Y., Kamenosono,H. & Uchida,T. (1998). High temperature measurements in well WD-1a and the thermal structure of the Kakkonda geothermal system,Japan. *Geothermics* **27**,591-607.

Kasai,K., Sakagawa,Y., Komatsu,R.,Sasaki,M., Akaku,K. & Uchida,T. (1998). The origin of hypersaline liquid in the Quaternary Kakkonda Granite, sampled from well WD-1a, Kakkonda geothermal system, Japan. *Geothermics* **27**,631-645.

Kimbara, K. (1983). Hydrothermal rock alteration and geothermal systems in the eastern Hachimantai geothermal area, Iwate Prefecture, northern Japan. *J. Japan. Assoc. Min. Petr. Econ. Geol.* **78**, 479-490 (in Japanese with English abstract).

Kodama,M. & Nakajima,T. (1988). Exploration and exploitation of the Kirishima geothermal field. *Journal of the Japan Geothermal Energy Association* **25**,201-230 (in Japanese with English abstract).

Komatsu,R. & Muramatsu,Y. (1994). Fluid inclusion study of the deep reservoir at the Kakkonda geothermal field,Japan. *Proc.16th NZ Geothermal Workshop* 1994,91-96.

Komatsu, R., Ikeuchi, K., Doi,N., Sasaki,M., Uchida, T. & Sasada, M. (1998). Characteristics of the Quaternary Kakkonda granite and geothermal system clarified by fluid inclusion study of deep investigation well, Kakkonda, Japan. *J. Geotherm. Res. Soc. Japan* **20**, 209- 224 (in Japanese with English abstract).

Moore, J.N. & Gunderson, R.P. (1995). Fluid inclusion and isotopic systematics of an evolving magmatic- hydrothermal stage. *Geochim. Cosmochim. Acta* **59**, 3887-3907.

Moore, J.N., Lutz, S.J., Renner, J.L., McCulloch, J. & Petty, S. (2000). Evolution of a volcanic-hosted vapor-dominated system: Petrologic and geochemical data from corehole T-8, Karaha-Telaga Bodas, Indonesia. *Geotherm. Res. Council Trans.* **24**, 24-27.

Muramatsu. Y. (1984). Fluid inclusion study in the Takinoue geothermal field, Iwate prefecture, Japan: An application to the estimate of the present underground temperature. *Proc.6th New Zealand Geothermal Workshop*, 21-25.

Muramatsu,Y. (1987). Distributions, paragenesis and fluid inclusions of hydrothermal minerals in the Kakkonda geothermal field, Northeastern Japan. *J. Miner. Petr. Econ. Geol.* **82**,216-229 (in Japanese with English abstract).

Muramatsu,Y. (1993). Considerations on application of fluid inclusion geothermometry for evaluation of geothermal well. *Geothermal Energy* **61**,90-103 (in Japanese with English abstract).

Muramatsu, Y. & Komatsu. R. (1996). Fluid evolution in the Kakkonda shallow geothermal reservoir, Iwate prefecture, Northeastern Japan: a fluid inclusion study. *J. Miner. Petr. Econ. Geol.* **91**, 145-161 (in Japanese with English abstract).

Muramatsu, Y. & Komatsu. R. (1999). Microthermometric evidence for the formation of Ca-rich hypersaline brine and CO_2-rich fluid in the Mori geothermal reservoir, Japan. *Resource Geol.* **49**,27-37.

Muramatsu, Y., Sawaki, T. & Arai, F.(2006). Geochemical study of fluid inclusions from the western upflow zone of the Matsukawa geothermal system, Japan. *Geothermics*, **35** 123-140.

Muramatsu, Y., Komatsu. R., Sawaki.T. & Sasaki.M. (1997). Gas composition of fluid inclusions from the Mori geothermal reservoir, Southwestern Hokkaido, Japan. *Resource Geol.* **47**,283-291.

Muramatsu, Y., Komatsu. R., Sawaki.T., Sasaki.M. & Yanagiya,S. (2000). Geochemical study of fluid inclusions in anhydrite from the Kakkonda geothermal system, northeast Japan. *Geochemical J.* **34**,175-193.

Nakamura, H. (1967). Geothermal exploration and subsurface structure of Matsukawa area. *J. Jap. Geotherm. Energy Assoc. (Chinetsu)*, **20**, 6-8 (in Japanese with English abstract).

New Energy and Industrial Technology Development Organization (1989). Report on geothermal development research, development of method for reservoir assessment.

Nitta,T.,Suga,S.,Tsukagoshi,S. & Adachi,M.(1987). Geothermal resources in the Okuaizu, Tohoku district, Japan. *J. Jap. Geotherm. Energy Assoc. (Chinetsu)*, **24**, 26-56 (in Japanese with English abstract).

Oakes,C.S., Bodnar,R.J. & Simonsin,J.M.(1990). The system $NaCl-CaCl_2-H_2O$: I. The ice liquidus at 1 atm total pressure. *Geochim. Cosmochim. Acta* **54**,603-610.

Ozeki, H., Fukuda, D. & Okumura, T. (2001). Historical changes of geothermal reservoir models at Matsukawa, Japan. *J. Jap. Geotherm. Energy Assoc. (Chinetsu)*, **167**, 339-368 (in Japanese with English abstract).

Potter II,R.W. & Clynne,M.A.(1978). The solubility of the noble gases He, Ne, Ar, Kr, and Xe in water up to the critical point. *J.Solution Cem.*, **7**,837-844.

Roedder, E. (1970). Application of an improved crushing microscope stage to studies of the gases in fluid inclusions. Schweiz. *Mineral.Petrogr.Mitt.*, **50** *(1):*41-58.

Roedder, E. (1984). Fluid inclusions: Vol.12. Reviews in Mineralogy. Washington, D.C., *Mineralogical Society of America.*

Roedder,E. & Coombs,D.S. (1967). Immiscibility in granitic melts indicated by fluid inclusions in ejected granite blocks from Ascension Island. *J.Petrol.***8**,417-451.

Sasada, M., Roedder, E. & Belkin.H.E. (1986). Fluid inclusions from drill hole DW-5, Hohi geothermal area, Japan: Evidence of procedure for estimating CO_2 content. *J. Volcanol. Geotherm. Res.* **30**, 231-251.

Sasada, M., Sawaki, T. & Takeno, N. (1992). Analysis of fluid inclusion gases from geothermal system, using a rapid-scanning quadrupole mass spectrometer. *Eur. J. Mineral.* **4**, 895-906.

Sasaki,M., Sasada,M., Fujimoto,K., Muramatsu,Y., Komatsu,R. & Sawaki,T. (1995). History of post-intrusive hydrothermal systems indicated by fluid inclusions occurring in the young granitic rocks at the Kakkonda and Nyuto geothermal systems, northern Honshu, Japan. *Resource Geol.* **45**,303-312 (in Japanese with English abstract).

Sasaki,M., Fujimoto,K., Sawaki,T., Tsukamoto,H., Muraoka,H., Sasada,M., Ohtani,T., Yagi,M., Kurosawa,M., Doi,N., Kato,O., Kasai,K., Komatsu,R. & Muramatsu,Y. (1998) Characterization of a magmatic/ meteoric transition zone at the Kakkonda geothermal system, northeast Japan. *Proc. 9th Int. Symp. Water-Rock Interaction, Taupo, New Zealand,*483-486.

Sawaki, T. & Muraoka,H. (2002). Fluid inclusion study on the wells MT-1 and MT-2 in the Mataloko geothermal system, *Indonesia.Bull Geol.Surv. Japan,* **53**,337-341.

Sawaki,T., Sasada,M., Sasaki,M. & Goko,K.(1997). Fluid inclusion study of the Kirishima geothermal system, Japan. *Geothermics,* **26**, 305-327.

Sawaki,T., Sasaki,M., Komatsu,R., Muramatsu,Y. & Sasada,M. (1999). Gas compositions of fluid inclusions from the shallow geothermal reservoir in the Kakkonda geothermal system, northeast Japan. *J. Geotherm. Res. Soc. Japan* **21**,127-141 (in Japanese with English abstract).

Shepherd,T., Rankin,A.H. and Alderton,D.H.M. (1985). Fluid inclusion studies. Blackie,USA:Chapman & Hall,New York.pp239.

Shimazu, M. & Yajima, J. (1973). Epidote and wairakite in drill cores at the Hachimantai geothermal area, northeastern Japan. *J. Japan. Assoc. Min. Petr. Econ. Geol.* **68** 363-371.

Sourirajan,S. & Kennedy,G.C.(1962). The system H_2O-NaCl at elevated temperatures and pressures. *Amer. Jour. Sci.* **260**,115-141.

Sterner,S.M., Hall,D.L. & Bodnar,R.J.(1988). Systematic fluid inclusions. Solubility relations in the system NaCl-KCl-H_2O under vapor-saturated conditions. *Geochim. Cosmochim. Acta* **52**,989-1006.

Sternfeld, J., Keskinen, M. & Blethen, R.(1983). Hydrothermal mineralization of a Clear Lake Geothermal Well, Lake County, California. *Geothermal Resour. Council Trans.* **7** 193-197.

Sumi, K. (1966). Hydrothermal rock alteration of the Matsukawa geothermal area, Iwate Prefecture. *Min. Geol.* **16**, 261-271 (in Japanese with English abstract).

Sumi, K. (1968). Structural control and time sequence of rock alteration in the Matsukawa geothermal area, with special reference to comparison with those of Wairakei. *J. Jap. Geotherm. Energy Assoc. (Chinetsu)*, **17**, 80-92 (in Japanese with English abstract).

Taguchi, S. & Hayashi, M. (1983). Past and present subsurface thermal structures of the Kirishima geothermal area, Japan. *Geothermal Resour. Council Trans.* **7**, 199-203.

Taguchi,S. & Nakamura,M.(1991). Subsurface thermal structure of the Hatchobaru geothermal system, Japan, determined by fluid inclusion study. *Geochemical J.* **25**,301-314.

Taguchi,S., Hayashi,M., Mimura,T.,Kinoshita,Y., Goko,K. & Abe,I.(1984). Fluid inclusion temperature of hydrothermal minerals from the Kirishima geothermal area, Kyushu, Japan. *Journal of the Japan Geothermal Energy Association* **21**,119-129(in Japanese with English abstract).

Takenaka,T. & Furuya,S. (1991). Geochemical model of the Takigami geothermal system, northeast Kyushu, Japan. *Geochem. J.* **25**,267-281.

Takenouchi, S. (1986). Fluid inclusions in anhydrite and calcite from the Nishiyama geothermal area, Fukushima. *J. Geotherm. Res. Soc. Japan* **8,** 191 (in Japanese with English abstract).

Takenouchi, S. (1990). Geothermal systems and hot spring-type gold deposits. *J. Jap. Geotherm. Energy Assoc. (Chinetsu)*, **27**, 23-33 (in Japanese with English abstract).

Takenouchi, S. & Kennedy, G.C. (1965). The solubility of carbon dioxide in NaCl solutions at high temperatures and pressures. *Amer.Jour.Sci.***263**, 445-454.

Takenouchi, S. & Shoji, T. (1984). Some geological methods applied to the exploration and evaluation of geothermal resources. *Research on Natural Energy* **8**, 481-486.

Truesdell, A.H.(1976). Summary of section III-Geochemical techniques in exploration. *Proc.2nd U.N.Symp. on the Development and Use of Geothermal Resources, San Francisco* 1975,1,1iii-1xxix.

Vanko, D.A., Bodnar, R.J. & Sterner, S.M. (1988) Synthetic fluid inclusions: VIII.Vapor-saturated halite solubility in part of the system $NaCl-CaCl_2-H_2O$, with application to fluid inclusions from oceanic hydrothermal systems. *Geochim. Cosmochim. Acta* **52**,2451-2456.

Yoshida, Y. & Ishizaki, Y. (1988). Geochemical model of the Matsukawa geothermal field. *Proceedings of the International Symposium on Geothermal Energy, Kumamoto and Beppu, Japan*, 128-131.

In: Energy Research Developments
Editors: K.F. Johnson and T.R. Veliotti
ISBN: 978-1-60692-680-2
© 2009 Nova Science Publishers, Inc.

Chapter 8

SIZE DEPENDENT INTERFACE ENERGY

Q. Jiang * *and H.M. Lu*

Key Laboratory of Automobile Materials (Jilin University),
Ministry of Education, and Department of Materials Science and Engineering,
Jilin University, Changchun, China

Abstract

As size of low dimensional materials decreases, which leads to dramatic increase of surface-to-volume ratio, their properties are essentially controlled by related interface energetic terms, such as interface energy and interface stress. Although such changes in behavior can be dominant effects in nanoscale structures, we still have remarkably little experience or intuition for the expected phenomena, especially the size dependent properties and their practical implication, except for electronic systems. In this contribution, the classic thermodynamics as a powerful traditional theoretical tool is used to model these energetic terms including different bulk interface energies and the corresponding size dependences. Among the above modeling a special emphasis on the size dependence of interface energy is carried out. The predictions of the established models without free parameters are in agreement with the experimental or other theoretical results of different kinds of low dimensional materials with different chemical bond natures.

Introduction

Scope

In this review, we shall focus attention of classic thermodynamic aspects to model the interface energy. For present purposes (and according to this author's preference) the examples have been divided according to those definitions based on the principle phenomenon involved. This contribution comprises six sections. Section 1 is entitled Introduction. After the introduction it concentrates on different interface energies and their

---

* E-mail address: jiangq@jlu.edu.cn. Fax: +86-431-5095876. (Author to whom all correspondence should be addressed.)

size dependence. The sequence from solid-liquid interface energy γ_{sl} (section 2), solid-solid interface energy γ_{ss} (section 3), solid-vapor interface energy γ_{sv} (section 4), and liquid-vapor interface energy γ_{lv} (section 5). The section 6 provides a summary and future prospect.

Overview

At materials sizes of nanosize range (interface size at least in one dimension is also in this size range), which are *above* the atomic scale and *below* the macroscopic scale, the corresponding materials properties could not be readily interpreted based on "classical" atomic or solution theories, and the regions of space involved were beyond the reach of existent experimental techniques [1]. Science to interpret these phenomena has taken a firm theoretical and experimental hold on the nature of matter at its two extremes: at the molecular, atomic, and subatomic levels, and in the area of bulk materials, including their physical strengths and weaknesses and their chemical and electrical properties. Between those two extremes lies the world of nanometer size range, or called as mesoscale, and even with the latest advanced techniques for studying the phases and the region between phases, a great many mysteries remain to be solved. That "region" of the physical world represents a bridge not only between chemical and physical phases, but also plays a vital but often unrecognized role in other areas of physics, chemistry, materials science, biology, medicine, engineering, and other disciplines [1].

"Interfacial" phenomena may be defined as those related to the interaction between one phase (solid or liquid) with another phase (solid, liquid, or gas) or a vacuum in the narrow region in which the transition from one phase to the other occurs. As will quickly become apparent, the two classes of phenomena are intimately related and often cannot be distinguished [1].

Our understanding of the nature of the interfacial region and the changes and transitions that occur in going from one chemical (or physical) phase to another has historically lagged behind that in many other scientific areas in terms of the development and implementation of both theoretical and practical concepts. Great strides were made in the theoretical understanding of interactions at interfaces in the late nineteenth and early twentieth centuries by thermodynamics [1]. Modern computational and analytical techniques made available in the last few years have led to significant advances toward a more complete understanding of the unique nature of interfaces and the interactions that result from their unique nature due to the rapid increase of computation ability/price ratio of computers [2]. The so-called computation materials science considers the interface properties from three different size scales [2]:

1. From atomic scale with the ab initio calculation based on the first principle, which considers many-body interaction behavior of several ten to hundreds molecules together;
2. From nanosize scale with molecular dynamics and Monte Carlo methods, which considers many-body interaction behavior of several thousands to several million molecules;

3. Engineering behavior of large-scale construction problems, or bulk materials using finite element methods where averaging constitutive laws are used to incorporate the microstructure.

However, because of the unusual and sometimes complex character of interfaces and associated phenomena, the development of fully satisfying theoretical models has been slow. There exists a great deal of controversy in many areas related to interfaces [1]. Also, the classic thermodynamics has more or less neglected to interpret the interface phenomena in modern science due to the appearance of computer simulation technique. Note that many present controversies in theories and experiments in interface phenomena are not bad, since it represents the fuel for continued fundamental and practical research. However, for the practitioner who needs to apply the fruits of fundamental research, such uncertainty can sometimes complicate attempts to solve practical interfacial problems [1].

Nanoscience and nanotechnology are rapidly developed field in materials science and engineering in recent years. As size of materials drops to nanometer size range, interface/volume ratio increases and thus interface effect on materials properties becomes evident. When the sample size, or grain size, or domain size becomes comparable with the specific physical length scale such as the mean free path, the domain size in ferromagnets or ferroelectrics, the coherence length of phonons, or the correlation length of a collective ground state as in superconductivity, then the corresponding physical phenomenon will be strongly affected [3].

The interface phenomena, which affect the materials behavior, are in nature produced by different energetic states of molecules on the interface in comparison with those within the materials. Since the molecular interactions on the interfaces differ from those on the interior of phases, the excess specific Gibbs free energy or interface energy for molecules of unit interface area exists, which is equal to the difference between the total Gibbs free energy of interface molecules and that within the phases per unit area. The interfaces and corresponding interface energies studied here are sometimes called as other ones. The following is several examples of them. When the interface is composed of the same solids, the interface is also called as grain boundary and the corresponding interface energy is named as grain boundary energy. Usually solid-vapor interface energy and liquid-vapor interface energy are considered to have the same size of solid-vacuum interface energy and that of liquid-vacuum interface energy since their differences are often at least one order smaller when the external pressure is ambient one although it is known that this difference just leads to the absorption of vapor on interfaces. The solid-vapor interface energy is also named as surface energy while the liquid-vapor interface energy is made as surface tension [4].

Although surface tension of liquid can mean both surface energy and surface stress where they have the same size and meaning, this identity fails for solids [5]. The former describes the reversible work per unit area to form a new solid surface, whereas the latter denotes the reversible work per unit area due to elastic deformation, which is equal to the derivative of γ with respect to the strain tangential to the surface.

All above phenomena change as materials dimension and broken symmetry vary. The dimensionality is dictated by the size in a certain direction of the object compared to an appropriate length scale for the phenomena being discussed. If only one dimension of the sample is small compared to this length parameter, then the object is two-dimensional. Similar definitions can be given for one-dimensional and zero-dimensional.

Although an increasing body of excellent computational materials science, the computer simulations remain many uncertainties in nanosize range. Thus, the classic thermodynamics still remain importance to model the above phenomena since the history of science demonstrate a belief in the universal nature of thermodynamics.

The laws of thermodynamics were formulated as phenomenological laws about observable and operationally defined quantities. One of the main points of classical thermodynamics is that the thermodynamic approach is applicable to macroscopic systems only, i.e. to systems containing great number of molecules, atoms, ions, or to black holes and the entire Universe [6]. From this point of view, a single small object, including a nanosized droplet and a nanocrystal, does not satisfy the definition of the macroscopic thermodynamic system. At the same time, an ensemble of many small objects will be a macroscopic system. Such an ensemble may be rather speculative or correspond to a real dispersed system (aerosol, micro-emulsion, composed material etc). However, if the system is monodispersed (a contemporary example is an ensemble of the same, in a well-defined approximation, working elements in micro- or nanoelectronics), the treatment may be reduced to the investigation of a single modeling ensemble element. Introducing adequately defined distribution functions, the polydispersed systems may be also replaced (at least in a first approximation) by a modeling monodispersed one. Thus, extension of thermodynamic methods to very small objects, including nanoparticles seems possible but faces many principal difficulties.

It is also noteworthy that, in the context of some modern experimental techniques such as the atomic force microscopy, the thermodynamic method is equally promising as a theoretical and empirical description of the experimentally investigated nanosystems. Indeed in many cases, experimental data on nanoparticle properties are rather scanty and contradictory. Thus if the particle is not entirely rigid, the experimental arrangement can have an effect on the object under investigation. It was demonstrated by Monte Carlo computer experiments [7] that the tip of the atomic force microscopy could have a noticeable effect on a nanodroplet in the space between the tip and the solid substrate. Hence difficulties of principle inevitably occur not only at the thermodynamic treatment but also at the experimental study of very small objects. However, no comprehensive overview of that field exists. This contribution aims to fill that gap, which is essentially based on the author's own recent works. It gives a review of modern approaches how the traditional thermodynamics treats these interface amounts. Particular emphasis is placed on the size dependence of these quantities at the nanometer size range.

Solid-Liquid Interface Energy

The Bulk Solid-Liquid Interface Energy γ_{sl0}

The solid-liquid interface free energy γ_{sl0}, which is defined as the reversible work required to form or to extend a unit area of interface between a crystal and its coexisting fluid plastically [8-13], is one of the fundamental materials properties. Many practically important processes and phenomena like crystal growth, homogeneous nucleation, surface melting and roughening transition are directly related to its value. Thus, a quantitative knowledge of γ_{sl0} values is necessary. However, direct measurements for γ_{sl0} are not at all easy even for elements in contrast to the case of interface energy of liquid-vapor γ_{lv0} [11-13]. Some attempts are made to obtain γ_{sl0} by theoretical approaches or computer simulations [8-17]. A widely used technique in an indirect way to determine γ_{sl0} value is nucleation experiments of undercooled liquid based on the classical nucleation theory (CNT), which was made firstly by Turnbull [9]. According to the CNT, the undercooled liquid crystallizes at nucleation temperature T_n with a critical nucleation size D_n induced by a localized structural and energetic fluctuation where the thermodynamic properties of nanometric aggregates of the newly nucleated phase are to be the same as those of the corresponding bulk one. Thus, the nucleus-liquid interface energy $\gamma_{sl}(D_n,T_n)$ at T_n is treated as the respective value for a planar interface γ'_{sl0} being temperature independent (this assumption is known as the capillarity approximation), which makes it possible to consider the Gibbs free-energy difference of a spherical nucleus in the liquid $\Delta G(D,T)$ as a sum of a volume term and an interface term [9],

$$\Delta G(D,T) = -(1/6)\pi D^3 g_m(T)/V + \pi D^2 \gamma'_{sl0} \qquad (2.1)$$

where V denotes the gram-atom volume and $g_m(T)$ is temperature-dependent Gibbs free energy difference between crystal and liquid. $g_m(T)$ for elements is assumed to be a linear function of T, namely contribution of specific heat difference ΔC_p between solid and liquid on $g_m(T)$ is negligible under small undercooling [9,11]. Thus, the corresponding $g_m(T)$ function is shown as [11],

$$g_m(T) = (1-T/T_m)H_m \qquad (2.2)$$

where H_m is the bulk melting enthalpy at bulk melting temperature T_m. With a consideration on homogeneous nucleation rate I_v, γ'_{sl0} in Eq. (2.1) can be determined, which as an empirical relationship is proportional to H_m [11],

$$\gamma'_{sl0} = c_1'hH_m/V \qquad (2.3)$$

where c_1' is a constant to be 0.45 for metals (especially close-packed metals) and 0.32 for nonmetallic elements, h is the atomic diameter. Note that $\gamma_{sl}(D_n,T_n) \approx \gamma'_{sl0}$ has been implied in Eq. (2.3) [11,18-19]. Moreover, the appropriate value of k increases noticeably for molecules having more asymmetry. Undoubtedly, Eq. (2.3) overlooks some important pieces of physics. In addition, the existence of c_1' to be determined also weakens the theoretical meaning of this equation, and makes it only be an empirical rule.

The most powerful methods at present available for experimentally measuring γ_{sl0} are to make direct use of the so-called Gibbs-Thomson expression [12-13,17]. This thermodynamic result shows that, if all other intensive variables (such as composition, pressure and strain energy) remain constant, a solid bounded by an element of interface having principle diameters of curvature D_1 and D_2 measured in the solid will be in equilibrium with its melt at a temperature $T_m(D_1,D_2)$, which is not equal to the liquidus temperature T_m in phase diagram. The expression may be written as,

$$T_m(D_1,D_2)/T_m = 1 - V\gamma_{sl0}(1/D_1 + 1/D_2)/H_m(T)$$
$$(2.4)$$

where $H_m(T)$ is temperature-dependent melting enthalpy.

Note that Eq. (2.4) is valid when D_1 and D_2 are large enough (e.g. 20 nm) [12-13]. Thus, when the appropriate physical constants are known, measurements of $T_m(D_1,D_2)$ for known values of D_1 and D_2 for a system at "equilibrium" will yield values of γ_{sl0} directly.

Fortunately, the size dependence of melting temperature of metallic nanocrystals $T_m(D)$ have been deduced as the following [20-21],

$$\frac{T_m(D)}{T_m} = \exp\left(-\frac{2S_{vib}}{3R}\frac{1}{D/D_0 - 1}\right) \qquad (2.5)$$

where S_{vib} denotes the vibrational component of the melting entropy S_m of bulk crystals at T_m, R is the ideal gas constant, D_0 shows a critical diameter at which all atoms of a particle are located on its surface. For low dimensional element crystals, D_0 depends on their dimension d and atomic diameter h through [20-21],

$$D_0 = 2(3-d)h \qquad (2.6)$$

where $d = 0$ for nanoparticles, $d = 1$ for nanowires and $d = 2$ for thin films. When $d = 0$, D has a normal meaning. For $d = 1$, D denotes the diameter of the nanowire. If $d = 2$, D is defined as the thickness of a thin film. Since a crystal is characterized by its long-range order, the smallest crystal should have at least a half of the atoms located within the crystal. Hence, the smallest value of D is $2D_0$ [20]. This estimation is consistent with experimental results for Bi film [22] and Pb nanowire in a carbon nanotube [23-24]. However, the parameter h must be redefined to adapt to the case of molecular crystals. For organic spherical molecules, h can be an averaged diameter of the molecules since the molecule for molecular crystals has a similar effect as the atom for metallic elements [25]. While for chain molecules, considering that the γ_{sl0} value states excess energy of interface molecules in a unit area, h thus may be defined as the mean size of a chain segment [25],

$$h = [mV/(nN_a)]^{1/3} \qquad (2.7)$$

where m and n denote the total atom number and the chain segment number of an organic molecule, respectively. N_a is Avogadro's constant.

S_m consists, at least, of three components: positional S_{pos}, vibrational S_{vib} and electronic S_{el} [26],

$$S_m = S_{vib} + S_{pos} + S_{el}. \qquad (2.8)$$

S_{pos} is given by [26],

$$S_{pos} = -R(x_A \ln x_A + x_B \ln x_B) \tag{2.9}$$

where x_A and x_B are molar fractions of crystals and vacancies, respectively. For a melting process, $x_A = 1/(1+\Delta V_m/V_s)$ and $x_B = 1-x_A$ where $\Delta V_m = V_l-V_s$ with subscripts of s and l for crystal and liquid phases, respectively. Note that Eq. (2.9) assumes that ΔV_m consists of vacancies with the same size of atoms. For metallic and organic crystals, the type of chemical connection does not vary during the melting transition. Thus, $S_{el} \approx 0$ [27], and

$$S_{vib} = S_m - S_{pos}. \tag{2.10-a}$$

For some semi-metals, $S_{el} \neq 0$, S_{vib} must be determined in a direct way, such as Mott's equation [28],

$$S_{vib} = 3R\ln(v_s/v_l) = (3/2)R\ln(\sigma_s/\sigma_l) \tag{2.10-b}$$

where v and σ denote characteristic vibration frequency and electrical conductivity. If the parameters in above equations are unavailable, the following equation can also be utilized as a first order approximation [26],

$$S_{vib} \approx S_m - R. \tag{2.10-c}$$

For organic crystals, ΔV on melting is small and S_{pos} is thus negligible as a first order approximation [25]. While for some organic crystals, one or more solid-state phase transformations closely precede the melting, which is thought to reduce the S_m values [29]. Herein, the cumulative entropy of fusion S^c_m should be introduced, which is defined as the summation of all the changes in entropy at all the transformation temperatures and melting temperature. Thus,

$$S_{vib} \approx S^c_m. \tag{2.10-d}$$

For semiconductors, the melting is accompanied by the semiconductor-to-metallic transition and the elements suffer contraction in volume rather than expansion for most of metals. Thus, S_{el} strongly contributes S_m and $S_{pos} \ll S_{el}$ and S_{pos} is thus neglected as a first order approximation [29]. Namely,

$$S_{vib} \approx S_m - S_{el}. \tag{2.11}$$

The above model, i.e. Eq. (2.5) has predicted the size-dependent melting for nanoparticles [30], for thin films [20], for metallic nanowires in carbon nanotubes [21], and the related size-dependent initial sintering temperature of metallic nanoparticles [31]. The available experimental evidences in a large size range from several nanometers to several hundred nanometers confirm the above-predicted results noted that the valid size range of Eq. (2.5) is from $2D_0$ to infinite although $T_m(D) \approx T_m$ when $D > 200$ nm. Thus, the approximate expression for the melting temperature of an object of small size in Eq. (2.5) could proceed from it to the limit of a large system.

When $D > 10D_0 \approx 20$ nm, in term of the mathematical relationship of $\exp(-x) \approx 1-x$ when x is small enough, Eq. (2.5) is rewritten as,

$$T_m(D)/T_m \approx 1-2D_0S_{vib}/(3RD).$$

For an actual γ_{sl0} measurement, the principal radii on the interface between the crystal and the liquid may be $D_1 = D_2$ ($d = 1$ and $D_0 = 4h$) or $D_1 = D$ and $D_2 = \infty$ ($d = 2$ and $D_0 = 2h$) [12]. At both cases, through comparing Eqs. (2.5) and (2.12) with Eq. (2.4), there is [32]

$$\gamma_{sl0} = 2hS_{vib}H_m(T)/(3VR). \tag{2.13}$$

Although $T_m(r)$ function is introduced in the course of the deduction of Eq. (2.13), γ_{sl0} determined by Eq. (2.13) can also be considered as the solid-liquid interface energy of bulk crystals. This is because the measured D values are in the magnitude of millimeter where the difference between γ_{sl0} and $\gamma_{sl}(D)$ is far smaller than the experimental errors.

$\gamma_{sl0}(T_m)$ for Elemental Crystals

To determine the generality of Eq. (2.13), the values of γ_{sl0} at T_m for different types of elemental crystals including true metals, semi-metals, and semiconductors are predicted firstly and shown in table 1. For comparison, the corresponding experimental γ_{exp} [12,33-40] and γ_{TS} based on the Turnbull-Spaepen relation [41] are also shown in table 1 noted that the Turnbull-Spaepen relation is read as [41],

$$\gamma_{TS} = 0.6 \frac{M^{1/3} H_m}{\rho_{liq}^{1/3} N_0^{1/3}} \tag{2.14}$$

with M being the atomic mass.

Although it is argued [46] that our model, i.e. Eq. (2.13) overestimates γ_{sl0} at T_m for metals typically by 50% to 100% in comparison with that obtained from CNT or the undercooling experiments of liquid droplets of elements, typically from the Turnbull's work [9], the values of our predictions γ_{sl0} are in agreements with those of γ_{exp} and γ_{TS} except for Si (the reason will be explained in the latter) and Mo where the reason remains unknown. As for Cr, the larger deviation between γ_{sl0} and γ_{TS} may result from lower S_m value, which is thought not to present its true value [29]. When the model predictions of 151, 215, 85, 48 and 88 mJ/m^2 for Al, Au, Sn, Pb and Bi are compared with the corresponding experimental values of 131~151 [33], 270±10, 62±10, 40±7 and 55~80 mJ/m^2 [12] derived from measurements of the depression of melting point noted that the theoretical data from Ref. [46] are 120, 151, 83, 71 and 95 mJ/m^2, it is obvious that our predictions correspond better to the experimental results than the latter especially for Al, Au, Pb and Bi. Thus, Eq. (2.13) can be employed to quantitatively estimate the γ_{sl0} values of metals.

Table 1. Comparisons of $\gamma_{sl0}(T_m)$ among the model predictions γ_{sl0} by Eq. (2.13), γ_{TS} by Eq. (2.14), and γ_{exp} [33-40]

	γ_{sl0}	γ_{TS}	γ_{exp}	h	ρ_{liq}	ρ_s	V_0	H_m	T_m	$\Delta V/V$	S_{vib}
		(mJ/m²)		(nm)	(g/cm³)		(cm³/g-atom)	(kJ/g-atom)	(K)	(%)	(J/g-atom K)
Ti	216	234		0.290	4.11		10.6	15.5	1943	3.2	6.80
V	364	354		0.262	5.56		8.8	20.9	2175	5.2	7.86
Cr	272	326		0.250	6.28	7.20	7.2	16.9	2130	7.2	6.08
Mn	243	246		0.273	5.73		7.4	12.1	1517	1.7	7.26
Fe	236	276	221	0.248	7.01		7.1	13.8	1809	3.4	6.47
Co	353	338		0.251	7.76		6.7	16.1	1768	3.5	7.85
Ni	421	367	326	0.249	7.91		6.6	17.2	1726	5.4	8.11
Cu	297	261	270	0.256	8.00		7.1	13.0	1358	4.2	7.75
Zn	137	122	141	0.267	6.58		9.2	7.3	693	4.3	8.60
Zr	192	208		0.318	6.24		14.1	16.9	2125	3.5	6.69
Nb	402	400		0.286	7.62	8.55	10.9	26.4	2740	5.5	7.73
Mo	626	533		0.273	9.00	10.2	9.4	32.0	2890	6.4	9.01
Tc	441	425		0.270	10.2	11.5	8.5	24.0	2473	5.8	7.74
Ru	431	434		0.265	10.7	12.2	8.3	24.0	2523	7.0	7.51
Rh	387	391		0.269	10.7	12.4	8.3	21.5	2236	8.9	7.41
Pd	311	304		0.275	10.5		8.9	17.6	1825	5.9	7.64
Ag	199	176		0.289	9.34		10.3	11.3	1234	3.8	7.82
Cd	92	81		0.298	8.02		13.1	6.1	594	3.8	8.87
Hf	341	310		0.313	11.9	13.1	13.7	24.0	2500	3.2	8.45
Ta	472	477		0.286	14.6	16.6	10.9	31.6	3287	6.7	7.60
W	511	582		0.274	16.2	19.3	9.5	35.4	3680	12	6.62
Re	541	581		0.274	18.0	21.0	8.8	33.2	3453	9.7	6.91
Os	525	572		0.268	19.2	22.4	8.5	31.8	3300	9.7	6.93
Ir	468	464		0.271	20.0	22.5	8.6	26.0	2716	5.5	7.67
Pt	336	333		0.278	19.0		9.1	19.6	2045	6.6	7.58
Au	215	201	270±10 190	0.288	17.4		10.2	12.5	1338	5.2	7.62
Hg	29	28		0.301	13.4		14.8	2.3	234	3.7	8.51
In	35	38		0.325	7.02		15.7	3.3	430	2.7	6.70
Tl	39	45		0.341	11.2		17.2	4.2	577	3.2	6.13
Pb	48	49	40±7	0.350	10.7		18.2	4.8	601	3.6	6.77
Se	72	78		0.232	3.99	4.80	16.5	6.7	494	13	10.2
Al	151	170	149	0.286	2.39		10.0	10.8	933		6.15
Sn	89	79	62±10	0.281	7.00		16.7	7.0	505		9.22
Ga	87	76		0.244	6.09		11.8	5.6	303		10.1
Ge	412	453		0.245	5.60		13.6	31.8	1210		9.40
Si	738	672	332	0.235	2.51		12.1	50.0	1685		10.3
Sb	185	200		0.290	6.48		18.2	19.7	904		7.80
Te	158	168		0.286	5.71		20.5	17.5	723		8.65
Bi	88	101	55~80	0.309	10.1		21.3	10.9	545		7.20
Li	31	38	30	0.304	0.51		13.1	3.0	454	2.2	5.77
Na	19	22	20	0.372	0.94		23.7	2.6	371	2.5	6.15
K	11	13		0.454	0.83		45.5	2.3	336	2.6	6.06
Rb	10	11		0.495	1.46		55.9	2.2	313	2.5	6.17
Cs	8	9		0.531	1.84		71.1	2.1	302	2.6	6.06
Be	287	303		0.223	1.69	1.85	5.0	12.2	1560	2.5	6.96
Mg	115	113		0.320	1.59		14.0	8.7	922	3.6	8.15
Ca	60	61		0.395	1.37		26.2	8.5	1112	4.1	6.24
Sr	56	58		0.430	2.37	2.60	33.7	8.3	1041	2.7	7.06
Ba	45	49		0.435	3.20	3.50	39.2	7.8	1002	2.4	6.94
Ref				43	43-44		43	43	43		45

S_{vib} is determined by Eq. (2.10-b) for Al and Sn, by Eq. (2.10-c) for Ga, by Eq. (2.11) for Ge, Si, Sb, Te and Bi. Since S_{el} of Si, Sb, Te and Bi are unknown, S_{el}/S_m for these semiconductors are assumed to be the same as that for Ge with $S_{el} = 16.8$ J/g-atom K [26]. S_{vib} for the rest elements is determined by Eq. (2.10-a).

$\gamma_{sl0}(T_m)$ for Organic Crystals

As shown in table 2, for 15 different organic molecular crystals, the predictions in terms of Eq. (2.13) partly correspond to the latest experimental results [13,47-49] but are smaller than the early experimental results [12,50-52].

Organic crystals as molecular crystals differ from metallic and ionic ones, whose chemical bonds are covalent within molecules but consist of van der Waals forces or hydrogen bonds among molecules. The former being responsible for stability of individual molecules, are much stronger than the latter, being primarily responsible for bulk properties of matter, such as γ_{sl0}. Because bond strengths of van der Waals forces or hydrogen bonds are weaker than those of metallic or ionic bonds, γ_{sl0} values of organic crystals are also smaller than those of metallic and ionic crystals, such as Pb with $\gamma_{sl0} \approx 40\pm7$ mJ/m^2 [γ_{sl0} value of Pb in fact is one of the smallest γ_{sl0} values among metallic or ionic crystals because its S_m and H_m values are small, which leads to a small γ_{sl0} value of Pb in terms of Eq. (2.13)] [12]. Moreover, γ_{sl0} values of compounds composed of full hydrogen bonds, such as H_2O with $\gamma_{sl0} \approx 25\sim45$ mJ/m^2, should also be larger than those of organic crystals since molecules of organic crystals have only partly hydrogen bonds while the other is van der Waals force [12]. Thus, γ_{sl0} values for organic crystals could be not more than 30~40 mJ/m^2.

It is known that organic molecules mostly consist of C-H bonds with intermolecular dispersion forces caused by relative movement between electrons and the atomic nucleus. Their relative movements change electron density within the molecule. Generally, the larger the number of electrons and the more diffuse the electron cloud in the molecule, the greater its dispersion forces. However, the forces hardly affect the net attraction applied to a unit area of interface, the size difference of γ_{sl0} values for organic molecules thus is smaller than those of metallic and ionic crystals (for other three kinds of intermolecular forces, orientation forces and induced forces, a similar discussion may be carried out).

Although measured γ_{sl0} values of organic crystals composed of chain molecules are much larger than the above limits shown in table 2 [50-52], their real values should be similar to those composed of spherical molecules since γ_{sl0} denotes excess energy of unit area where molecular weight has negligible effect on it. Even if chain molecules may contain one or more hydrogen bonds, γ_{sl0} values still vary little since most bonds of the molecules are van der Waals forces. This result also implies that anisotropy of γ_{sl0} of organic crystals is small, although this issue up to now is still debated [11-12,32].

Moreover, for a typical fcc crystal, the coordinate number (CN) or bond number decrease of molecules on a solid-liquid interface is usually 1~2 while that on a liquid-vapor interface is 3~4. If this bond number is proportional to the corresponding interface energy, with the note that the bond strength difference of molecules between solid and liquid states is only several percent, γ_{sl0} value must be lower than γ_{lv0} value, which is easy to measure with better measuring accuracy. Thus, γ_{lv0} value of the same substance is good reference as an upper limit on γ_{sl0} [25]. For organic crystals, the γ_{sl0} value may be around half γ_{lv0} as a rough estimation. Since $\gamma_{lv0} \approx 20\sim40$ mJ/m^2 shown in table 2, $\gamma_{sl0} < 20$ mJ/m^2, which is also smaller than the above-stated limits in terms of considerations of the bond strength of metallic, ionic, and hydrogen bonds.

Table 2. Comparisons of $\gamma_{sl0}(T_m)$ between our model predictions γ_{sl0} by Eq. (2.13) and available experimental results γ_{exp} [12-13,47-52] for ice and organic crystals where γ_{lv0} values are also listed as reference [36,53-55] and the other parameters come from Refs. [36,53-56]

	γ_{sl0}	γ_{exp} (mJ/m²)	γ_{lv0}	m	n	h (nm)	V (cm³/g-atom)	H_m (J/g-atom)	S_{vib} (J/g-atom K)
Ice	56	44±10 45±15		3	1	0.310	6.55	2006	7.34
Benzene	15.4	15.7 22±2	28.9	12	1	0.503	6.40	829	2.95
Naphthalene	18.6	8.2 61±11	28.8	18	2	0.452	6.18	1062	2.98
Ethylene dibromide	16.4	19.5 35±7	38.4	8	2	0.509	19.8	1368	5.81
Cis-decalin	8.5	11.6	32.2	28	2	0.503	5.48	515	2.24
Trans-decalin	9.6	18.4	29.9	28	2	0.509	5.67	516	2.59
Chlorobenzene	15.7	14.1	33.6	12	1	0.534	7.66	802	3.50
Cyclohexane	3.1	4.6	25.5	18	1	0.547	5.46	150	2.55
Heptane	10	17.1	20.1	23	7	0.302	5.04	616	3.34
Stearic acid	14.3	106-151 151±10 135-180	28.9	56	18	0.294	7.83	1010	2.95
Myristic acid	16.0	81 116±10	28.6	44	14	0.292	10.8	1031	3.15
Lauric acid	13.9	71±15 100±15	28.5	38	12	0.299	4.90	964	3.05
Pivalic acid	2.4	2.8 2.7±0.2		17	1	0.572	6.60	146	2.33
Carbontetrabromide	10.6	10~20		5	1	0.540	19.0	790	5.89
Succinonitrile	7.3	28±4 8.9 7.9±0.8	47	10	2	0.402	6.20	370	3.78
Diphenyl	14.4	24 50±10	34.5	22	2	0.505	7.06	905	2.78

On the basis of the above consideration, it is known that γ_{sl0} values of organic molecules must be smaller than those of the other types of crystals. Thus, literature values [50-52] in a size of $\gamma_{sl0} > 20$ mJ/m² may be questionable [25]. On the contrary, the predictions of Eq. (2.13) are in agreement with the principles of $\gamma_{sl0} < \gamma_{lv0}$ and the upper size limit of γ_{sl0} values. Thus, the disagreement between the predictions and the experimental results for the chain molecules [50-52] may hardly argue the incorrectness of Eq. (2.13). We believe that our predictions are at least qualitatively correct.

In table 2, although $\gamma_{sl0} < 20$ mJ/m² as analyzed for organic crystals in the above discussion, γ_{sl0} values of different organic crystals change from 0.62 mJ/m² for cyclohexane to 19.7 mJ/m² for ethylene dibromide, their sizes differ from about 30 times. This difference can be found in Eq. (2.13). The variants of γ_{sl0} in Eq. (2.13) are h, S_m/R, and H_m/V. Since h for different substances varies a little, the essential contributions to differences in γ_{sl0} are H_m/V, denoting the energetic difference between crystal and liquid, and S_m/R, showing the corresponding structural difference. The cited data in table 2 for H_m/V values are between 27 and 172 J/cm^3, while those for S_m/R values are in the range from 0.064 to 0.568. Since H_m/V and S_m/R for one substance do not simultaneously take the largest values induced by different T_m, with the known fact that $H_m = T_m S_m$, the real differences in γ_{sl0} values are smaller than the largest possible difference.

Let γ_{sl0} of Eq. (2.13) be expressed as $\gamma_{sl0} = c_1 h H_m/V$, there is $c_1 = 2S_{vib}/(3R)$. Because the sizes of S_{vib} are different for crystals with different types of bonds and almost follow the sequence of ionic bond, covalent bond, metallic bond, hydrogen bond, van der Waals force, c_1 is component dependent. As shown in table 2, the sizes of S_{vib} vary from 9.22 J/g-atom for Sn to 0.53 J/g-atom for Cis-decalin, which makes c_1 in the range from 0.74 to 0.04. This range is larger than that for c_1' [11]. The reason can be illustrated as the followings: On one hand, although γ'_{sl0} in Eq. (2.3) is considered as the bulk solid-liquid energy, $\gamma'_{sl0} \approx \gamma_{sl}(D_n, T_n)$ has been implied [11,18-19] (it will be further demonstrated in Section 2.3), which makes the maximum value of c_1' is smaller than 0.74; On the other hand, molecular crystals are not considered when Eq. (2.3) was proposed, thus the minimum value of c_1' is larger than 0.04.

$\gamma_{sl0}(T_m)$ for Intermetallic Compounds and Oxides

For intermetallic compounds and oxides listed in table 3, the predictions based on Eq. (2.13) also correspond to the available theoretical results [57-59] with the absolute deviation smaller than 6%. Although higher T_m and larger $\Delta V/V$ of these substances make S_{vib} comparable with those of elemental crystals, larger H_m and smaller V lead that their $\gamma_{sl0}(T_m)$ values are larger than those of most of elemental crystals.

Table 3. Comparison of $\gamma_{sl0}(T_m)$ between γ_{sl0} by Eq. (2.13) and other theoretical results γ' for intermetallic compounds and oxides 57-59]

	γ_{sl0}	γ'	h	ρ_s	V	H_m	T_m	$\Delta V/V$	S_{vib}
	(mJ/m^2)		(nm)	(g/cm^3)	(cm^3/g-atom)	(kJ/g-atom)	(K)	(%)	(J/g-atom K)
α-MoSi$_2$	651	620	0.277	6.27	8.1	28.2	2173	29	8.4
β-MoSi$_2$	538	509	0.460	6.32	8.0	22.9	2303	35	5.1
WO$_3$	233	241	0.193	7.2	8.1	17.9	1743	18	6.8
ZrO$_2$	491	500	0.223	5.89	7.0	29.1	2988	15	6.6
Ref			57,60-61	56-57		56	56-57	57,62	

The values of $\Delta V/V$ for MoSi$_2$ are calculated in terms of Eq. (5) of Ref. [57] and that of ZrO$_2$ is also taken from Ref. [57]. $\Delta V/V$ of WO$_3$ is unavailable and assumed to be the average of those of ZrO$_2$ and Al$_2$O$_3$ [62].

$\gamma_{sl0}(T_m)$ in Metals: fcc Versus bcc

The dependence of γ_{sl0} on crystal structure is crucial to understanding the role of metastable structures in nucleation pathways. In 1897, Ostwald [63] formulated his "step rule", which states that nucleation from the melt occurs to the phase with the lowest activation barrier, which is not necessarily the thermodynamically most stable bulk phase. In the case of the nucleation of fcc crystals, there is evidence that crystallization often proceeds first through the formation of bcc nuclei, which transform to fcc crystals later in the growth process. This phenomenon has been observed in experiments on metal alloys [64], in computer simulations of Lennard-Jones particles [65] and weakly charged collides [66], and in classical density-functional theory studies of nucleation in Lennard-Jones [67]. These results could be explained if γ_{sl0} for bcc crystals were significantly lower than those for fcc crystals in these systems as that would lead to substantially lower activation barriers [68]. Using a simple model of interfacial structure, Spaepen and Meyer [69] predicted that γ_{sl0} for bcc-melt interfaces should be about 20% lower than that for fcc-melt interfaces, based on packing considerations. In a recent paper, Sun et al. [34,70] determined γ_{sl0} for bcc- and fcc-melt interfaces for several models for iron by molecular-dynamics simulations, obtaining values of γ_{sl0} that were about 30~35% smaller for bcc-melt interfaces than those for fcc-melt interfaces.

However, it is also easy to understand the difference of γ_{sl0} between bcc- and fcc-interfaces in terms of Eq. (2.13). According to Goldschmidt premise for lattice contraction [71], h contracts 3% if the coordination number (CN) of the atom reduces from 12 to 8. Thus, the effect of the change of h on γ_{sl0} can be neglected as a first order approximation. Similarly, the effect of the change of V can also be neglected since the difference is only 1.3% [34]. Since ΔV values for bcc and fcc crystals are also very close, $\gamma^{bcc}/\gamma^{fcc} \approx (H_m^{bcc}/H_m^{fcc})^2/(T_m^{bcc}/T_m^{fcc})$ in terms of Eq. (2.13) with the simplification of $S_{vib} \approx H_m/T_m$. With the known values of $H_m^{bcc}/H_m^{fcc} = 0.715$ and $T_m^{bcc}/T_m^{fcc} = 0.805$ [34], $\gamma^{bcc}/\gamma^{fcc} \approx 0.64$, which is consistent with the reported values 30~35% [34,70].

The Size Dependence of Solid-liquid Interface Energy $\gamma_{sl}(D)$

For comparison, γ'_{sl0} values for elemental crystals Au, Al, Sn, Pb and Bi in terms of Eq. (2.3) are 132, 93, 33.6, 54.5 and 54.4 mJ/m^2 [11], which correspond to the lower limits of the corresponding experimental data for Sn, Pb and Bi [12] while are by far lower than the corresponding experimental data for Au and Al [12]. This disagreement results from the two approximations in the CNT: 1. The specific heat difference between solid and liquid ΔC_p is assumed to be zero. Namely, the influence of ΔC_p is neglected, which makes the neglect of temperature dependence of $H_m(T)$ in terms of $H_m(T) = g_m(T)-Tdg_m(T)/dT$ (Helmholtz function); 2. The nucleus-liquid interface energy $\gamma_{sl}(D_n,T_n)$ at T_n is treated as the value for a planar interface γ'_{sl0} and γ'_{sl0} is assumed to be temperature independent, i.e. the capillarity approximation [11]. The $g_m(T)$ functions have been improved theoretically and confirmed experimentally by Hoffman [72], Spaepen [73] and Singh [74]. Since the values of γ'_{sl0} in Eq. (2.3) are initially obtained for nuclei-liquid interface [11,18-19] while any nucleus during solidification is in nanometer size range, $\gamma_{sl}(D)$, not γ_{sl0}, has to be considered for theoretical prediction of actual nucleation and growth processes.

To determine the $\gamma_{sl}(D)$ function, we consider a compressible spherical particle, or a cube with cube side taken as D, immersed in the corresponding bulk liquid. According to Laplace-Young equation [66], we have

$$P = 2fA/(3V) = 4f/D \qquad (2.15)$$

where A is the surface area of the particle, P is the difference in pressure inside and outside of the particle and f is interface stress. Using the definition of compressibility $\kappa = -\Delta V/(VP)$, $\varepsilon = \Delta D/D = \Delta A/(2A) = \Delta V/(3V)$ under small strain and $A/V = 6/D$ where Δ denotes the difference,

$$\varepsilon = -4\kappa f/(3D). \qquad (2.16)$$

In terms of a scalar definition of f, there exists [5,75-76]

$$f = \partial G/\partial A = \partial(\gamma_{sl}A)/\partial A = \gamma_{sl}+A\partial\gamma_{sl}/\partial A \approx \gamma_{sl}+A\Delta\gamma_{sl}/\Delta A \qquad (2.17)$$

where $G = \gamma_{sl}A$ states the total excess Gibbs surface free energy, or $\Delta\gamma_{sl} = (\Delta A/A)(f-\gamma_{sl})$.

To find mathematical solutions of f and γ_{sl} or $\gamma_{sl}(D)$, two boundary conditions of $\gamma_{sl}(D)$ are needed. An understandable asymptotic limit is that as $D \to \infty$, $\gamma_{sl}(D) \to \gamma_{sl0}$. As $D \to \infty$, let

$$\Delta\gamma_{sl} = \gamma_{sl}(D)-\gamma_{sl0}. \qquad (2.18)$$

Substituting Eq. (2.18) into Eq. (2.17) and taking in mind that $V/A = D/6$ and $\Delta A/A = 2\varepsilon = -8\kappa f/(3D)$ in terms of Eq. (2.16), it reads [77],

$$\gamma_{sl}(D)/\gamma_{sl0} = [1-8\kappa f^2/(3\gamma_{sl0}D)]/[1-8\kappa f/(3D)]. \qquad (2.19)$$

Eq. (2.19) is consistent with general calculations of thermodynamics [15,78-79] and quantum chemistry [80] for particles.

To determine f, we assume that when almost all atoms of a low-dimensional crystal immersed in fluid is located on its surface with a diameter of D'_0, the crystal is indistinguishable from the surrounding fluid where the solid-liquid interface is at all diffuse. Note that the crystal is now similar to a cluster produced by an energetic fluctuation of the fluid. This assumption leads to a limit case: As $D \to D'_0$, $\gamma_{sl}(D) \to 0$ where D'_0 depends on the existence of curvature [81]. Note that when a crystal has plane surface, such as a film, the possible smallest value of D is $2h$. For any crystals with curved surfaces, such as a particle or a wire, the possible smallest D is $3h$ [77]. According to the above definition of D'_0, $hA/V = 1-V_i/V = 1-[(D'_0-2h)/D'_0]^{3-d} = 1$ where V_i is the interior volume of the crystal. The solution of the above equation is $D'_0 = 2h$. For the particle and the wire usually having a curved surface, the value of D cannot be smaller than $3h$. In this case, there is no exact solution of the equation. As a first order approximation, $D'_0 = 3h$ is taken. Note that this approximation does not lead to big error. For instance, for a spherical particle, $hA/V = 26/27 \approx 1$. In summary,

$$D'_0 = 3h \tag{2.20-a}$$

$$D'_0 = 2h \tag{2.20-b}$$

where Eq. (2.20-a) for a particle or a wire with a curved surface and Eq. (2.20-b) for a film with a plane surface. Now Eq. (2.19) can be rewritten as [77],

$$\gamma_{sl}(D)/\gamma_{sl0} = [1-D'_0/D]/[1-\gamma_{sl0}D'_0/(fD)] \tag{2.21}$$

with $8\kappa f^2/(3\gamma_{sl0}) = D'_0$ or $f = \pm[(3\gamma_{sl0}D'_0)/(8\kappa)]^{1/2}$. The different signs of f correspond to the tensile (+) and compressive (-) stress on the surface. (In our opinion, the possible physical background of the positive or negative f could be explained based on the following mechanism [77]: Atoms at the interface suffer a coordination number reduction, bond contracts spontaneously, which leads to the enhancement of the atomic binding energy and hence the tensile stress [82], or $f > 0$. This is the case of free nanoparticles. When the interface atoms of different elements are intermixing, CN or the bond strength may increase, such as alloying or compound formation. The alloying and chemical reaction may alternate the atomic valences, which may introduce repulsive stress among the ions. An alternative interpretation is that the element with lower T_m has lower bond strength within the layer than that on the interface, bond expansion of interface atoms is present, which should results in negative f.)

It is known that for most metals, f is one larger than γ_{sl0} according to theoretical and experimental results [77,83-86]. Some molecular dynamics works based on a hard sphere model show that f has the same magnitude of γ_{sl0} [87-88]. However, the hard sphere model itself may lead to this result where strain is absent, which co-exists with f [82]. Thus, Eq. (2.21) may be simplified as a first order approximation,

$$\gamma_{sl}(D)/\gamma_{sl0} \approx 1-D'_0/D. \tag{2.22}$$

Figure 1 shows an agreement between the model prediction of Eq. (2.22) and the computer simulation results [15] [Because the $\gamma_{sl}(D)/\gamma_{sl0}$ value of an unknown fcc crystal is cited for comparison, the simplified form of Eq. (2.22) is employed].

At the same time, $\gamma_{sl}(D)$ values of five organic nanocrystals have also been calculated in terms of Eq. (2.21) and shown in table 4. The model predictions are consistent with the experimental observation [13] with the note that there exists a size distribution of nanocrystals. As the size of crystals decreases, $\gamma_{sl}(D)$ decreases. At $D = 4$ nm, the decrease in $\gamma_{sl}(D)$ for different substances is different, which is mainly induced by different D'_0 or D'_0/D values. The drop of $\gamma_{sl}(D)$ values in comparison with the corresponding bulk values reaches 20~40% where the critical diameter of a nucleus may be near 4 nm. Thus, the energetic resistance for the nucleation procession in liquid may be smaller than what the CNT has estimated [77].

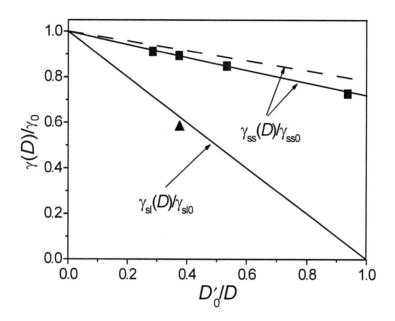

Figure 1. γ(D)/γ as a function of D'₀/D in terms of Eq. (2.22) and Eq. (3.4) with D'₀ = 3h. For $\gamma_{ss}(D)/\gamma_{ss0}$ function, the solid line and the segment line are obtained by use of negative and positive f, respectively. The symbols ■ and ▲ are the computer simulation results of $\gamma_{sl}(16h)/\gamma_{sl0} = 0.58$ for unknown fcc metal [15] and those for Cu [107] where γ_{ss0} =594 mJ/m².

Table 4. The comparison of $\gamma_{sl}(D)/\gamma_{sl0}$ values between the model predictions in terms of Eq. (2.21) with $D'_0 = 3h$ and the corresponding experimental results [13] where the experimental data of $\gamma_{sl}(D=4$ nm$)$ are obtained by measuring the slope of experimental data of melting temperature versus $1/D$ with two points of $D = 4$ nm and $D \approx 8.6$ nm in terms of Gibbs-Thomson equation

	$\gamma_{sl}(D)/\gamma_{sl0}$		κ
	Ref. [13]	Eq. (2.21)	(Mpa⁻¹ ×10⁻⁵)
Benzene	0.67	0.66	87
Naphthalene	0.68	0.69	≈ 87
Chlorobenzene	0.89	0.63	67
Heptane	0.63	0.80	134
Trans-decalin	0.60	0.68	≈ 87

Since κ values of crystals are not found, κ values of the corresponding liquid are used, which leads to minor error [53]. Note that κ values of naphthalene and *Trans*-decalin have been estimated as that of benzene.

The above agreement between Eq. (2.21) or Eq. (2.22) and experiments denotes that the energetic and structural differences between crystal and liquid decrease with size, which is proportional to surface/volume ratio with a $1/D$ relationship [27]. The success of model prediction for $\gamma_{sl}(D)$ values in return confirms again that H_m and S_{vib}, not H_m itself, determine the sizes of γ_{sl0} values as shown in Eq. (2.13) [32].

The Determination of Nucleus-liquid Interface Energy $\gamma_{sl}(D_n, T_n)$

As above-mentioned, several improved expressions for $g_m(T)$ function have been proposed through considering ΔC_p function below T_m and read as [72-74],

$$g_m^a(T) = \frac{7H_m T(T_m - T)}{T_m(T_m + 6T)}, \tag{2.23-a}$$

$$g_m^b(T) = \frac{2H_m T(T_m - T)}{T_m(T_m + T)}, \tag{2.23-b}$$

$$g_m^c(T) = \frac{H_m T(T_m - T)}{T_m^2} \tag{2.23-c}$$

where superscripts of a, b and c stand for metallic elements, ionic crystals and semiconductors, respectively. Eq. (2.23) predicts a steepest variation near T_m, and a much weaker temperature-dependence near the ideal glass transition temperature or isentropic temperature T_k, which can be determined in terms of the relationship of $dg_m(T)/dT = 0$ [89]. With these $g_m(T)$ functions, the respective $H_m(T)$ functions can also be determined in terms of $H_m(T) = g_m(T)-Tdg_m(T)/dT$ (Helmholtz function),

$$H_m^a(T) = 49H_m T^2 /(T_m + 6T)^2, \tag{2.24-a}$$

$$H_m^b(T) = 4H_m T^2 /(T_m + T)^2, \tag{2.24-b}$$

$$H_m^c(T) = H_m (T/T_m)^2. \tag{2.24-c}$$

Because $g_m(T)$ determined by ΔC_p while $\Delta C'_p$ between crystal and glass approaches to zero when $T \leq T_k$, the liquid must transform to glass [89]. Thus, Eq. (2.24) is valid only at $T > T_k$ where $T_k^a = (\sqrt{7} - 1)T_m /6$, $T_k^b = (\sqrt{2} - 1)T_m$ and $T_k^c = T_m /2$. Noted that $T > T_k$ is satisfied in undercooling experiments. Combining Eq. (2.13) and Eq. (2.24), the temperature-dependent $\gamma_{sl0}(T)$ functions can be expressed as,

$$\gamma_{sl0}^a(T) = \frac{2hH_m S_{vib}}{3RV}(\frac{7T}{T_m + 6T})^2,$$

$$\gamma_{sl0}^b(T) = \frac{2hH_m S_{vib}}{3RV}(\frac{2T}{T_m + T})^2, \tag{2.25-b}$$

$$\gamma_{sl0}^c(T) = \frac{2hH_mS_{vib}}{3RV}\left(\frac{T}{T_m}\right)^2.$$ (2.25-c)

$\gamma_{sl0}(T)$ in terms of Eq. (2.25) decreases as T drops. As $T \to T_m$, $\gamma_{sl0}^a \approx \gamma_{sl0}^b \approx \gamma_{sl0}^c$ due to the decreased effect of ΔC_p on $g_m(T)$. Although negative temperature dependence for γ_{sl0} has been considered [90], it differs from the usual understanding that differences of structure and surface state between crystal and liquid decrease with T [72-74].

Substituting Eq. (2.25) into Eq. (2.21), the integrated size- and temperature-dependent interface energy can be read as,

$$\gamma_{sl}^a(D,T) = \frac{2hH_mS_{vib}}{3RV}(1 - \frac{3h}{D})(\frac{7T}{T_m + 6T})^2$$ (2.26-a)

$$\gamma_{sl}^b(D,T) = \frac{2hH_mS_{vib}}{3RV}(1 - \frac{3h}{D})(\frac{2T}{T_m + T})^2$$ (2.26-b)

$$\gamma_{sl}^c(D,T) = \frac{2hH_mS_{vib}}{3RV}(1 - \frac{3h}{D})(\frac{T}{T_m})^2.$$

Substituting Eq. (2.23) into Eq. (2.1) and γ'_{sl0} is replaced by $\gamma_{sl}(D,T)$ in terms of Eq. (2.26), the critical size of nuclei D_n can be determined by letting $\partial \Delta G(D,T)/\partial r = 0$,

$$D_n^a/h = 2(A + \sqrt{A^2 - 3A\theta/2})/\theta$$

$$D_n^b/h = 2(4B + \sqrt{16B^2 - 18B\theta})/(3\theta),$$ (2.28-b)

$$D_n^c/h = 2(C + \sqrt{C^2 - 3C/2})$$ (2.27-c)

where $\theta = \dfrac{T_m - T}{T_m}$ is the degree of undercooling. $A = \dfrac{14S_{vib}}{3R}\dfrac{1-\theta}{7-6\theta}$, $B = \dfrac{S_{vib}}{R}\dfrac{1-\theta}{2-\theta}$ and $C = \dfrac{2S_{vib}}{3R}\dfrac{1-\theta}{\theta}$.

Substituting Eq. (2.27) into Eq. (2.26) with experimentally determined θ values, the interface energy $\gamma_{sl}(D_n,T_n)$ of the nucleus-liquid can be determined.

$\gamma_{sl}(D_n,T_n)$ for Metallic and Semiconductors Elements

Since metallic and semiconductors elements were firstly dealt with by Turnbull [11], the comparison between the predicted $\gamma_{sl}(D_n,T_n)$ values in terms of Eq. (2.26) and experimentally determined γ_{CNT} values in terms of the CNT [11] are firstly listed in table 5. A good agreement between them is shown. Thus, the obtained γ_{CNT} values in the CNT are not $\gamma_{sl0}(T_m)$ values, but $\gamma_{sl}(D_n,T_n)$ ones. $\gamma_{sl}(D_n,T_n) < \gamma_{sl0}(T_m)$ could be partly induced by different structures between nuclei and bulk crystals although it is difficult to determine the nucleus structure. The above argument can be further supported by considering the correlation of the incoherent interface energy between adjacent crystals or grain-boundary energy γ_{ss0} with $\gamma_{sl0}(T_m)$, which may be explained in detail in Section 3.1.

Table 5. The model predictions in terms of Eq. (2.26)
and the corresponding experimental results calculated from the
CNT [11] with relevant parameters listed in table 1

	$\gamma_{sl}(D_n,T_n)$	γ_{CNT}	θ_n	D_n/h
	(mJ/m^2)			
Co	248	234	0.187	11.0
Ni	283	255	0.185	11.6
Cu	200	177	0.174	12.0
Pd	211	209	0.182	10.8
Ag	137	126	0.184	11.2
Pt	243	240	0.181	11.2
Au	150	132	0.172	11.6
Pb	36	33	0.133	14.0
Mn	206	206	0.206	10.2
Fe	186	204	0.164	11.6
Hg	21	24	0.247	9.0
Al	110	93	0.14	12.0
Sn	62	55	0.208	12.0
Ga	61	54	0.25	10.8
Sb	109	101	0.15	12.6
Bi	45	54	0.166	9.6
Ge	201	181	0.184	11.0
Ref		11		

Similarly, although $\gamma_{sl0}(T_m)$ of pure Si was reported to be 352±38 and 332 mJ/m^2 [39], recent experiment has shown that $\gamma_{CNT} = 380$ mJ/m^2 at $\theta = 0.21$ [91]. Combining Eqs. (2.26-c) and (2.27-c), the real $\gamma_{sl0}(T_m)$ value of pure Si should be 680 mJ/m^2, which corresponds to $\gamma_{sl0}(T_m) = 738$ mJ/m^2 by Eq. (2.13) and $\gamma_{TS} = 672$ mJ/m^2 by Eq. (2.14).

As shown in Ref. [39], the $\gamma_{sl0}(T_e)$ values of Solid Al-liquid Al-Cu, Solid Al-liquid Al-Si and Solid Pb-liquid Pb-Sn systems are only a litter larger than the $\gamma_{sl0}(T_m)$ values of pure Al and Pb where T_e is eutectic temperature. However, $\gamma_{sl0}(T_e)$ of Solid Si-liquid Al-Si system is much smaller than $\gamma_{sl0}(T_m)$ of pure Si. The possible reason can be explained as the following: T_e/T_m for Al and Pb are about 0.90 and 0.76 while that for Si is 0.50. For the bulk crystals, $\gamma_{sl0}(T_e)/\gamma_{sl0}(T_m)$ for Al and Pb are 0.97 and 0.92 while that for Si is 0.25 in terms of Eqs. (16-a) and (16-c). Although it is inappropriate that Eqs. (2.26-a) and (2.26-c) for pure metals and semiconductors are directly applied to the eutectic system, it may be reflect some evident differences in $\gamma_{sl0}(T)$ between solid metal-liquid and solid semiconductor-liquid systems.

Combining our previous work [92], the model predictions for both T_n and $\gamma_{sl}(D_n,T_n)$ correspond well to Turnbull's experimental results although both $\gamma_{sl}(D,T)$ and $g_m(T)$ functions differ from those in the CNT. The possible reason may be the mutual compensation between $\gamma_{sl}(D,T)$ and $g_m(T)$ functions. However, there is about 30% difference in the value of D_n between CNT ($D_n \approx 8h$) and this model ($D_n \approx 11h$), which may result from neglecting of derivative of $\gamma_{sl}(D,T)$ with respect to D in the CNT. Although we cannot confirm the above difference from experiments due to the experimental difficulties, we believe that a little larger D_n is more reasonable.

According to Eq. (2.27), the size of D_n is decided by S_{vib} and θ due to the introduction of $\gamma_{sl}(D,T)$ function, and it increases with an increase in S_{vib} or a decrease in θ. Since S_{vib} values, which differ from the cases of S_m, are similar for different bond natures of elements while θ determined from experiments for distinct elements are also similar, D_c is in fact independent on the elemental types.

Let $\gamma_{sl}(D,T)$ of Eq. (2.26) is also expressed as $\gamma_{sl}(D,T) = c_2hH_m/V_m$, there are $c_2^a = \dfrac{2S_{vib}}{3R}(1-\dfrac{3h}{D})(\dfrac{7T}{T_m+6T})^2$ and $c_2^c = \dfrac{2S_{vib}}{3R}(1-\dfrac{3h}{D})(\dfrac{T}{T_m})^2$. The mean values of S_{vib}/R, D_n/h and θ for concerned metals and semiconductors listed in table 5 are 0.94, 11.2 and 0.187, and 0.98, 11.0 and 0.167, respectively, which leads to $c_2^a = 0.43$ and $c_2^c = 0.33$ at T_n and D_n. These values correspond well to the initial findings of $c_1'^a = 0.45$ and $c_1'^b = 0.32$ in Eq. (2.3) [11]. This correspondence confirms again that $\gamma_{CNT} \approx \gamma_{sl}(D_n,T_n)$.

If the temperature dependence of $H_m(T)$ is neglected, both γ_{sl0} and $\gamma_{sl}(D,T)$ increase in terms of Eqs. (2.25) and (2.26), which leads to $c_2^a = 0.46$ and $c_2^c = 0.48$. These are similar to $c_2 = 0.49 \pm 0.08$ for 22 metals and 4 semiconductors where $H_m(T) \approx H_m(T_m)$ is used [93]. However, this treatment will result in errors in analyzing the nucleation process due to the unreasonable assumption of $H_m(T) \approx H_m(T_m)$.

It seems from Eq. (2.3) that the size of h also affects γ'_{sl0} value. To accurately estimate the influence of h, all h values of elements should be unified to h' values where the elements with different structures have the same CN of 12. According to the Goldschmidt premise for lattice contraction [71], h contracts 3%, 4%, and 12% if CN of the atom reduces from 12 to 8, 6, and 4, respectively. The additional correlation between h and h' for the elements, which CN is not 4, 6 or 8, can be found in Ref. [94]. Replacing h in Eq. (2.3) with h', Eq. (2.3) is simplified as the following form [95],

$$\gamma'_{sl0} = c_3H_m/V \tag{2.28-a}$$

where $c_3 \approx 0.11 \pm 10\%$ nm is the slope except semi-metals Pb, Sn and Ga shown in figure 2 while Eq. (2.13) can also be simplified as the similar form [95],

$$\gamma_{sl0}(T_m) = c_4H_m/V \tag{2.28-b}$$

where $c_4 = 2S_{vib}h'/(3R) \approx 0.18 \pm 15\%$ nm is the slope except Fe, Al and Ga. Eq. (2.28) does not need to distinguish the bond natures of the elements with a unique c_3 or c_4 value rather than two different c_1' values for metals and semiconductors, respectively. Moreover, disappearance of h or h' in Eq. (2.28) implies that the unique parameter in deciding γ_{sl0} is H_m.

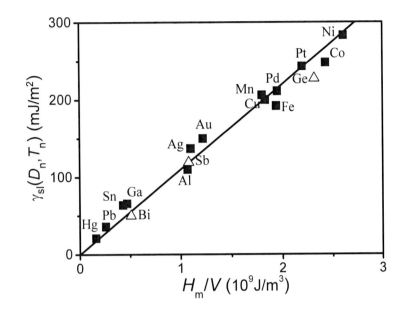

Figure 2. $\gamma_{sl}(D_n,T_n)$ as a function of H_m/V for a variety of elemental systems in terms of Eq. (2.28-a) (the solid line) where the symbols ■ and △ denote the $\gamma_{sl}(D_n,T_n)$ values of metals and semiconductors, respectively.

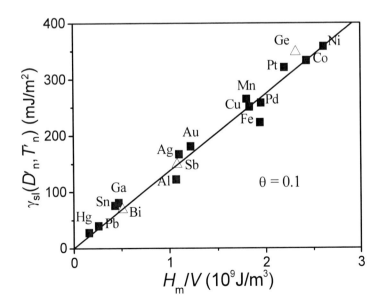

Figure 3. $\gamma_{sl}(D_n',T_n')$ as a function of H_m/V with $\theta = 0.1$ for a variety of elemental systems in terms of Eq. (2.29) (the solid line) where the symbols ■ and △ denote the $\gamma_{sl}(D_n',T_n')$ values of fcc and non-fcc elements, respectively.

Substituting Eq. (2.27) into Eq. (2.26) and plotting Eq. (2.26), the curve is linearly regressed as a function of θ, which leads to [95],

$$\gamma_{sl}(D_n',T_n') \approx (1.78-3.83\theta)H_m/V.$$

where T_n' is any nucleation temperature and D_n' is the corresponding radius of a nucleus. The standard deviations for 1.78 and 3.83 are 0.01 and 0.16, respectively. Eq. (2.29) indicates that Eq. (2.28) denotes two extreme cases where $\theta = \theta_n$ (the maximum undercooling, which is nearly a constant of 0.18±0.02 for the most elements [43,96]) and $\theta = 0$, respectively. Thus, at any θ value, the relationship of $\gamma_{sl}(D_n',T_n') \propto H_m/V$ exists always. As an example, figure 3 shows such a relationship at $\theta = 0.1$ where the slope is the linear function, which equals to 1.40 as indicated in Eq. (2.29).

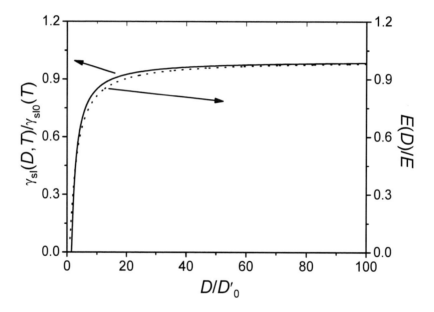

Figure 4. Comparison between $\gamma_{sl}(D,T)/\gamma_{sl0}(T)$ shown as solid line in terms of Eq. (2.22) and $E(D)/E$ shown as dot in terms of Eq. (2.30) where $D_0' = 2h$.

This linear relationship of Eq. (2.29) between γ_{sl} and H_m could be considered as that H_m is related to bond energy of crystalline atoms while γ_{sl} denotes the bond energy difference between surface atoms and interior atoms of a crystal. The behavior displayed by γ_{sl} has been found to be fruitfully compared to that of cohesive energy E_b [96]. It is known that cohesive energy determines the size of bond energy while E is also size dependent, which has been determined by [97],

$$E(D)/E_b = \left[1-\frac{1}{2D/h-1}\right]\exp\left(-\frac{2S_b}{3R}\frac{1}{2D/h-1}\right) \qquad (2.30)$$

where $S_b = E_b/T_b$ is the solid-vapor transition entropy of crystals with E_b being the bulk cohesive energy and T_b being the bulk boiling temperature. Comparing $\gamma_{sl}(D)/\gamma_{sl0}$ function in terms of Eq. (2.22) with $E(D)/E$ function in terms of Eq. (2.30) where $S_b \approx 12R$ is assumed except for Sb and Bi [42], a similar size dependence is found (see figure 4). This agreement indicates that the size dependence of $\gamma_{sl}(D)$ is originated from the size dependence of bond energy. In terms of Eq. (2.28) or Eq. (2.29), $\gamma_{sl}(D_n,T_n)/\gamma_{sl0}(T_m)$ is within 10% of the value of $E(D_n)/E_b$, which also confirms the validity of Eq. (2.28) or Eq. (2.29).

Another linear relationship between $\gamma_{sl}(D_n,T_n)$ or γ'_{sl0} and T_m with large scatter among different groups of elements [11] has been rejected as the basis for an empirical rule in favor of the correlation with H_m [98]. However, direct calculations for γ'_{sl0} values of transition fcc metals with hard-sphere systems have shown a strong correlation with T_m [88,98], which can also be extended to $\gamma_{sl0}(T_m)$ for all fcc metals. Eq. (2.28-b) can confirm this correlation since $H_m = T_m S_m$ and $c_4 S_m = 1.59 \pm 7\%$ nm J/mol K for all fcc metals listed in table 5. For other elements, $c_4 S_m$ values show large scatter in a range of 1.04 to 5.44 nm J/mol K due to the scatter of S_m values. This indicates a disagreement for a linear relationship between T_m and H_m. Thus, H_m values, but not T_m values, in general characterize $\gamma_{sl0}(T_m)$ value better.

Recently, it is thought [99] that Skapsky's assumption, e.g. $\gamma_{sl0}(T_m) = \gamma_{sv0}(T_m)-\gamma_{lv0}(T_m)$ [100], has been used in the derivation of Eq. (2.13) where subscript v denotes vapor phase. Because there is no nucleation barrier for surface melting, solids can be hardly superheated above T_m, which leads to $\gamma_{sl0}(T_m) \leq \gamma_{sv0}(T_m)-\gamma_{lv0}(T_m)$. Thus, $\gamma_{sl0}(T_m)$ in terms of Eq. (2.13) is the upper limit of γ_{sl}. However, it is not the case. As a monotone function of T and D in terms of Eq. (2.26), $\gamma_{sl}(D,T)$ takes its minimum at D_n and T_n (D_n and T_n are the smallest values possibly to be taken) and its maximum at $D \rightarrow \infty$ and $T \rightarrow T_m$. Combining the values of D_n/h and θ listed in table 5, Eq. (2.26) gives the both limits of γ_{sl} for elements as,

$$0.62 \leq 3V_m R\gamma_{sl}(D,T)/(2hS_{vib}H_m) \leq 1. \tag{2.31}$$

Eq. (2.31) shows the reason why $\gamma_{sl0}(T_m)$ in terms of Eq. (2.13) is the upper limit of γ_{sl} and overestimates $\gamma_{sl}(D_n,T_n)$ 50% to 100% [46].

$\gamma_{sl}(D_n,T_n)$ for Alkali Halides

In terms of Eqs. (2.26-b) and (2.27-b), the $\gamma_{sl}(D_n,T_n)$ values for alkali halides (ionic crystals) are also predicted and shown in table 6. For comparison, the corresponding γ_{CNT} values [101] are also listed.

As shown in table 6, the difference between $\gamma_{sl}(D_n,T_n)$ and γ_{CNT} is smaller than 20% except for CsF where the difference reaches to 33%. Considering that both $\gamma_{sl}(D_n,T_n)$ and γ_{CNT} values are very smaller, tiny difference in absolute values will make the deviation larger. Thus, it can also claim that an agreement between $\gamma_{sl}(D_n,T_n)$ and γ_{CNT} is found.

Moreover, for alkali halides, most of the D_n/h values are smaller while the corresponding θ_n values are larger than those of metallic and semiconductors elements, which leads to that $\gamma_{sl}(D_n,T_n)/\gamma_{sl0}$ values of alkali halides will be smaller than those of elements in terms of Eq. (2.26).

Table 6. Comparisons of $\gamma_{sl0}(T_m)$ between $\gamma_{sl}(D_n, T_n)$ by Eq. (2.13) and other theoretical results γ_{CNT} deduced from γ_{CNT} [101]

	γ_{sl} (D_n, T_n) (mJ/m²)	γ_{CNT}	h (nm)	V (cm³/g-atom)	T_m (K)	H_m (kJ/g-atom)	$\Delta V/V$ (%)	S_m (J/g-atom K)	S_{vib}	θ	D_n/h
LiF	101	124	0.125	4.9	1121	13.4	29.4	12.0	7.5	0.21	8.0
LiCl	72	86	0.257	10.0	883	10.0	26.2	11.3	7.1	0.21	8.0
LiBr	54	46	0.258	12.6	823	8.8	24.3	10.7	6.5	0.15	10.0
NaF	218	206	0.292	8.2	1269	16.8	27.4	13.2	8.9	0.22	9.0
NaCl	114	112	0.302	13.5	1074	14.1	25.0	13.1	8.9	0.23	9.0
NaBr	88	88	0.304	16.1	1020	13.1	22.4	12.8	8.8	0.23	9.0
KCl	86	82	0.328	18.8	1043	13.1	17.3	12.6	9.1	0.23	9.4
KBr	70	63	0.334	21.7	1007	12.6	16.6	12.6	9.1	0.23	9.4
KI	54	47	0.339	27.5	954	11.9	15.9	12.5	9.1	0.23	9.4
RbCl	61	57	0.326	22.0	996	11.8	14.3	11.9	8.7	0.23	9.0
CsF	61	46	0.288	18.5	976	11.3		11.6	8.8	0.23	9.0
CsCl	52	51	0.318	21.1	918	10.2	10.5	11.2	8.4	0.24	8.0
CsBr	64	66	0.329	24.0	909	11.8		13.0	10.2	0.26	9.0
CsI	81	69	0.342	28.1	900	12.9		14.2	11.4	0.23	12.0
Ref			44	56	29	29	44	29		101	

S_{vib} is determined by Eq. (2.10-a). $\Delta V/V$ for CsF, CsBr and CsI are unavailable and assumed to be equal to that of CsCl since the differences of $\Delta V/V$ between different alkalis halides of the same family are small.

Solid-Solid Interface Energy

The Bulk Solid-solid Interface Energy γ_{ss0}

For coherent or semi-coherent solid-solid interface, its interface energy can be determined according to some classic dislocation models [102]. While for incoherent solid-solid interface, which is the most case since even for semi-coherent interface, atomic diameter misfit on the interface must be smaller than 0.15-0.25, γ_{ss0} remains challenge.

It is well known that a liquid may be regarded as a solid with such a high concentration of dislocation cores that these are in contact everywhere [103]. Based on this consideration, solid-solid interface energy γ_{ss0} at T_m is considered to be twice of the solid-liquid interface energy γ_{sl} approximately [104], $\gamma_{ss0}(T_m) \approx 2\gamma_{sl0}(T_m)$. Combining with Eq. (2.13),

$$\gamma_{ss0}(T_m) \approx 4hS_{vib}H_m/(3VR). \qquad (3.1)$$

As shown in table 7, the $\gamma_{ss0}(T_m)$ values based on Eq. (3.1) for eleven elemental crystals (Cu, Ag, Au, Al, Ni, Co, Nb, Ta, Sn, Pb and Bi) and two organic crystals (Pivalic acid and Succinonitrile) correspond to the available theoretical values γ'_{ss0} [39,48-49,105-111]. In addition, the data of γ'_{ss0} and γ'_{sl0} for Pivalic acid and Succinonitrile [48-49] determined by the equilibrated grain boundary groove shapes also confirm the validity of $\gamma_{ss0} \approx 2\gamma_{sl0}$ with the absolute deviation smaller than 5.5%. Thus, Eq. (3.1) can be used to quantitatively calculate the γ_{ss0} values, at least for metals and organic crystals.

Table 7. Comparison of $\gamma_{ss0}(T_m)$ between γ_{ss0} by Eq. (3.1) and available theoretical or experimental results γ'_{ss0} for metals [39,105-111] and organic crystals [48-49]

System	γ_{ss0}	γ'_{ss0}	γ'_{sl0}
		(mJ/m^2)	
Cu	584	601	
Ag	398	392,375	
Au	430	400	
Al	302	300~380	
Ni	844	866,757	
Co	706	650	
Nb	804	760	
Ta	944	900	
Sn	158	160,164	
Pb	96	111.5±15.6	
Bi	176	140.5±14.1	
Pivalic acid	4.8	5.2±0.4	2.7±0.2
Succinonitrile	14.6	15.0±2.0	7.9±0.8

On the contrary, when γ_{CNT} values are used, e.g. 132, 126, 177, 255 and 54.4 mJ/m^2 [11] for Au, Ag, Cu, Ni and Bi, it is obvious that $2\gamma_{CNT}$ values are about 20~40% smaller than the corresponding theoretical or experimental values 400, 392, 601, 757 and 140.5±14.1 mJ/m^2 [105-107,109,111], which is unreasonable [11]. Moreover, this comparison supports again the claim that $\gamma'_{sl0} \approx \gamma_{sl}(r_c,T_n)$ and $\gamma'_{sl0} \neq \gamma_{sl0}(T_m)$.

The Size Dependence of Solid-solid Interface Energy $\gamma_{ss}(D)$

In the deduction of $\gamma_{ss}(D)$, the way is similar to the deduction of $\gamma_{sl}(D)$ while Eq. (2.15) and Eq. (2.16) must be modified due to different interface conditions [77]. We assume that f as a first order approximation keeps constant for both solid-liquid and solid-solid interface at T_m, which leads to the same strain on both sides of the interface when the grains are isotropic. However, elastic modulus of grain boundaries should be larger than that of the solid-liquid interface with less strain under the same stress. This is introduced by the fact that $A = 3V/(D)$ for solid-solid interface because two solid-liquid interfaces of particles combine to form one grain boundary with

$$P = 2fA/(3V) = 2f/D \tag{3.2}$$

and

$$\varepsilon = -2\kappa f/(3D). \tag{3.3}$$

Both equations indicate that the strain on the grain boundary is only a half of that on the solid-liquid interface and thus $\Delta A/A = 2\varepsilon = -4\kappa f/(3D)$. Now $\Delta\gamma_{ss} = \gamma_{ss}(D)-\gamma_{ss0} = (\Delta A/A)(f-\gamma_{ss})= -4\kappa f/(3D)(f-\gamma_{ss})$, or

$$\gamma_{ss}(D)/\gamma_{ss0} = [1-D'_0/(4D)]/[1-\gamma_{ss0}D'_0/(4fD)]. \tag{3.4}$$

Similar to the simplification of $\gamma_{sl}(D)$, Eq. (3.4) can also be simplified as

$$\gamma_{ss}(D)/\gamma_{ss0} = 1 - D'_0/(4D). \tag{3.5}$$

Figure 1 shows an agreement between the model predictions of Eq. (3.4) and the computer simulation results for Cu [112]. It is interesting that the use of a negative interface stress for $\gamma_{ss}(D)$ in Eq. (3.4) leads to a full agreement with the computer simulation results, which implies that $f < 0$.

If we compare Eq. (2.22) and Eq. (3.5), it can be found that $\gamma_{sl}(D)/\gamma_{sl0} < \gamma_{ss}(D)/\gamma_{ss0}$ which implies that the stiffer surrounding of particles brings out less decrease of the interface energy as D decreases. For the grain boundaries, even if when $D \rightarrow D'_0$, $\gamma_{ss}(D)/\gamma_{ss0} \approx 75\%$ while $\gamma_{sl}(D)/\gamma_{sl0} = 0$. Since when $D \approx 2D'_0$, the grains are no more stable and will transform to amorphous solids in terms of the computer simulation results [112], the smallest value of $\gamma_{ss}(D'_0)/\gamma_{ss0}$ could be about 85%.

Solid-Vapor Interface Energy or Surface Energy

The Bulk Surface Energy γ_{sv0}

The surface energy γ_{sv0} usually is defined as the difference between the free energy of the surface and that of the bulk or simply as the energy needed to split a solid in two along a plane, which is one of the basic quantities to understand the surface structure and phenomena [32,77,113]. Despite of its importance, γ_{sv0} value is difficult to determine experimentally. The most of these experiments are performed at high temperatures where surface tension of liquid is measured, which are extrapolated to zero Kelvin. This kind of experiments contains uncertainties of unknown magnitude [114-115] and corresponds to only γ_{sv0} value of an isotropic crystal [116]. Note that many published data determined by the contact angle of metal droplets or from peel tests disagree each other, which can be induced by the presence of impurities or by mechanical contributions, such as dislocation slip or the transfer of material across the boundary [117]. In addition, there are hardly the experimental data on the more open surfaces except for the classic measurements on Au, Pb and In [118] to our knowledge. Therefore, a theoretical determination of γ_{sv0} values especially for open surface is of vital importance.

During the last years there have been several attempts to calculate γ_{sv0} values of metals using either *ab initio* techniques [119-121] with tight-binding (TB) parameterizations [122] or semi-empirical methods [123]. γ_{sv0} values, work functions and relaxation for the whole series of bcc (A2) and fcc (A1) 4d transition metals have firstly been studied [119] using the full-potential (FP) linear muffin-tin orbital (LMTO) method in conjunction with the local-spin density approximation to the exchange-correlation potential [124-126]. In the same spirit, γ_{sv0} values and the work functions of the most elemental metals including the light actinides have been carried out by the Green's function with LMTO method [120-121,127-128]. Recently, the full-charge density (FCD) Green's function LMTO technique in the atomic-sphere approximation (ASA) with the generalized gradient approximation (GGA) has been utilized to construct a large database that contains γ_{sv0} values of low-index surfaces of 60 elements in the periodic table [126,129-132]. The results denote a mean deviation of 10% for the $4d$ transition metals from FP methods [133]. This database in conjunction with the pair-potential model [134] has been further extended to estimate the formation energy of mono-atomic steps on low-index surfaces for an ensemble of the A1 and A2 metals [135].

On the other side, the traditional broken-bond model is again suggested to estimate γ_{sv0} values of the transition metals and the noble metals with different facets [118,136-137]. The simplest approach to get a rough estimation of γ_{sv0} values at $T = 0$ K is to determine the broken bond number $Z_{(h\ k\ l)}$ for creating a surface area by cutting a crystal along certain crystallographic plane with a Miller index $(h\ k\ l)$. $Z_{(h\ k\ l)} = Z_B - Z_S$ where Z_S is CN of surface atoms and Z_B is the CN of the corresponding bulk one. Multiplying this number with the cohesion energy per bond E/Z_B for the non-spin-polarized atom at 0 K, γ_{sv} is determined by [138],

$$\gamma_{sv0} = (1-k_1)E_b/(N_a A_S) \qquad (4.1)$$

where $k_1 = Z_S/Z_B$ and A_S denotes the area of the two-dimensional unit cell of solid.

In Eq. (4.1), E_b is independent on crystalline structures as a first order approximation since energetic differences between solid structures are several orders of magnitude smaller than E_b for any structure when the bond type remains unaltered.

Since the bond strength becomes stronger for an atom with a smaller CN, this CN-bond-strength relation can be quantified using tight-binding approximation. In the second-moment approximation, the width of the local density of states on an atom scales with Z_S, leading to an energy gain to be proportional to the square root of Z_S due to the lowing of the occupied states [136]. Neglecting repulsive terms, the energy per nearest neighbor is then proportional to the square root of Z_S. By assuming that the total crystalline energy is a sum of contributions of all bonds of an atom, it follows that [136],

$$\gamma_{sv0} = [1-k_1^{1/2}]E_b/(N_a A_S). \qquad (4.2)$$

While Eq. (4.1) does not consider the variation of bond strength with CN, Eq. (4.2) is also not complete because only attractive forces are taken into account [119]. Namely, the former neglects while the latter overestimates the effect of relaxation on γ_{sv0} [139], which results in that Eq. (4.1) can only give a rough estimation of γ_{sv0} values while Eq. (4.2) is especially suitable for noble metals.

Although a direct utilization of Eq. (4.1) or Eq. (4.2) is reasonable, one of them cannot alone give satisfied predictions for γ_{sv0} values in comparison with the experimental and theoretical results [119]. To obtaining a more general formula, we arbitrarily assume that both of Eqs. (4.1) and (4.2) could make up the deficiency each other with the same weight to both formulae. Thus, γ_{sv0} values may be determined by an averaged effect of them without elaborate estimation on the relaxation energy [139],

$$\gamma_{sv0} = [2-k_1-k_1^{1/2}]E_b/(2N_a A_S). \qquad (4.3)$$

Eq. (4.3) implies that γ_{sv0} values still depend on the bond-broken rule although they are scaled by both of Z_S and the square root of Z_S.

In Eq. (4.3), Z_S can be determined according to the crystalline structure through determining $Z_{(h\,k\,l)}$ by a geometric consideration [140-141]. For an fcc or hcp structure, $Z_B = 12$; For a bcc lattice, although $Z_B = 8$ is taken according to the nearest-neighbor definition by some authors (probably the majority), others prefer to regard $Z_B = 14$ since the difference between the nearest neighbor bond length and the next-nearest neighbor bond length is small [142]. Here, the latter is accepted. By assuming that the total energy of a surface atom is the sum of contributions from both of the nearest neighbor and the next-nearest neighbor atoms, Eq. (4.3) should be rewritten for bcc metals after normalization [139],

$$\gamma_{sv0} = [(2-k_1-k_1^{1/2})+\varphi(2-k_1'-k_1'^{1/2})]E_b/[(2+2\varphi)N_aA_S] \tag{4.4}$$

where the superscript comma denotes the next-nearest CN on a surface and φ shows the total bond strength ratio between the next-nearest neighbor and the nearest neighbor [143]. To roughly estimate the size of φ, the Lennard-Jones (LJ) potential is utilized [143]. The potential is expressed as $u(r) = -4\eta[(\tau/r)^6-(\tau/r)^{12}]$ with η being the bond energy and τ insuring $du(r)/dr_{(r=h)} = 0$, i.e. $\tau = 2^{-1/6}h$ where r is the atomic distance. For bcc crystal, $h = 3^{1/2}a/2$ and $h' = a$, respectively. Let $r = a$, $\eta' = 2\eta/3$. Thus, $\varphi = [(2/3)\times6]/8 = 1/2$. Adding this value into Eq. (4.4) [139],

$$\gamma_{sv0} = [3-k_1-k_1^{1/2}-k_1'/2-(k_1'/4)^{1/2}]E_b/(3N_aA_S). \tag{4.5-a}$$

Note that the bonding of the LJ potential, which is utilized to justify φ value in Eq. (4.4), is different from the metallic bond in its nature. For instance in a LJ bonded system, the surface relaxation is outwards whilst in the transition metals it is inwards. However, this difference leads to only a second order error in our case and has been neglected.

The effect of next-nearest CN also occurs for simple cubic (sc) and diamond structure crystals because there are twice and thrice as many the second neighbors as the first neighbors, respectively. Similar to the above analysis, for sc crystals: $\eta' \approx \eta/4$ and $\varphi = [(1/4)\times12]/6 = 1/2$, which is the same for bcc and thus Eq. (4.5-a) can also hold for sc crystals. For diamond structure crystals, $\eta' \approx \eta/10$ and $\varphi = [(1/10)\times12]/4 = 3/10$. With this φ value, Eq. (4.4) is rewritten as [139],

$$\gamma_{sv0} = [26-10k_1-10k_1^{1/2}-3k_1'-(9k_1')^{1/2}]E_b/(26N_aA_S). \tag{4.5-b}$$

$Z_{(h\,k\,l)}$ can be determined by some known geometrical rules. For any surface of a fcc structure with $h \geq k \geq l$ [140-141],

$$Z_{(h\,k\,l)} = 2h+k \text{ for } h, k, l \text{ being odd} \tag{4.6-a}$$

$$Z_{(h\,k\,l)} = 4h+2k \text{ for the rest} \tag{4.6-b}$$

In a similar way, $Z_{(h\,k\,l)}$ for any surface of a bcc structure is determined with the consideration of the next-nearest CN [140],

$$Z_{(h\,k\,l)} = 2h+(h+k+l) \text{ for } h+k+l \text{ being even} \tag{4.7-a}$$

$$Z_{(h\,k\,l)} = 4h+2(h+k+l) \text{ for } h+k+l \text{ being odd and } h-k-l \geq 0 \tag{4.7-b}$$

$$Z_{(h\,k\,l)} = 2(h+k+l)+2(h+k+l) \text{ for } h+k+l \text{ being odd and } -h+k+l > 0 \tag{4.7-c}$$

where the 2nd item of the right-hand side of Eq. (4.7) denotes the broken bond number of the next-nearest neighbors.

For sc crystals, $Z_{(h\,k\,l)}$ values of the nearest and the next-nearest atoms are 1 and 4 for (100) surface as well as 2 and 5 for (110) surface, respectively. For diamond structure crystals, $Z_{(h\,k\,l)}$ values of the nearest and the next-nearest atoms are 1 and 6 for (110) surface.

For several surfaces of a hcp structure, $Z_{(h\,k\,i\,l)}$ is obtained by [141],

$$Z_{(h\,k\,i\,l)} = 4(h+k)+3l \text{ for } (0001) \tag{4.8-a}$$

$$Z_{(h\,k\,i\,l)} = 4(h+k)+(8h+4k)/3 \text{ for } (10\bar{1}0) \tag{4.8-b}$$

where the 1st item of the right-hand side of Eq. (4.8) denotes the average number of basal broken bonds while the 2nd item is that of non-basal broken bonds.

Table 8. A_S, Z_S and Z_B values for different surfaces and structures where a and c are lattice constants

Structure	Surface	A_S	Z_S	Z_B
Fcc	(111)	$3^{1/2}a^2/4$	9	12
	(100)	$a^2/2$	8	12
	(110)	$2^{1/2}a^2/2$	6	12
Bcc	(110)	$2^{1/2}a^2/2$	$Z_S = 6,\ Z'_S = 4$	$Z_B = 8,\ Z'_B = 6$
	(100)	a^2	$Z_S = 4,\ Z'_S = 4$	$Z_B = 8,\ Z'_B = 6$
	(111)	$3^{1/2}a^2$	$Z_S = 2,\ Z'_S = 0$	$Z_B = 8,\ Z'_B = 6$
Hcp	(0001)	$3^{1/2}a^2/2$	9	12
	(10$\bar{1}$0)	$(8/3)^{1/2}a^2$	16/3	12
Diamond	(110)	$2^{1/2}a^2/4$	$Z_S = 3,\ Z'_S = 6$	$Z_B = 4,\ Z'_B = 12$
SC	(100)	a^2	$Z_S = 5,\ Z'_S = 8$	$Z_B = 6,\ Z'_B = 12$
	(110)	$2^{1/2}a^2$	$Z_S = 4,\ Z'_S = 7$	$Z_B = 6,\ Z'_B = 12$

For bcc, the CN of surface and bulk metals are divided into the nearest and the next-nearest bonds. CN is determined by Eqs. (4.6) to (4.8). For hcp metals, an ideal c/a ratio (1.633) is assumed.

Table 9. Comparison of surface energies of fcc metals among the predicted values γ_{sv0} of Eq. (4.3), the FCD calculations γ'_{sv0} [116] and the experimental results γ''_{sv0} [114-115]. E_b and a are cited from Refs. [144] and [116,145]

	E_b (kJ/g-atom)	a (nm)	$(h\ k\ l)$	γ_{sv0} (J/m^2)	γ'_{sv0} (J/m^2)	γ''_{sv0} (J/m^2)
Cu	336	0.366	(111)	1.83	1.95	1.79, 1.83
			(100)	2.17	2.17	
			(110)	2.35	2.24	
Ag	284	0.418	(111)	1.20	1.17	1.25, 1.25
			(100)	1.40	1.20	
			(110)	1.51	1.24	
Au	368	0.420	(111)	1.52	1.28	1.51, 1.50
			(100)	1.80	1.63	
			(110)	1.94	1.70	
Ni	428	0.358	(111)	2.44	2.01	2.38, 2.45
			(100)	2.88	2.43	
			(110)	3.11	2.37	
Pd	376	0.385	(111)	1.85	1.92	2.00, 2.05
			(100)	2.15	2.33	
			(110)	2.35	2.23	
Pt	564	0.402	(111)	2.54	2.30	2.49, 2.48
			(100)	2.98	2.73	
			(110)	3.24	2.82	
Rh	554	0.387	(111)	2.70	2.47	2.66, 2.70
			(100)	3.15	2.80	
			(110)	3.41	2.90	
Ir	670	0.391	(111)	3.19	2.97	3.05, 3.00
			(100)	3.74	3.72	
			(110)	4.06	3.61	
Pb	196	0.511	(111)	0.55	0.32	0.59, 0.60
			(100)	0.64	0.38	
			(110)	0.70	0.45	
Al	327	0.405	(111)	1.45	1.20	1.14, 1.16
			(100)	1.68	1.35	
			(110)	1.84	1.27	
Ca	178	0.562	(111)	0.43	0.57	0.50, 0.49
			(100)	0.50	0.54	
			(110)	0.55	0.58	
Sr	166	0.617	(111)	0.33	0.43	0.42, 0.41
			(100)	0.39	0.41	
			(110)	0.43	0.43	
Mn[*]	282	0.353	(111)	1.65	3.10	1.54, 1.60
Ac	410	0.579	(111)	0.90	0.87	
			(100)	1.03	0.73	
			(110)	1.14	0.68	
Th	598	0.519	(111)	1.61	1.48	1.50
			(100)	1.85	1.47	
			(110)	2.36	1.45	

A_S and CN come from table 8. The symbol *, which has the same meaning in tables 10 and 11, denotes that when the low temperature equilibrium crystal structure has a lower symmetry than a close packing phase at high temperature or under a high pressure, the latter is utilized [116].

Table 10. Comparison of surface energies of crystals in bcc, sc and diamond structures among the predicted values γ_{sv0} based on Eq. (4.5), the FCD calculations γ'_{sv0} [116] and the experimental results γ''_{sv0} [114-115]

	E_b (kJ/g-atom)	a (nm)	$(h\,k\,l)$	γ_{sv0} (J/m^2)	γ'_{sv0} (J/m^2)	γ''_{sv0} (J/m^2)
Li	158	0.399	(110)	0.50	0.56	0.52,
			(100)	0.58	0.52	0.53
			(111)	0.72	0.59	
Na	107	0.420	(110)	0.29	0.25	0.26,
			(100)	0.34	0.26	0.26
			(111)	0.41	0.29	
K	90.1	0.530	(110)	0.16	0.14	0.13,
			(100)	0.18	0.14	0.15
			(111)	0.23	0.15	
Rb	82.2	0.571	(110)	0.12	0.10	0.12,
			(100)	0.15	0.11	0.11
			(111)	0.18	0.12	
Cs	77.6	0.626	(110)	0.10	0.08	0.10,
			(100)	0.12	0.09	0.10
			(111)	0.14	0.09	
Ba	183	0.503	(110)	0.36	0.38	0.38,
			(100)	0.41	0.35	0.37
			(111)	0.51	0.40	
Ra	160	0.537	(110)	0.27	0.30	
			(100)	0.32	0.29	
			(111)	0.40	0.32	
Eu	179	0.458	(110)	0.43	0.49	0.45,
			(100)	0.50	0.46	0.45
			(111)	0.61	0.52	
V	512	0.302	(110)	2.74	3.26	2.62,
			(100)	3.26	3.03	2.56
			(111)	4.04	3.54	
Cr*	395	0.285	(110)	2.39	3.51	2.35,
			(100)	2.83	3.98	2.30
			(111)	3.50	4.12	
Fe	413	0.286	(110)	2.52	2.43	2.42,
			(100)	2.92	2.22	2.48
			(111)	3.62	2.73	
Nb	730	0.376	(110)	2.58	2.69	2.66,
			(100)	2.99	2.86	2.70
			(111)	3.72	3.05	
Mo	658	0.317	(110)	3.20	3.45	2.91,
			(100)	3.81	3.84	3.00
			(111)	4.62	3.74	
Ta	782	0.335	(110)	3.40	3.08	2.90,
			(100)	4.05	3.10	3.15
			(111)	5.01	3.46	
W	859	0.358	(110)	3.36	4.01	3.27,
			(100)	3.90	4.64	3.68
			(111)	4.84	4.45	

Table 10. Continued

	E_b (kJ/g-atom)	a (nm)	$(h\,k\,l)$	γ_{sv0} (J/m^2)	γ'_{sv0} (J/m^2)	γ''_{sv0} (J/m^2)
Sb (sc*)	265	0.336	(100) (110)	0.66 0.77	0.61 0.66	0.60, 0.54
Bi (sc*)	210	0.326	(100) (110)	0.55 0.64	0.54 0.54	0.49, 0.49
Po (sc*)	144	0.334	(100) (110)	0.38 0.44	0.44 0.37	
Si (A4)	446	0.771	(110)	1.06		1.14
Ge (A4)	372	0.810	(110)	0.80		0.88

Table 8 shows some necessary parameters in the Eqs. (4.3) and (4.5). Tables 9~11 give the predicted γ_{sv0} values for fcc, bcc, hcp, diamond and sc structure crystals in terms of Eqs. (4.3) and (4.5) where two sets of experimental results γ''_{sv0} [114-115] and the first principle calculations γ'_{sv0} [116] are also shown. Note that the experimental results are not orientation-specific but are averaged values of isotropic crystals. Thus, they should be close to those of the most close-packed surface.

For both noble and transition metals, our predictions agree nicely with the experimental results and FCD calculations as shown in tables 9~11 although our predictions for transition metals have slightly larger deviations than those for the noble metals due to the fact that their d-bands are not fully filled and they present peaks at the Fermi level, which can slightly change from one surface orientation to the other and consequently the energy needed to break a bond changes also a little.

As shown in these tables, γ_{sv0} values of transition metals increase along an isoelectronic row where a heavier element has a larger γ_{sv0} value. This is because the d-level of a heavier element is higher in energy and the corresponding d-wave function with a stronger bonding is more extended. This is also true for elements in the same row in the Periodic table where a heavier element has more d-electrons [118]. An exception is in VA series where γ_{sv0} value of Nb is smaller than that of V possibly due to the rehybridization of Nb where Nb, whose d shell is less than the half-full, rehybridizes in the opposite direction, i.e., depletes their d_z^2 orbitals based on a charge density difference analysis [146-148].

γ_{sv0} values for sp metals except for Be are smaller than those for d-metals due to their bond nature of s- and p-electrons, which are more mobile than the localized d-electrons and therefore less energy is needed to break these bonds.

For fcc metals except Ca, Sr and Al, the mean-square root error χ between the predicted and the experimental results for the most close-packed (111) is about 7.5%. For aluminum, the degree of covalent Al-Al bonding increases or the nature of the bonding changes with reduced CN [149], which leads to model prediction deviating from the experimental results since our formula neglects the variation of bonding type. However, the reason of deviations for Ca and Sr is not clear.

Table 11. Comparison of surface energies of hcp metals among the predicted values γ_{sv0} of Eq. (4.3), the FCD calculations γ'_{sv0} [116] and the experimental results γ''_{sv0} [114-115]

	E_b (kJ/g-atom)	a (nm)	$(h\,k\,i\,l)$	γ_{sv0} (J/m²)	γ'_{sv0} (J/m²)	γ''_{sv0} (J/m²)
Be	320	0.222	(0001)	2.40	1.83	1.63,
			$(10\bar{1}\,0)$	2.88	2.13	2.70
Mg	145	0.320	(0001)	0.53	0.79	0.79,
			$(10\bar{1}\,0)$	0.65	0.78	0.76
Zn	130	0.268	(0001)	0.66	0.99	0.99,
		$(c/a=1.86)$	$(10\bar{1}\,0)$	0.72		0.99
Cd	112	0.306	(0001)	0.44	0.59	0.76,
		$(c/a=1.89)$	$(10\bar{1}\,0)$	0.47		0.74
Tl	182	0.371	(0001)	0.49	0.30	0.60,
			$(10\bar{1}\,0)$	0.60	0.35	0.58
Sc	376	0.330	(0001)	1.25	1.83	1.28
			$(10\bar{1}\,0)$	1.53	1.53	
Ti	468	0.295	(0001)	1.96	2.63	1.99,
			$(10\bar{1}\,0)$	2.39	2.52	2.10
Co	424	0.253	(0001)	2.42	2.78	2.52,
			$(10\bar{1}\,0)$	2.95	3.04	2.55
Y	422	0.355	(0001)	1.22	1.51	1.13
			$(10\bar{1}\,0)$	1.49	1.24	
Zr	603	0.325	(0001)	2.08	2.26	1.91,
			$(10\bar{1}\,0)$	2.54	2.11	2.00
Tc	661	0.274	(0001)	3.22	3.69	3.15
			$(10\bar{1}\,0)$	3.93	3.90	
Ru	650	0.272	(0001)	3.20	3.93	3.04,
			$(10\bar{1}\,0)$	3.90	4.24	3.05
La*	431	0.387	(0001)	1.05	1.12	1.02
			$(10\bar{1}\,0)$	1.28	0.92	
Lu	428	0.351	(0001)	1.27	1.60	1.23
			$(10\bar{1}\,0)$	1.55	1.42	
Hf	621	0.320	(0001)	2.22	2.47	2.19,
			$(10\bar{1}\,0)$	2.71	2.31	2.15
Re	775	0.276	(0001)	3.72	4.21	3.63,
			$(10\bar{1}\,0)$	4.54	4.63	3.60
Os	788	0.275	(0001)	3.80	4.57	3.44,
			$(10\bar{1}\,0)$	4.64	5.02	3.45

c/a ratios of Zn and Cd are cited from Ref. [116]. In FCD calculations, hcp structure has two $(10\bar{1}\,0)$ surfaces depending on the first interlayer distance d, $(10\bar{1}\,0)_A$ refers to the surface with $d_A = 0.29a$ while $(10\bar{1}\,0)_B$ denotes the surface with $d_B = 2d_A$. Here the results for $(10\bar{1}\,0)_A$ surface are present.

For hcp metals, $\chi \approx 10\%$ except Mg, Zn, Cd and Tl. For Cd and Tl, both of our predictions and FCD calculations deviate evidently from the experimental results. In the case of Zn and Cd, c/a ratios (1.86 and 1.89) are larger than the ideal value of 1.633. Thus, the nearest CN values will differ from the ideal condition, which should contribute the deviation of our prediction.

For sc metals, $\chi = 2.6\%$ where Sb and Bi with the rhombohedral structure are assumed to have slightly a distorted sc structure [116].

For bcc metals, $\chi = 10\%$. The smallest value of χ in all considered structures appears for diamond structure crystals with $\chi = 1.4\%$, which implies that the pure coherent bond does not change after a CN deduction.

Note that the temperature dependence of surface energy in our model is ignored although the experimental results listed in tables 9~11 are calculated at 0 K while the most lattice constants cited are measured at room temperature. This temperature effect decreases the prediction accuracy of our model and can be partly responsible for the disagreement with other experimental and theoretical results.

In the FCD calculations, there are often exceptions that the most close-packed surface does not have the lowest γ_{sv0} values or there exists a weak orientation-dependence [116]. These physically unacceptable results are fully avoided in our calculation. In addition, the anisotropy in our formula is perfectly considered. $\gamma_{sv}^{(100)} / \gamma_{sv}^{(111)} \approx 1.16$ and $\gamma_{sv}^{(110)} / \gamma_{sv}^{(111)} \approx 1.27$

for fcc metals as well as $\gamma_{sv}^{(0001)} / \gamma_{sv}^{(10\bar{1}0)} \approx 1.22$ for hcp metals, which are in agreement with the latest theoretical values of 1.15 and 1.22 [118,137]. In addition, $\gamma_{sv}^{(100)} / \gamma_{sv}^{(110)} \approx 1.16$ for sc and bcc metals, which is especially comparable with 1.14 for monovalent sp metals based on the jellium model [150].

If the experimental results are taken as reference, 60% of our predicted γ_{sv0} values of the most close-packed surfaces of 52 elements shown in tables 9~11 have better agreements with the experimental ones than those of the FCD calculations do while 20% of the FCD calculations are in reverse. Note that LDA is used in our formula while GGA is used in FCD. Recently, it has been shown that both methods need to be corrected due to the neglect of surface electron self-interactions where GGA is worse than LDA [146-148]. This is surprising because GGA is generally considered to be the superior method for energetic calculations [116].

For the transition metals and noble metals, the formula works better than for others as the greatest contribution to bonding is from the s-d interaction and the orbitals of the latter localize, which is more like a pair interaction. According to tables 9~11, the predicted γ_{sv0} values of divalent sp metals have bad correspondence with the experimental results since the many body (e.g. trimer) terms are here critical to understanding the cohesive energy. Thus, the used pair potentials physically may be not fully correct. It is possible that the background of our formula, namely the broken-bond model, is not universally applicable although the lattice constants used in Eqs. (4.3) and (4.5) have measuring error of about 2% [114,130].

According to the first principles calculations, the effect of relaxation on the calculated γ_{sv0} value of a particular crystalline facet may vary from 2 to 5% depending on the roughness [147,151]. The semi-empirical results indicate further that the surface relaxation typically affects the anisotropy by less than 2% [123]. Surface relaxations for vicinal surfaces have been studied mainly using semi-empirical methods due to the complexity arisen by the simultaneous relaxation of a large number of layers [118]. In our formula, the relaxation effect is simply considered by adding Eq. (4.2) into Eq. (4.1). According to tables 9~11, this measure leads to satisfactory results.

It should be noted that as a simple model without free parameter, the above formula supplies a new insight and another way for a general estimation of surface energy of elements. This success is difficult to achieve for present first principles calculation. Moreover, our model also supplies a basis of comparison and supplement for further theoretical and experimental considerations on γ_{sv0} values of elements.

Recently, Lodziana *et al.* have proposed that surface energy for θ-alumina is negative [152]. Their use of the term "negative surface energy" can and has caused confusion in the scientific community [153]. Before long, a clarification is present [153]: *In a single-component system, the Gibbs dividing surface can be located arbitrarily so that there is no 'surface excess' quantity, which is to say that because the solid is surrounded by its own vapor, there are no chemical effects to consider. For all stable solids then, energy input is required for creating new surface area and hence all clean solids have positive surface energies.*

The situation is slightly more complicated in a multicomponent system (such as θ-alumina + water), where chemical effects must be considered [153]. In such a system, the Gibbs dividing surface can be located such that there is no excess term for one component, but this only leads to non-zone excess quantities for the other components, which alters the solid's surface energy. Physically, this is a result of the interaction energies between the solid surface and the other components. Thus, in addition to reversible work for creation of new (clean) surface area, surface energy here also includes chemical interactions between the newly formed surface and the surroundings. Adsorption on solid surfaces is typically an exothermic process and it reduces the solid's surface energy. Others have indeed shown that chemical effects can lead to negative surface energies [154].

The Size-Dependent Surface Energy $\gamma_{sv}(D)$

The thermodynamic behavior of nanocrystals differs from that of the corresponding bulk materials mainly due to the additional energetic term of $\gamma_{sv}(D)A$—the product of the surface (or interfacial) excess free energy and the surface (or interfacial) area. This term becomes significant to change the thermal stability of the nanocrystals due to the large surface/volume ratio of nanocrystals or $A/V \propto 1/D$ [155-158]. When the surfaces of polymorphs of the same material possess different interfacial free energies, a change in phase stability can occur with decreasing D [159]. Despite of the fundamental thermodynamic importance of $\gamma_{sv}(D)$, few reliable experimental or theoretical values are available [25,116]. The effects of size and surrounding of nanocrystals on $\gamma_{sv}(D)$ are hardly studied [77,159-160].

However, in mesoscopic size range, the size-dependence of the liquid-vapor interface energy $\gamma_{lv}(D)$ was thermodynamically considered fifty years ago by Tolman and Buff, respectively [113,161]. The final form of the analytical equation is as follows [113],

$$\gamma_{lv}(D)/\gamma_{lv0} = 1-4\delta/D+... \tag{4.9}$$

where γ_{lv0} is the corresponding bulk value of $\gamma_{lv}(D)$, δ denotes a vertical distance from the surface of tension to the dividing surface where the superficial density of fluid vanishes [113]. As a first order approximation, although there is no direct experimental evidence to support Eq. (4.9), Eq. (4.9) should also be applicable to predicting $\gamma_{sv}(D)$ since the structural difference between solid and liquid is very small in comparison with that between solid and gas or between liquid and gas. In addition, it is unknown whether D in Eq. (4.9) can be extended from micron size to nanometer size. Hence, a theoretical determination of $\gamma_{sv}(D)$ is meaningful.

Although both the expressions of Eqs. (4.1) to (4.5) and the corresponding results are different, all of them indicate that,

$$\gamma_{sv} = kE_b/(N_aA_S) \qquad (4.10)$$

where $k < 1$ is a function of CN.

If the nanocrystals have the same structure of the corresponding bulk, k is size-independent. Thus, Eq. (4.10) may be extended to nanometer size as [162],

$$\gamma_{sv}(D) = kE(D)/(N_aA_S). \qquad (4.11)$$

Combining Eq. (4.11) with Eq. (2.30), there is [162],

$$\gamma_{sv}(D)/\gamma_{sv} = \left[1 - \frac{1}{2D/h-1}\right]\exp\left(-\frac{2S_b}{3R}\frac{1}{2D/h-1}\right). \qquad (4.12)$$

In terms of Eq. (4.12), comparisons of $\gamma_{sv}(D)$ of Be, Mg, Na, Al thin films and Au particles with different facets between model predictions and experimental and other theoretical results [163-166] are shown in figures 5 and 6 where the related parameters in Eq. (4.12) are listed in table 12. It is evident that our predictions are in agreement with the experimental values of Be and Mg (0001), and with other theoretical results for Na (110) and for three low-index surfaces of Au. The deviations in all comparisons are smaller than 5% except that for Al (110) with a deviation of about 10%.

As shown in the figures 5 and 6, $\gamma_{sv}(D)$ decreases with a decrease in size. This trend is expected since $E(D)$ of the nanocrystals increases as the size decreases [162]. In other words, $\gamma_{sv}(D)$ as an energetic difference between surface atoms and interior atoms decreases as energetic state of interior atoms increases.

Considering the mathematical relation of exp $(-x) \approx 1-x$ when x is small enough, Eq. (4.12) can be rewritten as,

$$\gamma_{sv}(D)/\gamma_{sv} \approx 1-S_bh/(3RD) \qquad (4.13)$$

Eq. (4.13) is in agreement with the general consideration that the decrease of the any size-dependent thermodynamic quantity is proportional to $1/D$ [27]. If $\gamma_{sv}(D)$ function of Eq. (4.13) and $\gamma_{lv}(D)$ function of Eq. (4.9) have the same size dependence, $\delta = S_bh/(12R) \approx h$ when $S_b \approx 12R$ as seen in table 12. Namely, the transition zone separating a solid phase and a vapor phase is only one atomic layer, which is an understandable result. This determined δ

value is expected since when the atomic distance is larger than h, the bond energy decreases dramatically. Thus, Eq. (4.9) can be rewritten as [162],

$$\gamma_{lv}(D)/\gamma_{lv0} \approx \gamma_{sv}(D)/\gamma_{sv0} \approx 1\text{-}4h/D. \tag{4.14}$$

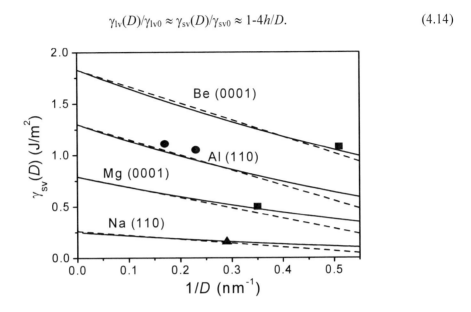

Figure 5. $\gamma_{sv}(D)$ as a function of $1/D$ in terms of Eq. (4.12) (solid lines) and Eq. (4.14) (segment lines) for nanocrystals Be, Mg, Na and Al with different facets. The symbols ■, ▲ and ● denote the experimental results for Be and Mg (0001) [163], the theoretical values for Na (110) [164] and Al (110) [165]. Note that the experiments on $\gamma_{sv}(D)$ and crystal equilibrium shapes with certain facets were performed at $T \approx T_m$. The equilibrium shape of a crystal follows Wulff construction [167].

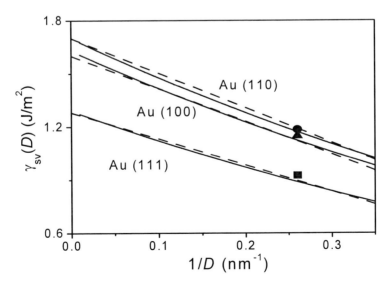

Figure 6. $\gamma_{sv}(D)$ as a function of $1/D$ for nanocrystals Au with different facets in terms of Eq. (4.12) (solid lines) and Eq. (4.14) (segment lines). The symbols ■, ▲ and ● denote the calculated results for (111), (100) and (110) facets in terms of a modified embedded-atom-method potential, respectively [166].

Table 12. Necessary parameters in Eq. (4.12). Note that γ_{sv0} values are cited from Ref. [116] while that of Al is cited from Ref. [118] since this value in Ref. [116] is doubtful where γ_{sv0} of Al (110) is smaller than that of Al (111), which is physically unacceptable

Element	Surface	D (nm)	h (nm)	E (kJ/mol)	T_b (K)	S_b (J/mol-K)	γ_{svo} (J/m²)
Be	(0001)	1.90	0.222	292.4	2745	106.5	1.83
Mg	(0001)	2.80	0.320	127.4	1363	93.5	0.79
Na	(110)	3.45	0.372	97.0	1156	83.9	0.26
Al	(110)	4.29 5.72	0.286	293.4	2793	105.0	1.30
	(111)						1.28
Au	(100)	3.80	0.288	334.4	3130	106.8	1.63
	(110)						1.70
Ref		163-166	42	42	42		

It is known that the surface energy ratio between different facets is a more important parameter in determining the crystalline shapes. Eq. (4.12) indicates that [162],

$$\frac{\gamma_{sv}^1(D)}{\gamma_{sv}^2(D)} = \frac{\gamma_{sv0}^1}{\gamma_{sv0}^2} \qquad (4.15)$$

where the superscripts 1 and 2 denote different facets. Eq. (4.15) implies that although the surface energy is size-dependent, the surface energy ratio between different facets is size-independent and is equal to the corresponding bulk ratio. Eq. (4.15) can also be compared with the theoretical results for Au where for example, $\gamma_{sv}^{(100)}/\gamma_{sv}^{(111)} \approx 1.24$ and $\gamma_{sv}^{(110)}/\gamma_{sv}^{(111)} \approx 1.28$ when $D = 3.8$ nm [166], which correspond well to the corresponding bulk ratios of 1.27 and 1.32 [116].

Noted that the structures of Be and Mg, Na, Al and Au belong to hcp, bcc and fcc structures, respectively. Owing to the above agreements shown in figures 5 and 6, the model should be applicable for all crystalline structures with different facets. Thus, Eq. (4.12) not only supplies a simple way to determine $\gamma_{sv}(D)$ values of different facets without any free parameter but also has an evident thermodynamic characteristic.

Liquid-Vapor Interface Energy or Surface Tension

The Bulk Surface Tension γ_{lv0} and its Temperature Coefficient γ'_{lv0}

The γ_{lv0} function and its temperature dependence are fundamental and important in the theory and practice of materials processing (e.g. crystal growth, welding and sintering), and its temperature coefficient $\gamma'_{lv0}(T) = d\gamma_{lv0}(T)/dT$ governs the well-known Marangoni convection on the surface of melt. There are several characteristics for the liquid surface. The first, the configuration of a liquid surface produced either by cleavage or by stretching is the same because the mobility of liquid molecules is high and the surface takes the equilibrium configuration of minimum energy. Thus, the work required for cleavage or surface stretching is the same when the adsorption is not taken into account, namely, $\gamma_{lv} = f_{lv}$. [5]. Second, because the liquid fails the elastic deformation resistance, $\gamma_{lv0}(T)$ called as surface tension equals surface stress when surface adsorption is not taken into account, which is defined as the reversible work per unit area involved in forming a new surface of a substance plastically [5]. Although early measurement methods of $\gamma_{lv0}(T)$ are sufficiently precise, there is still uncertainty regarding its absolute values and particularly regarding $\gamma'_{lv0}(T)$ function mainly due to the effect of impurities, which strongly changes the measured results. Therefore, considerable efforts have recently been directed towards the experimental determinations of $\gamma_{lv0}(T)$ and $\gamma'_{lv0}(T)$ of metals, and progress has been achieved with the advent of levitation processing and oscillating drop techniques [168-170]. However, such an experiment often suffers from the ambiguities in the interpretation of the resulting frequency spectra [168], it is also unlikely that experimental measurements will ever encompass all possible temperature ranges of interest and for all metals.

In contrast to the determination of $\gamma_{lv0}(T_m)$ value, $\gamma'_{lv0}(T_m)$ value is not well known experimentally even for elemental metallic liquids [171]. A recent analysis of existing data shows that this quantity is known with accuracy better than 50% for only 19 metals; For 28 metals, the accuracy is worse; For 18 metals (mainly refractory metals) there are no experimental results [171].

Computer stimulations with Monte Carlo or molecular dynamics methods are considered to be one of the reliable methods [172], in which γ_{lv0} can be calculated either using the mechanical expression for the surface stress, or from the viewpoint of the surface energy. Unfortunately, the former approach suffers from rather high fluctuation and statistical uncertainty, while the latter introduces additional complexity into performance. Thus, the demand of developing reliable prediction methods has never declined.

Semi-empirical predictions based on the correlation between the surface and bulk thermodynamic properties are always active [123,173-176]. Stephan firstly links γ_{lv0} to the heat of evaporation H_v' at $T = 0$ K [173],

$$\gamma_{lv0}(T_m) = c_5'H_v'/V^{2/3} \qquad (5.1)$$

with c_5' being an unknown constant. Since there is no suitable theoretical determination of c_5', Eq. (5.1) seems to be apparent only for transition metals [171,173]. Although Eq. (5.1) exists more than hundred years, the attempt to determine c_5' value theoretically is scarce.

On the other hand, $\gamma_{lv0}(T)$ of pure substances may be evaluated from values of the critical temperature T_c by the Eötvos or Guggenheim empirical equations [177],

$$\gamma_{lv0}(T)V^{2/3} = A+BT; \ T_c = -A/B \qquad (5.2\text{-}a)$$

or

$$\gamma_{lv0}(T)/\gamma_{lv0}(T_m) = (1-T/T_c)^\alpha \qquad (5.2\text{-}b)$$

where the exponent α is system dependent, e.g. 4/5 for strongly hydrogen bonded substances or 11/9 for H_2, N_2 and CO, etc [177]. However, α value for liquid metals is not known to our knowledge. Moreover, unlike organic fluid, T_c values of liquid metals are only available for alkali metals and mercury [42], which consumedly restricts the use of Eq. (5.2-a).

When $\gamma_{lv0}(T_m)$ and $\gamma'_{lv0}(T_m)$ values are known, under the assumption that $\gamma'_{lv0}(T)$ is nearly a constant being equal to $\gamma'_{lv0}(T_m)$, $\gamma_{lv0}(T)$ function is also calculated by [178],

$$\gamma_{lv0}(T) = \gamma_{lv0}(T_m) + \gamma'_{lv0}(T_m)(T-T_m). \tag{5.3}$$

However, Eq. (5.3) has not been strictly examined. Thus, both $\gamma_{lv0}(T)$ and $\gamma'_{lv0}(T)$ functions need to be further considered.

As stated in section 4.1, Eq. (4.4) can be used to calculate γ_{sv0} of elements [139]. Due to the structural similarity of liquid and solid at least near T_m, Eq. (4.4) for γ_{sv0} should give suggestions for γ_{lv0} modeling and analytical determination of c_s' value in Eq. (5.1).

About 60 years ago, noting that fusion has only a small effect on volume, cohesive forces, and specific heat of substance, Frenkel reached the conclusion that "the character of the heat motion in liquid bodies, at least near the crystallization point, remains fundamentally the same as in solid bodies, reducing mainly to small vibrations about certain equilibrium position" [179]. The very slight change in volume on melting is also thought to imply that the atoms in a liquid are tightly bound to one another like those in a crystalline solid [180]. Thus, the structural and energetic differences between solid and liquid are very small in comparison with those between solid and gas or between liquid and gas. Consequently, Eq. (4.4) can be extended to determine γ_{lv0} with several modifications: (i) Since $T \geq T_m$ for all concerned amounts, which is much higher than 0 K, E_b at $T = 0$ K should be replaced by $H_v(T)$ and A_S should be substituted by $A_L(T)$ where the subscript L denotes liquid; (ii) The influence of the molar excess surface entropy $S(T)$ should contribute γ_{lv0} due to the high temperature condition; (iii) The first coordination number of liquid is usually determined by integrating the radial distribution function (RDF) up to the first minimum while the distance of the second minimum of RDF is about twice of that, the effect of the next nearest-neighbors thus may be neglected in terms of the LJ potential, namely $\varphi \approx 0$. Thus, Eq. (4.4) can be rewritten for determining $\gamma_{lv0}(T)$ [181],

$$\gamma_{lv0}(T) = [m'H_v(T)-TS(T)]/[N_aA_L(T)] \tag{5.4}$$

with $m' = (2-k_1-k_1^{1/2})/2$.

Since metallic liquid is closely packed, the packing density of a random close packing $\eta = 0.637$ can be employed for the liquid [182], which leads to that the specific volume difference between solid with bcc structure and the corresponding liquid is only 0.2% [94,183-185]. It has been proposed that when an alloy liquid has no volume change upon crystallization, it can be supercooled to lower temperature [183]. If this rule is also applicable to elementary substance, the local order in the metallic liquid is very similar to the bcc-type short-range order (SRO) [184]. This is the case of liquid Zr [185]. Correspondingly, this consideration can also be applied on the surface structure of liquid metals with a similar (110) surface of bcc structure to assure the minimum of the surface energy [174]. As results, $k_1 = 3/4$ and $A_L = 8^{1/2}h^2/3$ with h being the atomic diameter. For any pure, isotropic, condensed material, $h = (6\eta V/\pi)^{1/3}$. V can be calculated from the atomic weight M and $\rho(T)$ by $V = M/[N_a\rho(T)]$. Thus, $A_L(T)$ can be determined as,

$$A_L(T) = \lambda\{M/[N_a\rho_L(T)]\}^{2/3} \tag{5.5}$$

with $\lambda = (8^{1/2}/3)(6\eta/\pi)^{2/3}$. $\rho_L(T)$ is given as $\rho_L(T) = \rho_L(T_m)+(d\rho_L/dT)(T-T_m)$ with $d\rho_L/dT$ being the temperature coefficient of liquid density since $d\rho_L/dT \approx d\rho_L(T_m)/dT$ for liquid metals at the temperature range of $T_m \sim 2T_m$. This range could up to $3T_m$ for Rb and Cs and $4T_m$ for Li and K [56].

It is known that $H_v(T)$ for most substance is zero at T_c and reaches the maximum at the triple point T_t where T_t is very close to T_m for metals [186]. Recently, an empirical equation $H_v(T)/H_v(T_m) = (1-t)^{it+j}$ has been proposed for liquids having a triple point where $t = (T-T_m)/(T_c-T_m)$, $i = 0.44$ and $j = -0.137$ [187]. In terms of the known T_m, T_b and T_c values for alkali metals (mercury is not involved in this work) [42], it is found that the temperature dependence of $H_v(T)$ at $T_m \leq T \leq T_b$ is very small (<2%) and thus negligible. Moreover, it is reported that $H_v(T_m)$ values of Cd, Cr, Mn and Zn are 101, 344, 234 and 115 kJ/mol while the corresponding $H_v(T_b)$ values are 100, 339, 226 and 119 kJ/mol [42,186]. Their differences are smaller than 3.5%. Thus, $H_v(T_m \leq T \leq T_b) \approx H_v(T_m) \approx H_v$ does not lead to large deviation and can be accepted as a first order approximation.

In Skapski's model [174], the main contribution to S results from the change of oscillation frequency v of atoms in the surface. Note that according to Lindemann criterion [188], the mean value of v reaches a certain value at T_m, which leads to a constant S value at T_m. It is known that H_v determines the size of bond strength of liquid atoms, which further determines the size of v. Since $H_v(T)$ varies little at $T_m \leq T \leq T_b$, the temperature dependence of $S(T)$ is thus negligible and $S(T) \approx S(T_m) \approx S$ can also be assumed.

With these considerations and inserting Eq. (5.5) into Eq. (5.4), $\gamma_{lv0}(T)$ at $T_m \leq T \leq T_b$ can be determined as [181],

$$\gamma_{lv0}(T) = \left[\frac{m'H_v - TS}{\lambda N_a^{1/3}}\right]\left(\frac{\rho_L(T)}{M}\right)^{2/3} \tag{5.6-a}$$

or

$$\frac{\gamma_{lv0}(T)}{\gamma_{lv0}(T_m)} = \left(1+p-p\frac{T}{T_m}\right)\left(1-q+q\frac{T}{T_m}\right)^{2/3} \tag{5.6-b}$$

where $p = 1/[m'H_v/(T_mS)-1]$ and $q = (d\rho_L/dT)[T_m/\rho_L(T_m)]$ are constants for certain metals. Deviation of Eq. (5.6-a) with respect to T leads to [181],

$$-\gamma'_{lv0}(T) = \frac{\gamma_{lv0}(T)}{T}\left[\frac{1}{m'H_v/(TS)-1} - \frac{2}{3}\frac{T}{\rho_L(T)}\frac{d\rho_L}{dT}\right]. \tag{5.7}$$

Determination of $\gamma_{lv0}(T_m)$ Values

Table 13 gives the comparison between the predicted $\gamma_{lv0}(T_m)$ values for 48 liquid metals in terms of Eq. (5.6-a) and the corresponding available mean values of experimental results $\gamma^e_{lv0}(T_m)$ [171,178,189]. These experimental data are mainly obtained by the maximum bubble pressure technique for low-melting-point oxidizable metals like Na, the sessile drop technique for moderate-melting-point metals like Cu, and the drop weight technique employed at the extremity of a pendant wire with electron bombardment heating for refractory metals like W [178].

It is found that $\phi = |\gamma_{lv0}(T_m)-\gamma^e_{lv0}(T_m)|/\gamma^e_{lv0}(T_m)$ for 40 elements from Cu to Ba (see table 13) is smaller than 10%. Note that although $\gamma^e_{lv0}(T_m) = 867$ mJ/m^2 was proposed for Al [171], several measurements suggested that the most data for $\gamma^e_{lv0}(T_m)$ of Al pertain to oxygen-saturated material and that for pure Al could be about 1070 mJ/m^2 [187,192]. If this result is taken, ϕ for Al will only be 3.6%. For the divalent metals Mg, Zn and Cd, the predictions are evidently smaller than $\gamma^e_{lv0}(T_m)$. According to Miedema and Boom [189], these three metals have exceptionally stable free atomic configuration, which is close to that of rare gas. Thus, smaller $\gamma_{lv0}(T_m)$ values in terms of Eq. (5.6-a) may be reasonable. Although ϕ values of Ta, Nb, Li, Be, and La range from 13% to 22%, the causes are unknown.

The above data imply that Eq. (5.6-a) is suitable not only for transition metals, but for all metals although the deviations from transition metals are slightly larger than those for other metals.

$\gamma_{lv0}(T_m)$ values of transition metals increase along an isoelectronic row where a heavier element has a larger $\gamma_{lv0}(T_m)$ value. This is because the d level of a heavier element is higher in energy and the corresponding d wave functions with stronger bonding are more extended. Two exceptions are Pd and Zr. For Pd, the full-filled d orbital drops the system energy in terms of Hund's rule, which makes that its H_v value only approaches that of Ni. Since V_m and T_m values of Pd are obviously larger than those of Ni, $\gamma_{lv0}(T_m)$ of Pd is thus smaller than that of Ni in terms of Eq. (5.6-a); For Zr, its H_v and V_m values approach those of Hf while its T_m value is obviously smaller than that of Hf, $\gamma_{lv0}(T_m)$ of Zr is thus larger than that of Hf in terms of Eq. (5.6-a). The reason of larger H_v value of Zr is unclear.

Table 13. Comparisons of γ_{lv0} for liquid metals between $\gamma_{lv0}(T_m)$ of Eq. (5.6-a) and the experimental results $\gamma^e_{lv0}(T_m)$ [171,178,189], as well as comparisons of γ'_{lv0} between $\gamma'_{lv0}(T_m)$ of Eq. (5.7) and the corresponding experimental or estimated results $\gamma'^e_{lv0}(T_m)$ [171,178,189-190]

	$\gamma_{lv0}(T_m)$	$\gamma^e_{lv0}(T_m)$	$-\gamma'_{lv0}(T_m)$	$-\gamma'^e_{lv0}(T_m)$	H_v	T_m	ρ_L	$d\rho_L/dT$
	(mJ/m^2)		(mJ/m^2-K)		(kJ/g-atom)	(K)	(kg/m^3)	(kg/m^3-K)
Cu*	1352	1355, 1310	0.21	0.19, 0.23	300	1358	8000	-0.801
Ag*	925	910, 925	0.18	0.17, 0.21	255	1234	9346	-0.907
Au*	1211	1138, 1145	0.18	0.19, 0.20	330	1338	17360	-1.500
Ni*	1810	1838, 1796	0.33	0.42, 0.35	378	1728	7905	-1.160
Pd	1467	1475, 1482	0.25	0.28, 0.28	380	1828	10490	-1.266
Pt*	1896	1746, 1860	0.31	0.29, 0.31	490	2045	19000	-2.900
Co*	1779	1830, 1881	0.30	0.37, 0.34	375	1768	7760	-0.988

Table 13. Continued

	$\gamma_{lv0}(T_m)$	$\gamma^c_{lv0}(T_m)$	$-\gamma'_{lv0}(T_m)$	$-\gamma'^e_{lv0}(T_m)$	H_v	T_m	ρ_L	$d\rho_L/dT$
	(mJ/m²)		(mJ/m²-K)		(kJ/g-atom)	(K)	(kg/m³)	(kg/m³-K)
Rh	2010	2000, 1970	0.26	0.30, 0.66	495	2237	10800	-0.896
Ir*	2241	2140, 2250	0.20	0.23, 0.25	560	2716	20000	-0.935
Fe*	1650	1830, 1855	0.26	0.23, 0.39	355	1811	7015	-0.883
Ru	2363	2180, 2250		0.31	580	2607	10900	
Os	2508	2500, 2500		0.23	630	3306	20100	
Mn	986	1152, 1100	0.21	0.20, 0.35	226	1519	5730	-0.700
Tc	2245	2350			550	2430	10300	
Re*	2755	2520, 2700	0.20	0.23	705	3459	18800	-0.800
Cr	1582	1628, 1642	0.19	0.20, 0.20	339	2180	6280	-0.300
Mo	2110	2250, 1915	0.21	0.20, 0.30	600	2896	9340	-0.743
W*	2676	2500, 2310	0.23	0.29, 0.21	800	3680	16200	-1.250
V*	1902	1855, 1900	0.23	0.19, 0.31	453	2175	5700	-0.531
Ti	1520	1525, 1500	0.27	0.26, 0.20	425	1941	4110	-0.702
Zr*	1669	1480, 1435	0.14	0.20, 0.17	580	2128	5800	-0.310
Hf*	1591	1630, 1490		0.21, 0.19	575	2506	11100	
Sc	895	939, 870		0.12, 0.12	318	1814	2846	
Y	899	872, 800		0.09, 0.09	380	1799	4243	
Ce	845	794, 740	0.09	0.07, 0.08	350	1068	6685	-0.227
Pr	782	743, 716	0.09	0.09, 0.08	330	1208	6611	-0.240
Nd	658	689, 687	0.10	0.09, 0.09	285	1297	6688	-0.528
Gd	690	664, 664		0.06, 0.06	305	1585	7140	
Th	1108	1006, 978		0.14	514	2028	10500	
U	1453	1550, 1552	0.15	0.14, 0.27	420	1405	17900	-1.031
Al*	1031	1070, 867	0.19	0.15, 0.16	283	933	2385	-0.280
Pb*	466	462, 457	0.12	0.11, 0.11	178	601	10678	-1.317
Tl*	439	461, 459	0.11	0.09, 0.11	165	577	11280	-1.430
Na*	215	200, 197	0.09	0.10, 0.09	98	371	927	-0.236
K*	110	112, 110	0.07	0.08, 0.07	79	337	827	-0.229
Rb	90	90, 85	0.06	0.07, 0.06	76	312	1437	-0.486
Cs*	73	69, 70	0.05	0.06, 0.05	69	302	1854	-0.638
Ca*	328	337, 366	0.09	0.11, 0.10	164	1115	1365	-0.221
Sr*	268	289, 286	0.08	0.08, 0.08	144	1050	2480	-0.262
Ba	231	226, 267	0.07	0.07, 0.07	150	1000	3321	-0.526
Mg	359	557, 583	0.14	0.15, 0.26	128	923	1590	-0.265
Zn	466	789, 815	0.18	0.25, 0.21	119	693	6575	-1.100
Cd	305	637, 642	0.13	0.20, 0.15	100	594	8020	-1.160
Ta*	2467	2180, 2010	0.22	0.25, 0.20	735	3290	15000	-1.147
Nb*	2335	2040, 1840	0.27	0.24, 0.18	690	2750	7830	-0.800
Li	465	404, 399	0.15	0.16, 0.15	137	454	525	-0.052
Be	1637	1350, 1320	0.24	0.29	297	1560	1690	-0.116
La	901	737, 728	0.09	0.11, 0.10	400	1193	5955	-0.237

S = 5.30 J/mol-K [176], $m' \approx 0.19$, H_v and T_m are cited from Ref. [42] while those of Hf come from Ref. [191] since it is unreasonable that H_v = 630 kJ/mol is larger than E = 621 kJ/mol [42,139]. ρ_L and $d\rho_L/dT$ are taken from Ref. [190]. Since the $d\rho_L/dT$ values of Rh, Ir, Re, Sr, Mo, Nb, Ta, W, V and Zr are unavailable in Ref. [190], Refs. [56] and [43] are employed. The asterisk * denotes that the accuracy on $\gamma'^e_{lv0}(T_m)$ is better than 50%.

$\gamma_{lv0}(T_m)$ values of sp metals except that of Be are smaller than those of d metals due to the bond nature of s and p electrons, which are more mobile than the localized d electrons. Moreover, in contrast to the transition metals, $\gamma_{lv0}(T_m)$ values of sp metals decrease along an isoelectronic row. This arises because the outmost $n's$ electrons (the number of period $n' = 2$-6) are progressively bound more loosely as they are screened from the nucleus by the increasing number of filled inner shells in the ionic core.

To find the similarity between Eq. (5.6-a) and Eq. (5.1), the prefactor c_5' in Eq. (5.1) can be determined by rearranging Eq. (5.6-a) at $T = T_m$ [181],

$$\gamma_{lv0}(T_m) = c_5 H_v / V_m^{2/3} \qquad (5.8)$$

with

$$c_5 = (m' - T_m S/H_v)/(\lambda N_a^{1/3}). \qquad (5.9)$$

Eq. (5.9) has determined c_5' value in Eq. (5.1) although Eq. (5.8) differs from Eq. (5.1) a little due to the difference of H_v and H_v', or $c_5 = c_5' H'_v / H_v$.

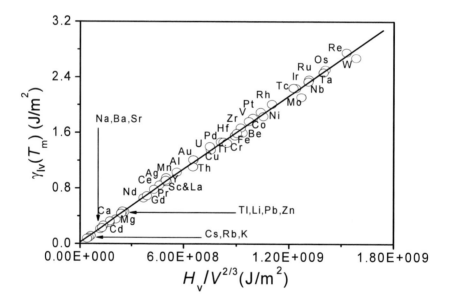

Figure 7. The plot of $\gamma_{lv0}(T_m)$ against $H_v/V^{2/3}$ for liquid metals in terms of Eq. (5.8) where the solid line is linearly regressed.

Figure 7 gives a plot of $\gamma_{lv0}(T_m)$ against $H_v/V^{2/3}$ for considered liquid metals in terms of Eq. (5.8) with a linearly regressed slope of $c_5 = 0.174 \times 10^{-8}$ mol$^{1/3}$ where the correlation coefficient of the fit is 0.998. All elements can thus be estimated by the same c_5 value, which implies that T_m/H_v is almost a constant (since S value has been taken as a constant according to Lindemann's criterion, $m' \approx 0.19$ and $\lambda \approx 1.08$) in terms of Eq. (5.9), which also confirms the correctness of Eq. (5.8). It is known that $H_v \propto H_m$ [176], and $T_m/H_m = 1/S_m$ where H_m and S_m are the melting enthalpy and melting entropy. Since S_m is almost a constant for metallic elements, c value as a constant is reasonable. In terms of the H'_v data listed in table 1 of Ref.

[189], it is found that H'_v/H_v is nearly a constant ($\approx$ 1.09) for transition metals. Thus, $c_5 = c_5'H'_v/H_v = 0.174 \times 10^{-8}$ mol$^{1/3}$ where $c_5' = 0.16 \times 10^{-8}$ mol$^{1/3}$ as determined [171,176]. In contrary to Eq. (5.1), which is considered to be only suitable for transition metals [171,176,189], both transition and non-transition metals are involved in figure 7. This improvement is only induced by the substitute of H'_v by $H_v(T_m)$. $H_v(T_m)$ describes the atom bonding of stable liquid and can be exactly measured while H'_v can be obtained only by extension of experimental results. In addition, since the difference between 0 K and T_m for transition metals are larger than that for non-transition metals, which leads to larger difference between $H_v(T_m)$ and H'_v for transition metals than for non-transition metals. This results in smaller suitability range of Eq. (5.1) than that of Eq. (5.8).

In terms of Eq. (5.6-a), the introduction of S drops the value of $\gamma_{lv0}(T)$. At T_m, the decreasing extents range from 8% (for La and Ce) to 20% (for Mg and Sr).

Determination of $\gamma'_{lv0}(T_m)$ Values

Table 13 also shows the comparison between $\gamma'_{lv0}(T_m)$ values of Eq. (5.7) and the available experimental or estimated results $\gamma'^e_{lv0}(T_m)$ [171,178,189-190]. A good agreement is also shown, which indicates that Eq. (5.7) provides a satisfactory description for $\gamma'_{lv0}(T_m)$.

Eq. (5.7) at $T = T_m$ can be written as,

$$-\gamma'_{lv0}(T_m) = (p-2q/3)\gamma_{lv0}(T_m)/T_m. \qquad (5.10)$$

In terms of the expressions for c and p, $p = m'/(c\lambda N_a^{1/3}) \approx 0.19$. Taking the mean value of -0.17 for $q = (d\rho_L/dT)[T_m/\rho_L(T_m)]$, there is the slope $\beta = p-2q/3 \approx 0.30$.

The relation between $-\gamma'_{lv0}(T_m)$ and $\gamma_{lv0}(T_m)/T_m$ for the Fourth, Fifth and Sixth periods are plotted in figure 8 in terms of Eq. (5.10) with the given slope $\beta = 0.30$ where $-\gamma'_{lv0}(T_m)$ functions increase almost linearly with increasing $\gamma_{lv0}(T_m)/T_m$ for the A family metals in the same period and the sequence is nearly the same as that in the Periodic table of the Elements although some deviations appears. It is understandable since their outmost electric configurations of $s+d$ electrons undergo nearly the same situation from the leftmost (IA metals) of one to the rightmost (VIIIA metals) of ten in these periods. The exceptions can be discussed as follows: (i) In the Fourth Period (from K to Ni), the anomalies of Mn and Cr are present where their $3d$ orbital is half-filled; (ii) Similarly, the appearance of the full $4d$ orbital also results in the anomaly of Pd of the Fifth Period (from Rb to Pd). On the contrary, the occurrence of half full $4d$ orbital in Mo does not change the sequence; half full $5d$ orbital of Re in the Sixth Period (from Cs to Pt) also does not change it. These may be explained as the following: In terms of Hund's rule, the half and the full fillings of orbital usually lead to the drop of the system energy and the effect of full filling is more effective. For example, the H_v values of Cr and Mn are evidently smaller than the neighbor elements V and Fe as shown in table 13. It is also applicable to Pd in comparison with Rh (Ag is not involved because it is B family metal). While the increase of electronic shell decreases the effect of electric configuration, the H_v value of Mo (Re) is thus in between those of Nb (W) and Tc (Os). Since $\gamma_{lv}(T_m)$ is proportional to the total energetic level of the system H_v in terms of Eq. (5.7), the abnormity only happens in Cr, Mn and Pd.

Note that Tc and the lanthanide elements are not involved in figure 8 because $\gamma'_{\mathrm{lv0}}(T_\mathrm{m})$ value of Tc is absent while those of the lanthanide elements are abnormally small possibly due to the effect of f electrons on $\gamma'_{\mathrm{lv0}}(T_\mathrm{m})$ value. Since the $s+d$ electrons of most lanthanide elements remain constant, their $\gamma'_{\mathrm{lv}}(T_\mathrm{m})$ values hardly change as shown in table 13, and the corresponding β values thus approach zero. In other words, f electrons hardly work as valence electrons.

When elements with empty or full filled electrons of second outmost sub-shell are considered, only the outermost electron layer (s electron layer) is valence electrons, two groups of elements exist, namely $s = 1$ (IA and IB metals) and $s = 2$ (IIA and IIB metals). There still exists a linear correlation between $\gamma'_{\mathrm{lv0}}(T_\mathrm{m})$ and $\gamma_{\mathrm{lv0}}(T_\mathrm{m})/T_\mathrm{m}$ as shown in figure 9. When figure 8 and figure 9 are compared, the β value of metals in groups IIA an IIB with $s = 2$ is similar to that of Eq. (5.10). However, the β value of metals in groups IA an IB with $s = 1$ is 30% smaller than that when the period number n' remains constant.

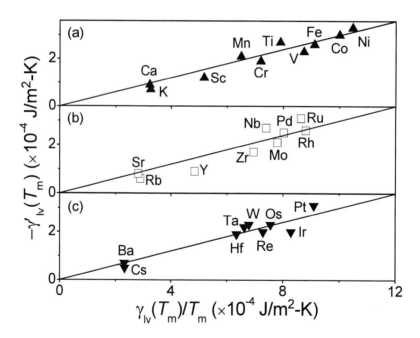

Figure 8. $-\gamma'_{\mathrm{lv0}}(T_\mathrm{m})$ as a function of $\gamma_{\mathrm{lv0}}(T_\mathrm{m})/T_\mathrm{m}$ for (a) the fourth (7), (b) the fifth ($\triangledown$) and (c) the sixth (B) periods A family metals where the solid lines are determined by Eq. (5.10).

As shown in figures 8 and 9, all elements with the sub-shell of $(n'-1)d$ are located on the right of the figure while all elements with the sub-shell of $(n'-1)p$ are found on the left of the figure. When the sub-shell is $(n'-1)s$, the elements are located in the middle or right of the figure. In addition, as n' increases, $-\gamma'_{\mathrm{lv0}}(T_\mathrm{m})$ values decreases linearly. Thus, the electron orbital movements and valence electron contributions of sub-shells are different for different orbitals and different n values as temperature increases. From thermodynamic point of view, the difference of β between figure 9a and figure 8 is induced by different p values in Eq. (5.10) since the corresponding $H_\mathrm{v}/T_\mathrm{m}$ value is systematically larger than that of the transition metals.

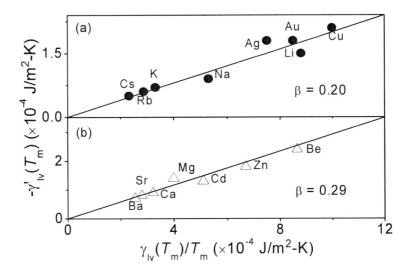

Figure 9. $-\gamma'_{lv0}(T_m)$ as a function of $\gamma_{lv0}(T_m)/T_m$ for (a) IA and IB (,) as well as (b) IIA and IIB (8) metals.

Noted that $\gamma_{lv}(T_m) = 359, 466, 305$ mJ/m^2 for Mg, Zn and Cd based on Eq. (5.6-a) are employed in plotting figures 7 and 9, the existences of these linear correlations imply that these calculated values may be reasonable.

In terms of Eq. (5.7), the contribution of S on $\gamma'_{lv0}(T_m)$ value ranges from 36% (for metals like Nb and Pt with larger γ_{lv0} values) to 78% (for metals like Cd, Li and Sr with smaller γ_{lv0} values). Neither S nor $d\rho_L/dT$ is thus negligible.

Estimation of $\gamma_{lv0}(T)$ and $\gamma'_{lv0}(T)$ Functions

Figure 10a shows the comparison of $\gamma_{lv0}(T)/\gamma_{lv0}(T_m)$ between the predictions of Eq. (5.6-b), Eq. (5.3) and the available experimental results for transition metals Ni, Co, Re and W with good agreement [168,193-195]. The experimental data for liquid Ni, Co, Re and W in the temperature ranges of 1573~1893, 1541~1943, 2800~3600 and 3360~3700 K respectively correspond to the undercooling of 155, 227, 659 and 320 K as well as the overheating of 165, 175, 141 and 20 K [168,193-195]. The plots in terms of Eq. (5.3) and Eq. (5.6-b) are nearly identical for Co, Re and W. Although it seems that the difference between Eq. (5.3) and Eq. (5.6-b) is large for Ni, its actual value is only about 1%. The agreements shown in figure 11a not only confirm the validities of Eqs. (5.3) and (5.6-b), but also indicate that the linear correlation between $\gamma_{lv0}(T)$ and T exists at T near T_m (including $T < T_m$ and $T \geq T_m$), namely, Eqs. (5.3) and (5.6-b) are also applicable for supercooled liquid metals.

Figure 10b shows the comparisons of $\gamma_{lv0}(T)/\gamma_{lv0}(T_m)$ between the predictions of Eq. (5.6-b) and the available experimental results for non-transition metals Na, K, Rb and Cs [196-197]. It is found that the differences between the predictions and the experimental data are smaller than 5% where the experimental data encompass the largest temperature range (from T_m to $3.5T_m$) to the best of our knowledge. Here, the approximately linear relation between $\gamma_{lv0}(T)$ and T is present again, which implies that Eq. (5.3) is a good approximation of Eq. (5.6-b).

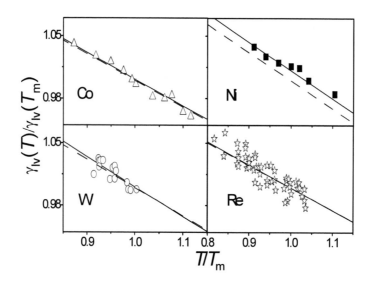

Figure 10a. Comparison of $\gamma_{lvo}(T)/\gamma_{lvo}(T_m)$ function between the predictions of Eq. (5.6-b) (the solid line), Eq. (5.3) (the two-point segment) and available experimental data for transition metals Ni (!), Co (8), Re (ψ) and W (–) [168,193-195].

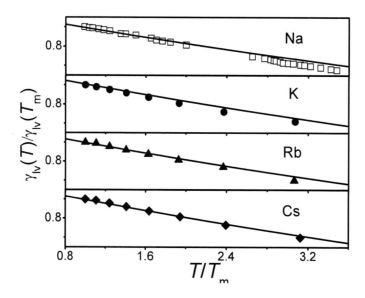

Figure 10b. Comparison of $\gamma_{lvo}(T)/\gamma_{lvo}(T_m)$ function between the predictions of Eq. (5.6-b) (the solid line) and available experimental results for non-transition alkali metals Na ($\forall$), K (,), Rb (7) and Cs (Λ) [196-197].

Substituting the expression $\rho_L(T) = \rho_L(T_m)+(d\rho_L/dT)(T-T_m)$ into Eq. (5.6-b), there is

$$\gamma_{lv0}(T) = \left[\frac{m'H_v - TS}{\lambda N_a^{1/3}}\right]\left[\frac{1+u(T-T_m)}{M/\rho_L(T_m)}\right]^{2/3}$$ with $u = (d\rho_L/dT)/\rho_L(T_m)$. Since $(d\rho_L/dT)/\rho_L(T_m)$ is

nearly a constant being equal to -10^{-4} K^{-1} [171,176], while $(T-T_m)$ values concerned in figure 10 are smaller than 800 K, $u(T-T_m)$ is smaller than 0.1. Considering the mathematical relation of $(1+x)^{2/3} \approx 1+2x/3$ when x is small enough (e.g. $x < 0.1$), Eq. (5.6-b) can be rewritten as $\gamma_{lv}(T) \approx (m'H_v-TS)[1+2u(T-T_m)/3]/(\lambda N_a^{1/3}V^{2/3})$. Because $(m'H_v-TS)[1+2u(T-T_m)/3] = v+(T-T_m)\{-S+(2uv/3)[1-S(T-T_m)/v]\}$ with $v = m'H_v-T_mS$, there is

$$\gamma_{lv}(T) \approx \{v+(T-T_m)\{-S+(2uv/3)[1-S(T-T_m)/v]\}\}/(\lambda N_a^{1/3}V^{2/3}). \qquad (5.11)$$

In terms of m', H_v, T_m and S values listed in table 13, $S(T-T_m)/v < 0.1$ for the most metals at $T-T_m < 800$ K and thus negligible as a first order approximation. Thus, Eq. (5.11) can be simplified as $\gamma_{lv}(T) \approx [v+(-S+2uv/3)(T-T_m)]/(\lambda N_a^{1/3}V^{2/3})$. Since $v/(\lambda N_a^{1/3}V^{2/3}) = \gamma_{lv0}(T_m)$ and $(-S+2uv/3)/(\lambda N_a^{1/3}V^{2/3}) = \gamma'_{lv0}(T_m)$, the agreement between Eq. (5.3) and Eq. (5.6-b) shown in figure 10a is not only understandable, but also inevitable.

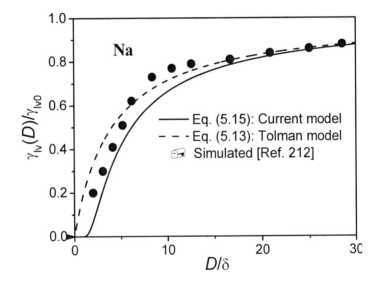

Figure 11. Comparison of the D/δ ($\delta = h$) dependence of $\gamma_{lv}(D)/\gamma_{lv0}$ described by various models and computational results for Na [212].

Since $\gamma_{lv0}(T)$ is related to the bond strength of atoms and its derivative corresponds to the bond strength change or electron orbitals change as temperature varies, thus, as long as the substance remains liquid and T has a temperature distance from T_c, electron orbitals of liquids change linearly with temperature. Deviation of Eq. (5.11) with respect to T leads to [181],

$$\gamma'_{lv0}(T) \approx \gamma'_{lv0}(T_m)+e(T-T_m) \qquad (5.12)$$

with $e = -4uS/(3\lambda N_a^{1/3}V_m^{2/3})$ being the temperature coefficient of $\gamma'_{lv0}(T)$.

Eq. (5.12) indicates that $\gamma'_{lv0}(T)$ is positively proportional to T since $e > 0$ induced by $u < 0$. This tendency against T is in reverse to that of $\gamma_{lv0}(T)$ because the energetic difference between liquid and gas drops as T increases. Although Eq. (5.12) ultimately gives positive temperature dependence, it should be indicated that e has only a secondary effect and can be neglected as a first order approximation. For instance, $e \approx 3.0 \times 10^{-5}$ and 1.6×10^{-5} mJ/m^2-K^2 for

Ni and V. Even $T-T_m = 1000$ K, $\gamma'_{lv0}(T)/\gamma'_{lv0}(T_m)$ values of Ni and V are only 1.09 and 1.08, or the error range is smaller than 10%.

Note that the simulated results based on the Monte Carlo method in conjunction with the embedded-atom method respectively show 20~60% underestimations for Al, Ni, Cu, Ag and Au, and 20% overestimation for Co when they are compared with the experimental data [198-200]. Thus, computer simulation methods for estimation of $\gamma_{lv0}(T)$ values of metals need to be further improved, which indicates that the above theoretical method is a powerful and even unique tool at present to determine $\gamma_{lv0}(T)$ function with good accuracy.

The Size Dependence of Surface Tension $\gamma_{lv}(D)$

A method of dividing surface pioneered by Gibbs defines that γ_{lv} with a given bulk value γ_{lv0} depends on pressure P, T and the composition of the two coexisting bulk phases [201]. However, when the liquid-vapor interface is curved, γ_{lv} is a function of D of the droplet, $\gamma_{lv}(D)$. Guggenheim suggested that the $\gamma_{lv}(D)$ would change when D falls below 100 nm based on statistical mechanical considerations [202].

A half-century ago, Tolman extended the idea of Gibbs and showed that if the radius R_s of the surface of tension of the droplet did not coincide with the equimolar radius R_e, the surface tension must vary with droplet size [161]. Moreover, Tolman proposed that the two surfaces must, in general, be distinct from each other. Tolman estimated the Tolman's length δ, or the separation between the equimolar surface and the surface of tension, $\delta = R_e-R_s$ [203]. He assumed that δ could be taken as a constant in the nanometer region, and derived the equation [161],

$$\gamma_{lv}(D)/\gamma_{lv0} = 1/(1+4\delta/D). \tag{5.13}$$

Kirkwood and Buff developed a general theory based on statistical mechanics for the interfacial phenomena and confirmed the validity of Tolman's approach [204]. For a sufficiently large droplet, Eq. (5.13) may be expanded into power series. Neglecting all the terms above the first order, the asymptotic form was obtained, which has been illustrated as Eq. (4.9) [113]. Values for $\gamma_{lv}(D)/\gamma_{lv0}$ determined by Eqs. (4.9) and (5.13) are close to each other at $D/\delta \geq 20$.

Tolman predicted that $\gamma_{lv}(D)$ should decrease with decreasing particle size [161], indicating the positive δ. The asymptotic Tolman's length in the limit of $D\rightarrow\infty$, δ_∞, is independent of the choice of the dividing surface [205], and $\delta_\infty = h$ [161]. However, δ was also predicted to be negative by a rigorous thermodynamic derivation [206], which would lead to an increase of $\gamma_{lv}(D)$ when the size is decreased. It is generally assumed that $\delta > 0$ for spherical droplets and $\delta < 0$ for bubbles in a liquid [207-209]. This consideration can also be simply translated as that the δ value is always positive while D may be positive for the droplets but negative for the bubbles. In addition to the uncertainty in the sign of δ, the validity of Eq. (5.13) is considered to be questionable for very small particles [210].

It is known that for a planar interface at the melting point, $\gamma_{sv0}/\gamma_{lv0} = w$ for metallic elements with $w = 1.18\pm0.03$ [114]. Note that in the derivation of $\gamma_{sv}(D)$, it is assumed that the nanocrystal has the same structure of the corresponding bulk [162]. Since the structure and energy differences between solid and liquid are small in comparison with that between solid and gas or between liquid and gas, the above expression for the bulk may be extended to nanometer size with the same form [181],

$$\gamma_{sv}(D)/\gamma_{lv}(D) = w. \tag{5.14}$$

Combining Eqs. (4.12) and (5.14) as well as the expression for the bulk, there is [211],

$$\frac{\gamma_{lv}(D)}{\gamma_{lv0}} = \left[1 - \frac{1}{2D/h - 1}\right]\exp\left(-\frac{2S_b}{3R}\frac{1}{2D/h - 1}\right). \tag{5.15}$$

In terms of Eq. (5.15), comparisons of $\gamma_{lv}(D)/\gamma_{lv0}$ for Na and Al droplets between the model predictions and the computer simulation results [212] are shown in figures 11 and 12 where the parameters involved in Eq. (5.15) are listed in table 14. It is evident that the model predictions are in agreement with the computer simulation results for Na and Al. This agreement in return confirms the validity of the assumption in Eq. (5.14). As a comparison, the predictions of Eq. (4.9) with $\delta = \delta_\infty = h$ are also shown in these two figures.

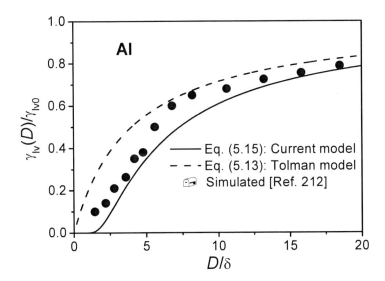

Figure 12. D/δ ($\delta = h$) dependence of $\gamma_{lv}(D)/\gamma_{lv0}$ for Al [212].

Although Eq. (5.15) is deduced in light of the relation between γ_{sv0} and the broken bond number of surface atoms for metals [162], this relation should be also applicable to other types of materials. Figure 13 shows $\gamma_{lv}(D)$ function of water droplets in terms of Eq. (5.15). The model prediction also corresponds well to the computer simulation results [54]. Note that the definition for h in this case is redefined as O-H bond length [213]. Similarly, the prediction of similarly with $\delta = h$ is also shown in figure 13.

Table 14. Several necessary parameters used in Eqs. (5.15) and (5.16)

	h (nm)	γ_{lv0} (mJ/m^2)	E_b (kJ/g-atom)	T_b (K)	S_b (J/g-atom K)
Na	0.372		97.7	1156	84.5
Al	0.286		293.4	2792	105
H$_2$O	0.096	75	13.6	373	36.5
Ref	42,213	214	42,54	42,54	

As shown in figures 11~13, $\gamma_{lv}(D)$ decreases with size, following the trend of $\gamma_{sv}(D)$ and $E(D)$ [97,162]. This is because $\gamma_{lv}(D)$ as an energetic difference between surface molecules and interior molecules of droplets decreases as energetic state of the interior molecules increases more quickly than that of the surface molecules. Note that, prediction of Eq. (5.15) provides the same or better accuracy of Eq. (5.13).

Although δ is assumed to be a constant as required by the derivation of Eqs. (4.9) and (5.13), several applications of statistical thermodynamics have indicated that δ depends strongly on D [209,215-216]. Since the results of these treatments are based on rather complex numerical calculations, it would be difficult to express $\delta(D)$ analytically. Fortunately, Eq. (5.15) can be used to satisfy this requirement.

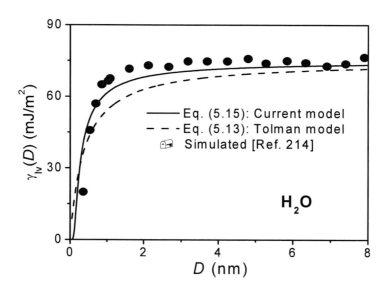

Figure 13. D dependence of $\gamma_{lv}(D)/\gamma_{lv0}$ with $\delta = h$ for water [214].

Substituting Eq. (5.15) into Eq. (5.13) rather than Eq. (4.9) because the latter is an approximation of Eq. (5.13) and will lead to error when $D/\delta \leq 10$ [211],

$$\delta(D) = \frac{D}{4}\left[\exp\left(\frac{2S_b}{3R}\frac{1}{2D/h-1}\right)\Big/\left(1-\frac{1}{2D/h-1}\right)-1\right]. \qquad (5.16)$$

$\delta(D)/h$ determined by Eq. (5.16) for Na, Al and water droplets are shown as a function of D/h in figure 14. It is observed that the Tolman's length is positive for these droplets and decreases when the size is increased, being consistent with statistical thermodynamics [209,215-216], computer simulations [217-218] and other approaches [219] for Lennard-Jones fluids. However, there is an obvious difference between our model predictions and others, namely, $\delta(D)$ remains positive among the whole size range in this model while it will decrease to a negative limiting value for the planar interface in the others.

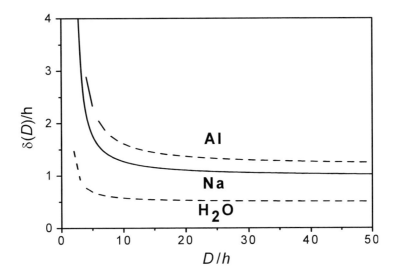

Figure 14. $\delta(D)/h$ as a function of D/h in terms of Eq. (5.16) for Na, Al and water droplets.

The value of δ in Eq. (5.16) is on the verge of infinitude when D reaches its lower limit h [162]. When D is sufficiently large, considering the mathematical relation of exp $(-x) \approx 1-x$ when x is small enough (e.g. $x < 0.1$), the minimal value δ_{min} in terms of Eq. (5.16) can be written as $\delta_{min} = \delta_\infty = hS_b/(12R)$, or

$$hS_b/(12R) < \delta \qquad (5.17)$$

$S_b \approx 12R$ for metallic elements as shown in table 14 leads to $\delta_\infty \approx h$ for Na and Al as indicated by Tolman [161] while $\delta'_\infty \approx 3h/8$ for water due to $S_b \approx 9R/2$. This is the reason that the differences between the model predictions in terms of Eqs. (5.13) and (5.15) appear at $D/h \leq 10$ for Na and Al while at $D/h \geq 20$ for water. Thus, the size dependence of $\delta(D)$ strongly depends on the value of S_b. Eq. (5.17) also implies that the decrease of the bond strength leads to the diffusion of the liquid-vapor interface. The corresponding physical picture is that the energetic difference of the molecule on the liquid surface and that in the vapor decreases as the bond strength weakens. Thus, the liquid-vapor interface transition zone becomes narrow.

Summary and Further Prospects

Several analytical models based on classic thermodynamics, free of adjustable parameters, are presented to quantitatively determine the bulk and size dependent interface energies as well as related interface stresses. The establishments of these models are of vital importance in quantitatively studying some basic problems of materials, such as solidification [92] and epitaxial growth [220]. With the high-speed developments of nanomaterials and nanotechnology, the models for the size dependences of interface energetic terms provide the clue to explore the fantastic properties of nanomaterials and, in particular, the continuous changes of thermodynamic properties in larger size range (consisting of $10^2 \sim 10^{23}$ atoms), which is beyond the reach of the computer simulations. Moreover, the size range of material referred in these works is the focus of the study in current or forthcoming nanotechnology, such as so-called "65 nm technique" where 65 nm denotes the line width of the integrate circuit.

The successes of the above classic thermodynamics in the mesoscopic and microscopic size ranges of materials not only further enrich the classical thermodynamic theory, but also offer powerful, irreplaceable and unfailing theoretical guidance for the development of materials science under the condition that the computer simulations play more and more important role.

However, since thermodynamics can only describe the statistical behavior of large numbers of molecules, it cannot depict the action of single molecule. When the size of low-dimensional materials decreases to a certain value, namely D_0 or D'_0, quantum effect becomes prominent and the classical thermodynamics is helpless. In addition, when clusters are small enough, the crystalline structure of bulk materials failed while new structures arise. Thus, the energy band theory and the electron theory should be included to discuss the limit cases, and this combination will provide new approaches for the classical thermodynamics to investigate the problems of nanomaterials.

The models considered in this work for surface energies and surface tension only are applicable for the elements. When they are extended to compounds, electronic structures are different in different surfaces. If the models are further used to predict the solid solutions and different substances, the chemical interaction parameters must be considered although it is known that these parameters are also size-dependent [221]. All of these are future works, which are important for establishments of binary nano-alloy phase diagrams.

Acknowledgement

The financial supports from the NNSFC under Grants Nos. 59671010, 59931030, 50071023, 50025101, from the Trans-Century Training Program Foundation for the Talents by the Ministry of Education of China, from National Key Basic Research and Development Program under Grant No. 2004CB619301, and from Project 985-Automotive Engineering of Jilin University are acknowledged.

References

[1] Myers, D. *Surfaces, interfaces, and colloids, principles and applications*; 2nd edition, John Wiley: New York, 1999.

[2] Plummer, E. W.; Ismail Matzdorf, R.; Melechko, A. V.; Pierce, J. P.; Zhang, J. D. *Surf. Sci.* 2002, 500, 1.

[3] Raabe, D. *Computational materials science: the simulation of materials, microstructures and properties*; Wiley-VCH: Weinheim, 1998.

[4] Lu, H. M.; Jiang, Q. *Phys. Rev. Lett.* 2004, 92, 179601.

[5] Cammarata, R. C.; Sieradzki, K. *Annu. Rev. Mater. Sci.* 1994, 24, 215.

[6] Tolman, R. C. *Relativity Thermodynamics and Cosmology*, Clarendon Press: Oxford, 1969.

[7] Samsonov, V. M.; Bazulev, A. N.; Sdobnyakov, N. Yu. *Central. Euro. J. Phys.* 2003, 3, 474 and reference therein.

[8] Christian, J. W. *The theory of transformation in metals and alloys, Part I Equilibrium and General kinetic Theory*; 2nd edition, Pergamon Press: London, 1975.

[9] Turnbull, D.; Fisher, J. C. *J. Chem. Phys.* 1949, 17, 71.

[10] Turnbull, D.; Cech, R. E. *J. Appl. Phys.* 1950, 21, 804.

[11] Turnbull, D. *J. Appl. Phys.* 1950, 21, 1022.

[12] Jones, D. R. H. *J. Mater. Sci.* 1974, 9, 1.

[13] Jackson, C. L.; McKenna, G. B. *J. Chem. Phys.* 1990, 93, 9002.

[14] Cahn, J. W.; Larche, F. *Acta Metall.* 1984, 32, 1915.

[15] Broughton, J. Q.; Gilmer, G. H. *J. Chem. Phys.* 1986, 84, 5759.

[16] Spaepen, F. *Mater. Sci. Eng. A* 1994, 178, 15.

[17] Jackson, C. L.; McKennam G. B. *Chem. Mater.* 1996, 8, 2128.

[18] Rusanov, A. I. *Fazovye Ravnovesiya i Poverkhnostnye Yavlenia*; Izd. Khimya: Leningrad, 1967.

[19] Forkin. V. M.; Zanotto, E. D. *J. Non-Cryst. Solids* 2000, 265, 105.

[20] Jiang, Q.; Tong, H. Y.; Hsu, D. T.; Okuyama, K.; Shi, F. G. *Thin Solid Films* 1998, 312, 357.

[21] Jiang, Q.; Aya, N.; Shi, F. G. *Appl. Phys. A* 1997, 64, 627.

[22] Mitch, M. G.; Chase, S. J.; Fortner, J.; Yu, R. Q.; Lannin, J. S. *Phys. Rev. Lett.* 1991, 67, 875.

[23] Iijima, S. *Nature* 1991, 354, 56.

[24] Ajayan, P. M.; Iijima, S. *Nature* 1993, 361, 333.

[25] Wen, Z.; Zhao, M.; Jiang, Q. *J. Phys. Chem. B* 2002, 106, 4266.

[26] Regel', A. R.; Glazov, V. M. *Semiconductors* 1995, 29, 405.

[27] Jiang, Q.; Shi, H. X.; Zhao, M. *J. Chem. Phys.* 1999, 111, 2176.

[28] Mott, N. F. *Proc. R. Soc. A* 1934, 146, 465.

[29] Tiwari, G. P.; Juneja, J. M.; Iijima, Y. *J. Mater. Sci.* 2004, 39, 1535.

[30] Shi, F. G. *J. Mater. Res.* 1994, 9, 1307.

[31] Jiang, Q.; Shi, F. G. *J. Mater. Sci. Technol.* 1998, 14, 171.

[32] Jiang, Q.; Shi, H. X.; Zhao, M. *Acta Mater.* 1999, 47, 2109.

[33] Zuinson, J. F.; Brun, M.; Eyraud, C. *J. Chem. Phys.* 1979, 76, 857.

[34] Sun, D. Y.; Asta, M.; Hoyt, J. *J. Phys. Rev. B* 2004, 69, 174103.

[35] Hoyt, J. J.; Asta, M.; Karma, A. *Phys. Rev. Lett.* 2001, 86, 5530.

[36] Hoyt, J. J.; Asta, M.; Karma, A. *Mater. Sci. Eng. R.* 2003, 41, 121.

[37] Cortella, L.; Vinet, B. *Phil. Mag. B* 1995, 71, 11.

[38] Morris, J. R. *Phys. Rev. B* 2002, 66, 144104.

[39] Gündüz, M.; Hunt, J. D. *Acta Mater.* 1985, 33, 1651.

[40] Taylor, J. W. *Phil. Mag.* 1955, 46, 857.

[41] Spaepen, F.; Turnbull, D. Rapidly Quenched metals, *2nd International Conference*, edited by Grant, N. J.; Giessen, B. C.; MIT Press: Cambridge, 1976, pp 205-229.

[42] *http://www.webelements.com/*

[43] Vinet, B.; Magnusson, L.; Fredriksson, H.; Desré, P. J. *J. Colloid Interf. Sci.* 2002, 255, 363.

[44] Ubbelohde, A. R. *Melting and Crystal Structure*, Clarendon Press: Oxford, 1965, p 137, 146 and pp 171-173.

[45] Ubbelohde, A. R. *The Molten State of Matter: Melting and Crystal Structure*; John Wiley: Chichester, 1978, pp 239-241.

[46] Jones, H. *Mater. Lett.* 2002, 53, 364.

[47] Rabinstein, E. B.; Glicksman, M. E. *J. Cryst. Growth* 1991, 112, 84.

[48] Bayender, B.; Maraşh, N.; Çadırlı, E.; Şişman, H.; Gündüz, M. *J. Cryst. Growth* 1998, 194, 119.

[49] Maraşh, N.; Keşlioğlu, K.; Arslan, B. *J. Cryst. Growth* 2003, 247, 613.

[50] Sill, R. C.; Skapski, A. S. *J. Chem. Phys.* 1956, 24, 644.

[51] Skapski, A. S.; Billups, R.; Rooney, A. *J. Chem. Phys.* 1957, 26, 1350.

[52] Skapski, A. S.; Billups, R.; Casavant, D. *J. Chem. Phys.* 1959, 31, 1431.

[53] Dean, J. A. *Lange's Handbook of Chemistry*, 13th edition, McGraw-Hill Book: New York, 1985.

[54] Dean, J. A. *Handbook of Organic Chemistry*; McGraw-Hill Book: New York, 1987.

[55] Hunten, K. W.; Mass, O. *J. Am. Chem. Soc.* 1929, 51, 153.

[56] Weast, R. C. *Handbook of Chemistry and Physics*; 69th edition, CRC Press Inc: Cleveland 1988-1989.

[57] Fan, X. B.; Ishigaki, T. *J. Cryst. Growth* 1997, 171, 166.

[58] Boulova, M.; Lucazeau, G. *J. Solid State Chem.* 2002, 167, 425.

[59] Bouvier, P.; Godlewski, J.; Lucazeau, G. *J. Nuclear Mater.* 2002, 300, 118.

[60] Woodward, P. M.; Sleight, A. W.; Vogt, T. *J. Solid State Chem.* 1997, 131, 9.

[61] Bouvier, P.; Djurado, E.; Lucazeau, G.; Le Bihan, T. *Phys. Rev. B* 2000, 62, 8731.

[62] Ansell, S.; Krishnan, S.; Weber, J. K. R.; Felten, J. J.; Nordine, P. C. *Phys. Rev. Lett.* 1997, 78, 464.

[63] Ostward, W. *Z. Phys. Chem.* (Leipzig) 1897, 22, 289.

[64] Notthoff, C.; Feuerbacher, B.; Franz, H.; Herlach, D. M.; Holland-Moritz, D. *Phys. Rev. Lett.* 2001, 86, 1038.

[65] ten Wolde, P. R.; Ruiz-ontero, M. J.; Frenkel, D. *Phys. Rev. Lett.* 1995, 75, 2714.

[66] Auer, S.; Frenkel, D. *J. Phys.: Condens. Matter.* 2002, 14, 7667.

[67] Shen, Y. C.; Oxtoby, D. W. *Phys. Rev. Lett.* 1996, 77, 3585.

[68] Davidchack, R. L.; Laird, B. B. *Phys. Rev. Lett.* 2005, 94, 086102.

[69] Spaepen, F.; Meyer, R. B. *Scr. Metall.* 1976, 10, 257.

[70] Sun, D. Y.; Asta, M.; Hoyt, J. J.; Mendelev, M. I.; Srolovitz, D. *J. Phys. Rev. B* 2004, 69, 020102(R).

[71] Sun, C. Q.; Li, S.; Tay, B. K.; Chen, T. P. *Acta Mater*. 2002, 50, 4687.

[72] Hoffman, J. D. *J. Chem. Phys*. 1958, 29, 1192.

[73] Thompson, C. V.; Spaepen, F. *Acta Met*. 1979, 27, 1855.

[74] Singh, H. B.; Holz, A. *Solid State Comm*. 1983, 45, 985.

[75] Weissmüller, J.; Cahn, J. W. *Acta Mater*. 1997, 45, 1899.

[76] Spaepen, F. *Acta Mater*. 2000, 48, 31.

[77] Jiang, Q.; Zhao, D. S.; Zhao, M. *Acta Mater*. 2001, 49, 3143.

[78] Streitz, F. H.; Cammarata, R. C.; Sieradzki, K. *Phys. Rev. B* 1994, 49, 10699.

[79] Cammarata, R. C.; Sieradzki, K.; Spaepen, F. *J. Appl. Phys*. 2000, 87, 1227.

[80] Müller, H.; Opitz, Ch.; Strickert, K.; Skala, L. *Z. Phys. Chemie. Leipzig*. 1987, 268, 625.

[81] Nix, W. D.; Gao, H. *Acta Mater*. 1998, 39, 1653.

[82] Needs, R. J.; Godfrey, M. *J. Phys. Rev. B* 1990, 42, 10933.

[83] Needs, R. J.; Godfrey, M. J.; Mansfield, M. *Surf. Sci*. 1991, 242, 215.

[84] Mansfield, M.; Needs, R. *J. Phys. Rev. B* 1991, 43, 8829.

[85] Streitz, F. H.; Cammarata, R. C.; Sieradzki, K. *Phys. Rev. B* 1994, 49, 10707.

[86] Davidchack, R. L.; Laird, B. B. *J. Chem. Phys*. 1998, 108, 9452.

[87] Davidchack, R. L.; Laird, B. B. *Phys. Rev. Lett*. 2000, 85, 4751.

[88] Jiang, Q.; Liang, L. H.; Zhao, D. S. *J. Phys. Chem. B* 2001, 105, 6275.

[89] Jiang, Q.; Zhao, M.; Xu, X. Y. *Phil. Mag. B* 1997, 76, 1.

[90] Broughton, J. Q.; Gilmer, G. H. *Acta Metall*. 1983, 31, 845.

[91] Shao, Y.; Spaepen, F. *J. Appl. Phys*. 1996, 79, 2981.

[92] Jiang, Q.; Zhou, X. H.; Zhao, M. *J. Chem. Phys*. 2002, 117, 10269.

[93] Kelton, K. F. *Solid State Physics* 1991, 45, 75.

[94] Lu, H. M.; Jiang, Q. *Phys. Stat. Sol. B* 2004, 241, 2472.

[95] Lu, H. M.; Wen, Z.; Jiang, Q. submitted to *Surf. Sci*.

[96] Skapski, A. S. *Acta Met*. 1956, 4, 583.

[97] Jiang, Q.; Li, J. C.; Chi, B. Q. *Chem. Phys. Lett*. 2002, 366, 551.

[98] Laird, B. B. *J. Chem. Phys*. 2001, 115, 2887.

[99] Digilov, R. M. *Surf. Sci*. 2004, 555, 68.

[100] Skapski, A. S. *Acta Met*. 1956, 4, 576.

[101] Luo, S. N. et al, *Phys. Rev. B* 2003, 68, 134206.

[102] Hirth, J. P.; Lothe, J. *Theory of Dislocations*; John Wiley: New York, 1982.

[103] Kuhlmann-Wilsdorf, D. *Phys. Rev*. 1965, A140, 1599.

[104] Kotzél, A.; Kuhlmann-Wilsdorf, D. *Appl. Phys. Lett*. 1966, 9, 96.

[105] Hilliard, J. E.; Cohen, M.; Averbach, B. L. *Acta Met*. 1959, 8, 26.

[106] Lange, W.; Hassner, A.; Mischer, G. *Phys. Stat. Sol*. 1964, 5, 63.

[107] Robinson, J. T.; Peterson, N. L. *Acta Met*. 1973, 21, 1181.

[108] Maraşh, N.; Hunt, J. D. *Acta Mater*. 1996, 44, 1085.

[109] Surholt, T.; Herzig, Chr. *Acta Mater*. 1997, 45, 3817.

[110] Lewis, A. C.; Josell, D.; Weihs, T. P. *Scr. Mater*. 2003, 48, 1079.

[111] Erol, M.; Maraşh, N.; Keşlioğlu, K.; Gündüz, M. *Scr. Mater*. 2004, 51, 131.

[112] Wang, J.; Wolf, D.; Philpot, S. R.; Gleiter, H. *Phil. Mag. A* 1996, 73, 517.

[113] Buff, F. P. *J. Chem. Phys*. 1951, 19, 1591.

[114] Tyson, W. R.; Miller, W. A. *Surf. Sci*. 1977, 62, 267.

[115] de Boer, F. R.; Boom, R.; Mattens, W. C. M.; Miedema, A. R.; Niessen, A. K. *Cohesion in Metals*; North-Holland Publishing Company: Amsterdam, 1998, Vol 1.

[116] Vitos, L.; Ruban, A. V.; Skriver, H. L.; Kollár, J. *Surf. Sci.* 1998, 411, 186.

[117] Mattsson, A. E.; Jennison, D. R. *Surf. Sci.* 2002, 520, L611.

[118] Galanakis, I.; Papanikolaou, N.; Dederichs, P. H. *Surf. Sci.* 2002, 511, 1.

[119] Methfessel, M.; Hennig, D.; Scheffler, M. *Phys. Rev. B* 1992, 46, 4816.

[120] Skriver, H. L.; Rosengaard, N. M. *Phys. Rev. B* 1992, 46, 7157.

[121] Kollár, J.; Vitos, L.; Skriver, H. L. *Phys. Rev. B* 1994, 49, 11288.

[122] Mehl, M. J.; Papaconstantopoulos, D. A. *Phys. Rev. B* 1996, 54, 4519.

[123] Rodríguez, A. M.; Bozzolo, G.; Ferrante, J. *Surf. Sci.* 1993, 289, 100.

[124] von Barth, U.; Hedin, L. *J. Phys. C* 1972, 5, 1629.

[125] Hohenberg, P.; Kohn, W. *Phys. Rev.* 1964, 136, B864.

[126] Kohn, W.; Sham, L. J. *Phys. Rev.* 1965, 140, A1133.

[127] Skriver, H. L.; Rosengaard, N. M. *Phys. Rev. B* 1991, 43, 9538.

[128] Alden, M.; Skriver, H. L.; Mirbt, S.; Johansson, B. *Phys. Rev. Lett.* 1992, 69, 2296.

[129] Vitos, L.; Kollár, J.; Skriver, H. L. *Phys. Rev. B* 1997, 55, 13521.

[130] Andersen, O. K.; Jepsen, O. *Phys. Rev. Lett.* 1984, 53, 2571.

[131] Lambrecht, W. R. L.; Andersen, O. K. *Phys. Rev. B* 1986, 34, 2439.

[132] Perdew, J. P.; Burke, K.; Ernzerhof, M. *Phys. Rev. Lett.* 1996, 77, 3865.

[133] Vitos, L.; Kollár, J.; Skriver, H. L. *Phys. Rev. B* 1994, 49, 16694.

[134] Moriarty, J. A.; Phillips, R. *Phys. Rev. Lett.* 1991, 66, 3036.

[135] Vitos, L.; Skriver, H. L.; Kollár, J. *Surf. Sci.* 1999, 425, 212.

[136] Desjonquères, M. C.; Spanjaard, D. *Concepts in Surface Physics, Springer Series in Surface;* Springer-Verlag: Berlin, Heidelberg, 1993, Vol 30.

[137] Galanakis, I.; Bihlmayer, G.; Bellini, V.; Papanikolaou, N.; Zeller, R.; Blügel, S.; Dederichs, P. H. *Europhys. Lett.* 2002, 58, 751.

[138] Haiss, W. *Rep. Prog. Phys.* 2001, 64, 591.

[139] Jiang, Q.; Lu, H. M.; Zhao, M. *J. Phys.: Condens. Matter* 2004, 16, 521.

[140] Mackenzie, J. K.; Moore, A. J. W.; Nicholas, J. F. *J. Phys. Chem. Solids* 1962, 23, 185.

[141] Mackenzie, J. K.; Nicholas, J. F. *J. Phys. Chem. Solids* 1962, 23, 197.

[142] Frank, F. C.; Kasper, J. S. *Acta Cryst.* 1958, 11, 184.

[143] Baskes, M. I. *Phys. Rev. Lett.* 1999, 83, 2592.

[144] Kittel, C. *Introduction to solid state physics*; 5th edition, John Wiley: New York, 1976, p 74.

[145] King, H. W. In *Physical Metallurgy*; Cahn, R. W.; North-Holland Publishing Company: Amsterdam, 1970, p 60.

[146] Mattsson, A. E.; Jennison, D. R. *Surf. Sci.* 2002, 520, L611.

[147] Mattsson, T. R.; Mattsson, A. E. *Phys. Rev. B* 2002, 66, 214110.

[148] Mattsson, A. E.; Kohn, W. *J. Chem. Phys.* 2001, 115, 3441.

[149] Mansfield, M.; Needs, R. *J. Phys. Rev. B* 1991, 43, 8829.

[150] Lang, N. D.; Kohn, W. *Phys. Rev. B* 1970, 1, 4555.

[151] Feibelman, P. *J. Phys. Rev. B* 1992, 46, 2532.

[152] Lodziana, Z.; Topsoe, N. Y.; Norskov, J. K. *Nature Mater.* 2004, 3, 289.

[153] Mathur, A.; Sharma, P.; Cammarata, R. C. *Nature Mater.* 2005, 4, 186.

[154] Gumbsch, P.; Daw, M. S. *Phys. Rev. B* 1991, 44, 3934.

[155] McHale, J. M.; Auroux, A.; Perotta, A. J.; Navrotsky, A. *Science* 1997, 277, 788.

[156] Jacobs, K.; Zaziski, D.; Scher, E. C.; Herhold, A. B.; Alivisatos, A. P. *Science* 2001, 293, 1803.

[157] Beenakker, C. W. J.; van Houten, H. *Sol. State Phys*. 1991, 44, 1.

[158] Bezryadin, A.; Dekker, C.; Schmid, G. *Appl. Phys. Lett*. 1997, 71, 1273.

[159] Zhang, H. Z.; Gilbert, B.; Huang, F.; Banfield, J. F. *Nature* 2003, 424, 1025.

[160] Zhang, H. Z.; Penn, R. L.; Hamers, R. J.; Banfield, J. F. *J. Phys. Chem. B* 1999, 103, 4656.

[161] Tolman, R. C. *J. Chem. Phys*. 1949, 17, 333.

[162] Lu, H. M.; Jiang, Q. *J. Phys. Chem. B* 2004, 108, 5617.

[163] Wachowicz, E.; Kiejna, A. *J. Phys.: Condens. Matter* 2001, 13, 10767.

[164] Plieth, W. *J. Surf. Sci*. 1985, 156, 530.

[165] Kiejna, A.; Peisert, J.; Scharoch, P. *Surf. Sci*. 1999, 432, 54.

[166] Shim, J. H.; Lee, B. J.; Cho, Y. W. *Surf. Sci*. 2002, 512, 262.

[167] Wei, S. Q.; Chou, M. Y. *Phys. Rev. B* 1994, 50, 4859.

[168] Sauerland, S.; Lohöfer, G.; Egry, I. *J. Non-Crystal Solids* 1993, 156-158, 833.

[169] Egry, I.; Lohöfer, G.; Sauerland, S. *Int. J. Thermophys*. 1993, 14, 573.

[170] Mills, K. C.; Brooks, R. F. *Mater. Sci. Eng. A* 1994, 178, 77.

[171] Eustathopoulos, N.; Ricci E.; Drevet, B. *Techniques de I'Ingénieur* 1998, M67, 1.

[172] Miyazaki, J.; Barker, J. A.; Pound, G. M. *J. Chem. Phys*. 1976, 64, 3364.

[173] Stephan, J. *J. Ann. Phys*. 1886, 29, 655.

[174] Skapski, A. *J. Chem. Phys*. 1948, 16, 386.

[175] Tegetmeier, A.; Cröll, A.; Benz, K. W. *J. Cryst. Growth* 1994, 141, 451.

[176] Eustathopoulos, N.; Drevet, B.; Ricci, E. *J. Cryst. Growth* 1998, 191, 268.

[177] Rebelo, L. P. N.; Lopes, J. N. C.; Esperança, J. M. S. S.; Filipe, E. *J. Phys. Chem. B* 2005, 109, 6040.

[178] Keene, B. *J. Int. Mater. Rev*. 1993, 38, 157.

[179] Wallace, D. C. *Phys. Rev. E* 1997, 56, 4179.

[180] Han, J. H.; Kim, D. Y. *Acta Mater*. 2003, 51, 5439.

[181] Lu, H. M.; Jiang, Q. submitted to *J. Phys. Chem. B*.

[182] Bernal, J. D.; Mason, J. *Nature* 1960, 188, 910.

[183] Shen, T. D.; Harms, U.; Schwarz, R. B. *Appl. Phys. Lett*. 2003, 83, 4512.

[184] Kresse, G.; Hafner, *J. Phys. Rev. B* 1993, 48, 13115.

[185] Jakse, N.; Pasturel, A. *Phys. Rev. Lett*. 2003, 91, 195501.

[186] Maftoon-Azad, L.; Boushehri, A. *Int. J. Thermophys*. 2004, 25, 893.

[187] Kuz, V. A.; Meyra, A. G.; Zarragoicoechea, G. *J. Thermochimica Acta* 2004, 423, 43.

[188] Lindemann, F. A. *Z. Phys*. 1910, 11, 609.

[189] Miedema, A. R.; Boom, R. *Z. Metallkd*. 1978, 69, 183.

[190] Smithells, C. J. *Metals Reference Book*; 5th edition, Butterworths: London & Boston, 1976, pp 944-946.

[191] *Table of Periodic Properties of the Elements*; Sargent-Welch Scientific Company: Illinois, 1980, p 1.

[192] Kalazhokov, K. K.; Kalazhokov, Z. K.; Khokonov, K. B. *Tech. Phys*. 2003, 48, 272.

[193] Wang, H. P.; Yao, W. J.; Cao, C. D.; Wei, B. *Appl. Phys. Lett*. 2004, 85, 3414.

[194] Ishikawa, T.; Paradis, P. F.; Yoda, S. *Appl. Phys. Lett*. 2004, 85, 5866.

[195] Paradis, P. F.; Ishikawa, T.; Fujii, R.; Yoda, S. *Appl. Phys. Lett*. 2005, 86, 041901.

[196] Goldman, J. H. *J. Nucl. Mater*. 1984, 126, 86.

[197] Ghatee, M. H.; Boushehri, A. *High Temp-High Press.* 1994, 26, 507.

[198] Chen, M.; Yang, C.; Guo, Z. Y. *Mater. Sci. Eng. A* 2000, 292, 203.

[199] Webb III, E. B.; Grest, G. S. *Phys. Rev. Lett.* 2001, 86, 2066.

[200] Yao, W. J.; Han, X. J.; Chen, M.; Wei B.; Guo, Z. Y. *J. Phys.: Condens. Matter* 2002, 14, 7479.

[201] Gibbs, J. W. *The Collected Works*; Longmans, Green and Company: New York, 1928, Vol 1.

[202] Guggengeim, E. A. *Trans. Faraday Soc.* 1940, 36, 407.

[203] Tolman, R. C. *J. Chem. Phys.* 1949, 17, 118.

[204] Kirkwood, J. G.; Buff, F. P. *J. Chem. Phys.* 1949, 17, 338.

[205] Bykov, T. V.; Zeng, X. C. *J. Phys. Chem. B* 2001, 105, 11586.

[206] Giessen, E. V.; Blokhuis, E. M.; Bukman, D. J. *J. Chem. Phys.* 1998, 108, 1148.

[207] Rowlinson, J. C.; Widom, B. *Molecular Theory of Capillarity*; Clarendon Press: Oxford, 1982.

[208] Fenelonov, V. B.; Kodenyov, G. G.; Kostrovsky, V. G. *J. Phys. Chem. B* 2001, 105, 1050.

[209] Talanquer, V.; Oxtoby, D. W. *J. Phys. Chem.* 1995, 99, 2865.

[210] Koga, K.; Zeng, X. C.; Shchekin, A. K. *J. Chem. Phys.* 1998, 109, 4063.

[211] Lu, H. M.; Jiang, Q. *Langmuir* 2005, 21, 779.

[212] Samsonov, V. M.; Shcherbakov, L. M.; Novoselov, A. R.; Lebedev, A. V. *Colloid Surface A* 1999, 160, 117.

[213] Dewar, M. J. S.; Zoebisch, E. G.; Healy, E. F.; Stewart, J. J. P. *J. Am. Chem. Soc.* 1985, 107, 3902.

[214] Samsonov, V. M.; Bazulev, A. N.; Sdobnyakov, N. Y. *Dokl. Phys. Chem.* 2003, 389, 83.

[215] McGraw, R.; Laaksonen, A. *J. Chem. Phys.* 1997, 106, 5284.

[216] Koga, K.; Zeng, X. C. *J. Chem. Phys.* 1999, 110, 3466.

[217] Nijmeijer, M. J.; Bruin, C.; van Woerkom, A. B.; Bakker, A. F. *J. Chem. Phys.* 1992, 96, 565.

[218] Haye, M. J.; Bruin, C. *J. Chem. Phys.* 1994, 100, 556.

[219] Bartell, L. S. *J. Phys. Chem. B* 2001, 105, 11615.

[220] Li, J. C.; Liu, W.; Jiang, Q. *Acta Mater.* 2005, 53, 1067.

[221] Liang, L. H.; Liu, D., Jiang, Q. *Nanotechnology* 2003, 14, 438.

In: Energy Research Developments
Editors: K.F. Johnson and T.R. Veliotti

ISBN: 978-1-60692-680-2
© 2009 Nova Science Publishers, Inc.

Chapter 9

SUSTAINABLE USE OF ENVIRONMENTAL RESOURCES: OPTIMIZATION OF LOGISTICS OPERATIONS

Riccardo Minciardi[1,2,3], Michela Robba[1,2,3] and Roberto Sacile[1,2,3]

[1] DIST-Department of Communication, Computer and System Sciences
[2] CIMA-Interuniversity Center of Research in Environmental Monitoring
[3] E.R.-Laboratory for Energy production from Renewable Resources

Abstract

The use of natural resources is necessary for human life and activities. Finding strategies that can satisfy energy and material demands, be sustainable for the environment, and preserve natural resources from depletion, seems to be a crucial point for the modern society.

This chapter deals with the role of Decision Support Systems (DSS) for environmental systems planning and management, with specific attention to the use of forest biomasses for energy production, and energy and material recovery from urban solid waste. The two mentioned issues are treated in two different sections to highlight the legislative and technological differences beyond the "treatment" and the "possibilities of use" of biomasses and solid wastes. Then, after a general introduction about how to build Environmental DSSs, decision models (characterized by decision variables, objectives, and constraints), and about the state of the art, the chapter is divided in two main sections: energy production from renewable resources, and material and energy recovery from solid urban waste.

The first one is focused on forest biomass exploitation and includes a detailed mathematical description of the formalized decision model and its solution applied to a real case study. The DSS described in the present study is organized in three modules (GIS, data management system, optimization) and is innovative, with respect to the current literature, because of the design of the whole architecture and of the formalization of the different decision models. In particular, the optimization module includes three sub-modules: strategic planning, tactical planning, and operational management. In this chapter, an example related to strategic and tactical planning is given.

The second one regards a more qualitative discussion about multi-objective decision problems for solid waste management in urban areas. In this case, the structure of DSSs able to help decision makers of a municipality in the development of incineration, disposal, treatment and recycling integrated programs, taking into account all possible economic costs, technical, normative, and environmental issues, is described. Finally, conclusions and future developments about EDSS for material recovery and energy production are reported.

1. Introduction

The use of natural resources is necessary for human life and activities. Finding strategies that can satisfy energy and material demands, be sustainable for the environment, and preserve natural resources from depletion, seems to be a crucial point for the modern society.

This chapter is about the role of Decision Support Systems (DSS) for environmental systems planning and management, with specific attention to material recovery and energy production within a general waste management context. Specific DSS models and related case studies are introduced and discussed as regards the use of wood biomasses (with specific reference to forest biomasses and scraps coming from wood industries and public green maintenance) for energy production in a rural context, and energy and material recovery from the planning and management of municipal solid waste in an urbanised context. The two mentioned issues are treated in two different sections to highlight the legislative and technological differences beyond the "treatment" and the "possibilities of use" of biomasses and solid wastes.

Specifically, the chapter is organised as follows. In this introduction, waste management is introduced as the originating problem for which material recovery and energy production represent the most important opportunities to face the problem. Specifically, the European Union (EU) strategies to cope with waste once it has been generated, which are greenhouse neutral energy production, recycling, and final optimal disposal, are briefly examined, since they do represent the general framework on which the decisional problems of this chapter are based on. The methodological possibilities to support these decisional problems are also introduced, with specific reference to DSS, the related information management aspects and the mathematical modelling of the decision. The state of the art as regards DSS for the two main issues on which this chapter is based, that are energy production from biomass and municipal solid waste management, is examined.

Two applicative sections follow, dealing with DSS for energy production from renewable resources, and for material and energy recovery from municipal urban waste. The DSS models and applications are based both on original work, and on previous works to which the authors contributed to, and to which the readers may refer for further details.

The first applicative section is focused on wood biomass use to produce greenhouse neutral energy, which, above all in rural territories, may represent an important solution not only to reduce waste, but also to improve the economic, social and environmental characteristics of a territory. Specifically, the possibility to use different kinds of biomasses has been considered: scraps coming from "wood-industries", scraps coming from public green, and forest biomasses (according to the available biomass reported in the Forest Management Plans). The whole supply chain is detailed in terms of strategic planning and tactical planning, and the connections with efficient manufacturing processes planning and management are highlighted. The second applicative section regards the formalization of multi-objective optimization problems for municipal solid waste management in urban areas. In this case, the DSSs can help decision makers of a municipality in the development of incineration, disposal, treatment and recycling integrated programs, taking into account all possible economic costs, technical, normative, and environmental issues. Great efforts are dedicated to the definition of suitable strategies for material recovery and energy production. The general aim is the one of showing how to build decision models applied to Solid Waste Management, with different degrees of complexity and different objectives. A multi-objective procedure is described and applied to a real case study.

Finally, conclusions and future developments about DSS for material recovery and energy production are reported.

1.1. Sustainable Use of Resources

Waste production is increasing all over the world. Total waste generation in EU is 1300 million tonnes per year. Construction and demolition and manufacturing industries generate half of total waste, while an important contribution, 14%, is also given by municipal waste. In addition, there is an evident relation between economic growth and waste generation. This overall increasing trend goes against the general policy target of waste prevention, and both policies and strategies are requested to face this problem.

The cornerstones of the EU's strategy to coping with waste (figure 1) are to: prevent waste in the first place; recycle waste; turn waste into a 'greenhouse neutral' energy source; optimise the final disposal of waste, including its transport.

Preventing waste requires a deep analysis of waste production processes both as regard industrial activities and general human habits. This activity is of fundamental importance to achieve a reduction of waste production, and, in case of industrial activity is often connected to quality assurance management, for example according to the International Organization for Standardization (ISO) 14001 and the Eco-Management Audit Scheme (EMAS). Rather than defining a single general strategy, preventing waste should be specialised for each single industrial activity, including human living as regards municipal waste. Due to the objective of the book, these aspects are not discussed in this chapter and the reader is recommended to look at (Ville Niutanen, Jouni Korhonen, 2003; Wenk, 2004; Broman, 2000; Del Breo and Beatriz, 2003; Van Cooten, 2005) for further details.

The fact from which the discussion in this chapter starts is that industrial and human activities produce waste, that in this specific context is referred to as "gross waste". One of the main goals of waste management is to reduce the quantity of "gross waste" to a lower (in a utopian zero emission world, zero) quantity of waste. This hopefully lower quantity is hereinafter referred as "net waste", and it is obtained by transforming part of the "gross waste" in other products (mainly energy and recovered materials). Specifically, this chapter is centred on material recovery and energy production, so the last three of the four above mentioned European Union strategies related to waste management ("greenhouse neutral" energy production (GNEP), recycling (REC), and optimising final disposal (OFD)) will be analysed and discussed in detail, and decision processes and systems to support them will be proposed.

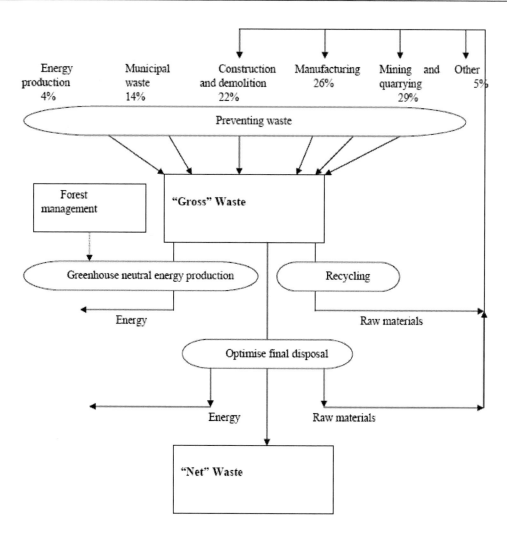

Figure 1. The waste management problem: prevention, material and energy recovery, disposal.

1.2. The Role of Environmental Decision Support Systems

In the recent literature, several papers can be found about Environmental Decision Support Systems (EDSS), their definition, characteristics, complexity, implementation, etc. One of the most recent papers on this subject (Matthies et al. (2005)) states that an EDSS often consists of various coupled environmental models, databases and assessment tools, which are integrated under a graphical user interface (GUI), often realized by using spatial data management functionalities provided by geographical information systems (GIS). Denzer (2005) states that Environmental Information Systems (EIS) and EDSS are major building blocks in environmental management and science today. They are used at all levels of public bodies (community, state, national and international level), in science, in management and as information platforms towards the public. EIS and EDSS are usually said to have certain characteristics, which distinguish them from standard information systems, e.g. information complexity in time and space or uncompleteness or fuzziness of data items.

By the very nature of the complex tasks involved, different methodologies can be an option while developing new systems, for instance modelling, decision theoretic approaches, artificial intelligence, geographical analysis, statistics and many more. A GIS can play the role of an EDSS whenever a territorial aspect is taken into account and if it is enhanced with a decision support module. So a GIS itself is often taken into account as a decision making system, and, in this case, is often referred to as GIDSS. In general, there may be some misunderstanding on what a GIS is, so it seems to be worthwhile to explain its role as an EDSS related to territorial problems.

A GIS, as many other modern information systems, is a complex collection of information processes allowed by a great number of hardware, software and communication technologies, with which the users, characterized by different competences and different objectives, can interact. In addition, a GIS is the fundamental tool to define, model and implement many information/knowledge/decisional class of problems. The continuous growing of the anthropisation of the territory more and more needs an accurate monitoring action of the environment, in order to evaluate and plan the necessary interventions for the protection of the environmental quality and of the health of the population. In general, a GIS aims at filling the information gap between the status of the natural or anthropic environment (in a wide meaning of the term, such as, for example, a forest, a river, an expanse of sea, but also a city, its traffics or an industrial district) and a set of persons generally defined as "decision makers" who can select the most appropriate actions for the environmental protection.

A GIS can be taken into account as a layered information system. Each layer, which may be more or less distributed, can offer a reliable service to the higher layers, and the link between adjacent layers is guaranteed by an adequate telematics communication.

The first level, from bottom to top, can be defined as the "environmental monitoring layer", whose task is to acquire information from the environment by the use of sensors, in a broad sense, including in this definition, for example, rain gauges, anemometer, cameras for map data acquisition from satellites, people collecting water samples and related water quality analysis processes ... In the "environmental monitoring" layer, sensors and related software for data acquisition and communication transfer are present and distributed in space, providing quite different content and formats of heterogeneous data.

The second level can be defined as the "DataBase Management System (DBMS) layer". In this layer, data coming from telematics connections or from storage supports are modelled and stored in one or more databases. The most frequently used data models are based on the relational paradigm, and these data are often centralized, such as for example at local or regional authorities. From a software architecture and applications point of view many are the trends in this layer, such as for example, client server web based architectures, java applications, open software applications, object oriented databases ... In general, a common approach to data modelling in this layer is to define as much as possible a relational data architecture, leaving map and cartographic data to specific proprietary data models related to GIS software for efficiency reasons.

The third level can be defined as a "data analysis, synthesis and processing layer", and it can be generally divided into two sub-components, of which at least one is always present in a GIS: the "environmental modelling sub-layer" and the "GIS software sub-layer". In the environmental modelling sub-layer, software tools are introduced in relation to models, generally mathematically formulated, describing the behaviour of an environmental system when physical, chemical or biological conditions change. The aim of these models is to produce "simulations", whose results are likely to be stored in the DBMS layer, either to predict future behaviours of the monitored environmental system, or structural changes, when new anthropic or environmental scenarios are supposed. Data that are necessary to the simulations to calibrate the system on the basis of historical data or to start calculations on the basis of real-time data are generally acquired from the previous DBMS layer. The second sub-layer is the true GIS software allowing the graphic visualisation of the information contained in the DBMS layer, giving relevance to the geographic characteristics of the data, while providing a user-friendly interface, which allows to show different information aspects (vegetation, population density, rainfall data etc.) at a customable scale. Both the sub-layers, complementarily, can provide the decision maker with an essential tool to synthesize the great amount of data collected and stored by the previous layers.

In the conventional GIS, working for example in public information systems or supporting for example environmental management systems of industrial companies, only the previous layers are generally present. As a matter of fact, several authors underline the importance of GIS as a decisional system (for example, Longley et al., 2001), and that the application of specific decision analysis techniques is more and more felt (Malczewski, 1999). Following this trend, several commercial GIS software are starting to add a new decisional layer, that is a DSS layer, providing aid in decisions, such as for example in vehicle routing problems. In truth, in environmental management, there is a wealth of problems which needs aid in the decisional processes, and which are an interesting and moderately explored field for researchers in environmental engineer.

In this chapter, the decisional aspects are explained in more detail, and the following sections are focused on the DSS layer. In fact, the DSS layer might be represented by a series of possible approaches: such as a knowledge based approach, where, for example, the experience of an expert to face the problem is captured with a set of rules to be fired whenever a problem is posed; or a black box approach, whenever the problem appears so complex that even no expert exist to explain how to face the problem, and the way to solve it can be done only by specific automatic "learning" algorithms (such as in artificial neural networks) on the base of past experiences. In this chapter, and specifically in sections 2 and 3, the DSS layer is always represented under a specific mathematical modelling formulation, that is under a mathematical programming approach. The objective, or the objectives, are always formulated as mathematical functions to be maximised (minimised), subject to a set of constraints again formulated through mathematical functions.

1.2.1. Forest Biomass Use for Energy Production

The consequences of the use of traditional fossil fuels has led, in the last years, the European Union to promote and encourage the development and the use of renewable energies rather than the traditional ones. Energy production from biomass exploitation, instead of the one from fossil fuels, does not involve the CO_2 increase in the atmosphere, contributing in this way to the duties assumed by the European Community during the International Conference of Kyoto (1997): as a matter of fact, the same quantity of CO_2 that is formed when burning the biomass is taken away from the atmosphere during the vegetation growing, through the photosynthesis process. This cyclic process results in a net inflow of CO_2 in the atmosphere equal to zero.

Without any doubt, the energy production from biomass represents an important part within an energy plan based on renewable resources (Basosi et al. (1999), Pari (2001), Hall and Scarse (1998), Varala et al. (1999)). Indeed, the White Book of the European Union *"Energy for the future: renewable resources –White Book for a strategy and an action plan of the Community"* (Com (97)599 def. del 26/11/97), that represents the proposal of an action plan for the development of renewable resources, states that the main contribution to energy production should be furnished by biomasses, and, in second instance, by the other renewable sources.

According to the guidelines in the White Book of the European Union, it has been presented in Italy at the National Conference "Energy and Environment" (Rome, 25-28 November 1998), the "White Book for Renewable resources (Rome, 1999) where specific issues like social impact, technical, juridical and legislative frameworks, incentives connected to biomass exploitation have been investigated with the aim of doubling the energy production from renewable sources within year 2010.

An extensive literature reports biomass utilization experience in different territorial realities (among others, Ediger and Kentel, 1999; Martinot, 1998; Ushiyma, 1999; Basosi et al., 1999). Quantitative analysis about strategies for renewable energy sources from biomass has been performed either evaluating the potential resources of bio-energy in different kind of countries (Hall and Scrase, 1998) or matching the woody biomass demand and supply by the forest industries in Europe (Kuiper et al., 1998).

Decision support systems (DSS's) have been proposed to help biomass management for energy supply at a regional level. Nagel J. has proposed a methodology (Nagel. J., 2000a), tested in the state of Brandenburg, Germany (Nagel. J., 2000b), to determine an economic energy supply structure based on biomass. The problem is formulated as a mixed-integer linear optimization using the dynamical evaluation of economic efficiency, and with 1-0 conditions to solve the question whether to build or not a heating system, a heating plant or a co-generation plant. Nagel's works focus on many aspects such as the user typology that can benefit from biomass use for energy supply, on the dimension and typology of heating plants, and on the sensibility of the decision with respect to fuel costs. Among the conclusions of these two works, it was assessed that using biomass in individual plants is already economic for some consumers, although an attempt should be made to reduce the biogenic fuel prices. In addition, since biomass can help CO_2 emissions, an economy effort should be dedicated to establish CO_2 taxes or state subsidies for biomass-fired energy conversion plants or by changing the payment for electricity produced by biomass.

Another decision support system called AUHDSS for bio-energy application, with special reference to harvesting wood for energy from conventional forestry and short rotation forestry has been recently described. Such a system concerns the calculation of delivery cost of wood fuel from conventional forest in the UK (Mitchell C.P., 2000). In this work, an exhaustive review of topics related to the problem is given as well as an exhaustive list of computer models of bioenergy systems. Always from the same research group, other DSS's have been proposed for biomass management: CDSS (Coppice Decision Support System), a spreadsheet model that can be used to model the costs of growing short rotation coppice under UK conditions (Mitchell, 1995); CHDSS (Coppice Decision Support System), modeling the supply chain from the standing coppice crop through harvesting, storage and transport (Mitchell, 2000). The previous DSS's, as well as other models, were linked together to produce BITES, now presented in an extended spread-sheet based format called BEAM (Mitchell, 2000), that is an integrated biomass to electricity model.

The territorial evaluation, involving geographical (Noon et al, 1996, Graham et al., 2000), environmental (Nagel. J., 2000b) and socio-economic (Varela et al., 1999) characteristics of the region are also very important aspects in the decision modeling of biomass management. In this respect, Geographic Information Systems (GIS) based approaches have been recently proposed. Noon and Daly, 1996, have proposed a GIS-based Biomass Resource Assessment, Version One, called BRAVO. BRAVO was defined as a computer-based DSS to assist the Tennessee Valley Authority in estimating the costs for supplying wood fuel to any one of its 12 coal-fired power plants. In BRAVO, the GIS platform allows the efficient analysis of transportation networks so that accurate estimates of hauling distances and costs can be determined. In a subsequent work (Graham et al., 2000) the previous work was extended under several aspects, one of which was the estimation of the costs and of the environmental implications of supplying specified amounts of energy crop feedstock across a state, considering where energy crops could be grown, the spatial variability in their yield, and transportation costs.

Forest biomass exploitation for energy production should be helpful for increasing the value of the territory through a specific attention to the forest ecosystem, to good hydro geological conditions, and to social and economical issues. In fact, these improvements of the environment are possible because, when building a Decision Support System for forest biomass exploitation, it is necessary to take into account different issues (energy production, preservation of the environment, social consequences, etc.).

The problem of the effects on the environment from human activities foresees a decisional process that should take to the conclusion if an intervention is feasible or not, if the choices for the development are sustainable, and then, in the realization phase, if the technical solutions are compatible with the environment. The role of Environmental Impact Assessment (EIA) in the decisional process is linked to the foresee and to the estimation of the effects of relevant actions on the environment, and it furnishes useful information for the final decision. The EIA procedure has the positive aspect of considering these effects before they can come out, because it is a technical-administrative procedure, performed on the interventions that modify the natural environment.

In this study, some aspects that are inside the EIA have been taken into account and integrated into the DSS. Besides, the study has been implemented taking into account European and International guidelines regarding energy production from renewable resources, and the objectives of the Energy Policy in the Liguria Region that have the aim of augmenting the energy efficiency, of reducing pollutant emissions, and reaching the percentage of 7% of energy demand for the energy produced by renewable resources. Specifically, Liguria region has evaluated the possibility to reach in the next 10 years the objective of producing an installed power for forest biomass equal to 150 MW. Biomasses can be used to produce electric and/or thermal power in spite of traditional fuels. The main techniques to convert biomasses in energy are combustion, gasification, and pyrolisis. The combustion represents the biomass oxidation under air excess, which products are hot gas that are used to produce superheated steam from which electricity is obtained. In the gasification process, the biomass is partly oxidized through sub-stechiometric oxygen quantities. The pyrolisis process is characterized by a thermal degradation of the material in absence of oxygen.

The favourable attitude of Liguria Region towards the plants that exploit forest biomasses for the production of energy, has led to the definition of the base criterions for the layout of the screening report that the plant proponent has to furnish to the public administration and the regional office.

The motivations that have led the regional administration to intervene in the case of this typology of plants are clearly express in the D.G.R. n. 965 of 05/09/2002; in the Deliberation of the Regional Authority, it is clearly expressed that "... the plants that exploit the sources of renewable energy (...) are more easily compatible than other plants with the surrounding territory, they increase the local possibilities of territorial equilibrium and the control of the territory, they favour reforestation, with notable improvements for the defence of the ground and for the hydro geological conditions;... ". The objective that wants to be achieved through the emanation of this deliberation is "... to point out some criterions that can simplify the elaboration of the screening report to make easier (...) in the context of the technical fulfilments to be observed as required by the article 4 of the new technical norms for the screening report."

The points necessary for the EIA have been taken into account when formalizing the DSS presented in this chapter.

In the following sections, a methodology, based on the concept of Decision Support System, and case study are proposed. The use DSS and of Territorial Informative Systems (SIT) for the verification of the sustainability of plants turned to the production of biomass energy, and for the definition of strategies of planning and management is considered innovative and promising both from a scientific point of view and for practical application (Najel (2000), Mitchell (2000), Graham et al. (2001), Voivontas et al. (2001), Freppaz et al. (2004)). For the creation of the DSS, different decision models have been defined adapting and integrating a previously formulated model (Freppaz et to the, 2004) to the case of study, and formalizing other, more general, decision models.

1.2.2. Solid Waste Management in Urban Areas

The need of defining strategies that can be sustainable for the environment and the ecosystem is one of the main priorities of European Union in the last years. Two areas of great importance when it comes to sustainable development are the waste management system and the energy system (Holmen and Henning, 2004). Indeed, when waste is energy recycled, it replaces other fuels. When waste is the material recycled, it replaces virgin material and also saves energy because the production processes that use recovered material are less energy-intensive than those that use virgin material.

Wastes represent a big problem for the society because they are produced in high quantities and because, when disposed, they are a danger for the environment. Moreover, European legislation recommends the development of local integrated management plans, which give priority to prevention, waste reduction and recovery, and allow using landfill only for the disposal of refuses that cannot be recovered. In Italian municipalities, the new legislation, the rapid increase of solid waste production and the frequent landfill closings have encouraged the development of incineration and recycling programs. The definition of such programs must take into account an integration of economic, environmental, social and technical considerations. That is a quite hard task, as it is necessary to properly take into account economic, technical, and normative aspects, paying particular attention to environmental issues. Moreover, public/private sectors consideration and public participation considerations make the problem even more complex.

Chang et al. (2005) assess that installing material recovery facilities (MRFs) in a solid waste management system could be a feasible alternative to achieve sustainable development goals in urban areas. They formalize a decision model for the optimal site selection and capacity planning for a MRF in conjunction with an optimal shipping strategy of solid waste streams in a multi-district urban region. Screening of material recovery and disposal capacity alternatives can be achieved in terms of economic feasibility, technology limitation, recycling potential, and site availability. The optimization objectives include economic impacts characterized by recycling income and cost components for waste management, while the constraint set consists of mass balance, capacity limitation, recycling limitation, scale economy, conditionality, and relevant screening constraints.

A fundamental difficulty in planning a Municipal Solid Waste (MSW) management system is the necessity of taking simultaneously into account conflicting objectives. It is really difficult for planners and regulators to develop a sustainable approach to waste management and to integrate strategies aiming at producing the best practicable and environmentally sustainable option. To formalize these strategies, in the last two decades, considerable research efforts have been directed towards the development of economic-based optimization models for MSW flow allocation. Several examples of mathematical programming models have been developed for MSW management planning, such as, for example, in Chang and Chang (1998), D'Antonio and Fabbricino (1998), Daskalopoulos et al. (1998), Fiorucci et al. (2003), Badran and El-Haggar (2005).

However, an approach to the waste management problem merely based on economic considerations cannot be considered as completely satisfactory, and a wide set of possible improvements can be pursued. Above all, modeling the environmental impact of solid waste management requires modeling and analyzing a quite heterogeneous set of subsystems, which, in turn, are affected by the decisions concerning solid waste management. Examples of such subsystems are the atmospheric pollution model, the city traffic system, the sanitary landfill, etc. In this respect, Tsiliyannis (1999) has discussed the main environmental problems related to MSW management, and in particular those concerning pollutant releases.

Another possible approach is based on life cycle assessment (Finnveden, 1999; Barton et al., 1998) and Finnveden (1999) has discussed some methodological issues arising in this case. More recently, Costi et al. (2003) have proposed a decisional strategy that takes into account the environmental impact of MSW in the constraints of the model.

Clearly, the necessity of taking into account economic, technical, normative aspects, paying particular attention to environmental problems (which usually cannot be dealt with by economical quantifications only) is more and more felt. Such a reason has led several authors to propose multi-criteria decision approaches that, in some cases, allow a formal representation of uncertainty or imprecise information. Recently, several authors have proposed a number of models and tools based on outranking approaches for multiple criteria decision making (MCDM) and multiattribute rating techniques applied to MSW management. Such approaches have paid a special attention to the different aspects (economic, technical, normative, environmental) of the decision process. Among others, the following methodologies have been proposed: Electre III [Hokkanen and Salminen, 1997], and DEA ranking techniques [Sarkis, 2000].

Other works have proposed fuzzy multi-objective formalizations [Chang N.B. et al., 1997; Chang N.B. et al., 2000]. Chang and Chen (1997) highlight how the common objective of minimizing the present value of the overall cost/benefit can be extended to deal explicitly with environmental considerations, such as air pollution, traffic flow limitation, and leaching and noise impacts.

Shekdar and Mistry (2001) have proposed an interactive goal programming model of multi-objective planning of the system. The system incorporates the activities involved from waste collection to final disposal of waste. Specifically, it includes processes associated with waste collection, transportation, resource recovery, treatment and disposal. The considered strategy regards the maximization of energy recovery, the maximization of material recovery, the minimization of expenditure, the limitation on the land filling capacity. It is important to note that there can be variations in the priority assigned to a particular goal by the management while managing the system. The Shekdar and Mistry's software procedure uses a goal programming method to allow each of the objective functions to be solved without necessarily providing a relationship between them. The model can be run interactively by setting the goals in any priority order. Generally speaking, there are different procedures of interactive multiple objective programming, and Gardiner and Steuer (1994) showed how these procedures can be unified into a single algorithm. As concerns environmental management, which is often formulated as a multi-objective problem, the reference point methodology (Wierzbicki et al, 2002) has proposed as an appropriate approach.

2. Supply Chain Optimization for Forest Biomass Use for Energy Production

The developed DSS, that is a generalization and the further research activity presented in a previous work (Freppaz et al., 2004), is based on the integration of Geographic Information Systems tools, a relational database, and decision models (in terms of decision variables, objectives, and constraints). There is one main general goal that the DSS application wishes to achieve: the definition of a clear path of the decisional processes on "how-to" make a biomass supply chain (BSC) effective on a territory, in order to create sustainable, economically attractive strategies. The BSC organization can be taken into account as a manufacturing process with specific attention to environmental impacts, actors participation, and costs minimization. Modern manufacturing requires a deep horizontal integration, for example among customers and suppliers, supported mainly by Internet/Intranet technologies. In this respect, the availability of the Supply-Chain Operations Reference-model (SCOR) (figure 2.1) recently had a major impact on business system planning. This model breaks down Supply Chain Management (SCM) into four main processes:

PLAN, related to typical production planning activities, such as the definition of the Master Production Schedule;
SOURCE, related to activities typically associated with the management of providers and of the inventory;
DELIVER, related to the management of customers and the distribution of products;
MAKE, related to production processes.

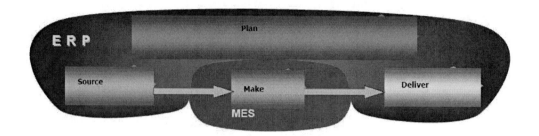

Figure 2.1. The Supply-Chain Operations Reference (SCOR) model (redrawn from (Supply Chain Council, 1996)).

With respect to the SCOR model, ERP (Enterprise Resource Planning) products can address the requirements of the PLAN, SOURCE and DELIVER processes, while MES (Manufacturing Execution Systems), jointly with control layer components, is the primary component of the MAKE process.

The DSS described in the present study is organized in three modules (GIS, data management system, optimization) and is innovative, with respect to the current literature, because of the design of the whole architecture and of the formalization of the different decision models. In particular, the optimiza tion module includes three sub-modules: strategic planning, tactical planning, and operational management. The necessity to introduce different decision levels derives from the different time scales to be considered and the different decisions to be taken. Strategic planning level decisions refer to plant sizing and location, to the choice of the technology to be adopted, to the degree of exploitation of forest resources. Tactical planning level decisions refer to a medium-short term horizon, and are generally considered within a discrete-time setting, assuming that the decisions of the strategic planning level have been established (i.e., plant capacity and location are fixed). In this case, a dynamical formalization of vegetation growth models and carbon sequestration models is needed. Finally, the operational level requires the explicit modelling of the supply-chain, as an ordered sequence of the operations that have to be performed from biomass collection to energy conversion.

In the following, the strategic planning (static decision problem) and the tactical planning (dynamic decision problem) levels are described. The developed DSS for sustainability problems has the main objective to define the sustainability of different plant capacity scenarios on a territory inside the Savona District. The formalization of the static and the dynamic optimization problem is here presented in two different sub-sections.

2.1. Strategic Planning: Formalization of the Decision Problem

The cost function, C [€], that should be minimized (the decision variables - represented in bold - are the biomass u_i [m^3 $year^{-1}$] collected in the i-th parcel, per unit time) is composed by two terms:

$$C = C_T + C_R \qquad (2.1)$$

where:

- C_T represents the biomass transportation cost (per unit time);
- C_R is the biomass collection cost (per unit time).

Transportation costs may be expressed as

$$C_T = C_{TR} \sum_{i=1}^{N} d_i MV_i \boldsymbol{u}_i \qquad (2.2)$$

where:

$C_{TR} \; [\text{€} \; Km^{-1} Kg^{-1}]$ is the unit transportation cost;
$d_i \; [Km]$ is the distance of the i-th parcel from the plant;
$MV_i \; [Kg \; m^{-3}]$ is biomass density of the i-th parcel.
Collection costs may be expressed as

$$C_R = \sum_{i=1}^{N} Cr_i MV_i \boldsymbol{u}_i \qquad (2.3)$$

where $Cr_i \; [\text{€} \; Kg^{-1}]$ is the unit collection cost for parcel i.

Such costs are strongly influenced from the characteristics of the territory: every parcel has its collection costs, depending on the viability conditions, the presence of infrastructures, the terrain accessibility and slope. In order to take into account such aspects, unit collection costs have been grouped and defined considering four levels about biomass collection: from an easy level to a non-practicable level.

Thus, the overall *cost function* that has to be minimized is

$$C = C_{TR} \sum_{i=1}^{N} d_i MV_i \mathbf{u}_i + \sum_{i=1}^{N} Cr_i MV_i \mathbf{u}_i \qquad (2.4)$$

The minimization of the cost function (2.4) has to be carried out under constraints relevant to biomass collection and to the biomass flow entering the plant.

Limits on biomass collection: the possibility of biomass collection on a territory must respect the forest regulations inside the Forest Plans which contains all the guidelines about the methods of forest management, treatment, and use. From the analysis of such documents and data, it is possible to consider a medium "cut turnover system" for coppice and high forest equal to 20 and 100 years, respectively, corresponding to a percentage of annual cutting of 5% and 1%.

In every parcel, known the biomass quantity x_i, in m^3, the quantity that may be collected is

$$u_i \le \alpha_i \cdot x_i \quad i=1,...,N \tag{2.5}$$

$$u_i \ge 0 \quad i=1,...,N \tag{2.6}$$

where α_i is the maximum percentage of biomass quantity that is fixed by law. These coefficients assume a value equal to 0.05 for coppice and 0.01 for high forest, according to the local regulations.

Constraints on biomass flow: the biomass quantity entering the considered plant should be equal to the plant capacity. This yields the following constraint

$$\frac{PCI}{3600 \cdot 8000} \sum_{i=1}^{N} MV_i u_i = \frac{CAP}{\eta_e} \tag{2.7}$$

where:

PCI is the low heating value assumed constant for not-treated biomasses, assuming a medium humidity of 30-35%;

η_e is the plant electric energy efficiency;

8000 are the functioning hours in a year;

3600 is a conversion factor (hours in seconds).

The overall problem (cost function and constraints) so formalized is a linear programming problem and can be solved by commonly used optimization tools.

2.2. Tactical Planning: Formalization of the Decision Problem

Referring to a given plant, as in the previous subsection, and considering the biomass growth dynamics, the objective is now to plan collection in the first five years of the plant. This means that the decision variables, namely u^t_i, are now functions of the parcels and of time discrete values. Moreover, state variables x^t_i [$m^3 \cdot ha^{-1}$] have to be introduced, in order to represent the available biomass in parcel i at time t. Apart this, the two optimization problems do not present substantial differences in their formulation.

The *cost function* in this case is expressed as:

$$C = C_{TR} \sum_{t=0}^{T-1} \sum_{i=1}^{N} d_i MV_i u^t_i + \sum_{t=0}^{T-1} \sum_{i=1}^{N} Cr_i MV_i u^t_i \tag{2.8}$$

where T is the length of the optimization horizon. Similarly, the constraints must be written for any time interval, that is

$$u_i^t \leq \alpha_i \cdot x_i^t \quad i=1,\dots,N \ t=0,\dots,T\text{-}1 \tag{2.9}$$

$$u_i^t \geq 0 \quad i=1,\dots,N \ t=0,\dots,T\text{-}1 \tag{2.10}$$

$$\frac{PCI}{3600 \cdot 8000} \sum_{i=1}^{N} MV_i u_i^t = \frac{CAP}{\eta_e} \quad t=0,\dots,T\text{-}1 \tag{2.11}$$

Besides, respect to the static model, another constraints has to be introduced: that representing the dynamics of biomass growth. To describe such a phenomenon the mathematical model adopted by Berryman (1981), Begon and Mortimer (1981), and Bernetti (1998) has been used. This model expresses the describes the dynamics of the growth as

$$\dot{x} = b_o x - b_1 x^2 \tag{2.12}$$

where values of b_0 and b_1 value are different for each type of biomass. However, the quantity of biomass in a parcel depends on the growth dynamics, as well as on the quantity of biomass collected, that is

$$\dot{x} = b_0 x - b_1 x^2 - u \tag{2.13}$$

The above continuous-time model can be discretized as follows

$$\mathbf{x}(t + \Delta t) = [1 + \Delta t b_0]\mathbf{x}(t) - b_1 \Delta t \mathbf{x}^2(t) - \Delta t \mathbf{u}(t) \tag{2.14}$$

Thus, selecting a unitary discretization interval, namely $\Delta t = 1$, and taking into account all the parcels, the following dynamic constraints arise

$$\mathbf{x}_{t+1}^i = (1 + b_{0,i})\mathbf{x}_t^i - b_{1,i}(\mathbf{x}_t^i)^2 - \mathbf{u}_t^i \tag{2.15}$$

$$t=0,\dots(T\text{-}1) \ i=1,\dots,N$$

2.3. The Case Study

The DSS has been applied to a territory within the Savona District, where it was necessary to evaluate the environmental impact and the sustainability of a plant producing only electric energy to be built in a specific area. Actually, from a purely economic point of view, a plant that produces energy from biomasses is not particularly advantageous, above all if the production is limited to electric energy. Regarding the environmental benefits, it is not easy to quantify them in monetary terms. However, it is possible to affirm that, for the considered study area, the installation of this type of activity on the territory would produce a whole series of benefits that range from the production of renewable source energy, to the constant and programmed maintenance of the territory, and to the improvement of the occupational level of the area under concern.

The study area is represented by a territory belonging to two different Mountain Communities: Val Bormida and Val Pollupice. Such territory has a high tree density index (with respect to Italian tree densities), and has some industrial activities linked to wood-use and producing scraps that can be burnt in the plant.

The first step has been the analysis of the specific areas, within the considered territory, that should be exploited to collect biomasses. Through GIS tools the parcels centroids have been identified, and different data have been associated (area, slope, biomass quantity and growth parameters). Figure 2.2 reports the communes that are included in the study. The whole area considered is partially in Val Bormida (Bardineto, Bormida, Calizzano, Mallare, Murialdo, Origlia, Pallare), and partially in Val Pollupice (Calice Ligure, Magliolo, Orco Feglino, Rialto). The overall area of the communes in the considered territory is about 32000 *hectares*. In figure 2.2a, the forest parcels in the considered territory are indicated. However, not all the forest parcels can be considered. For example in figure 2.2b the areas under legislative constraints (because of natural parks and protected areas) have been eliminated. The area of the remaining forest parcels is about 20000 *hectares*.

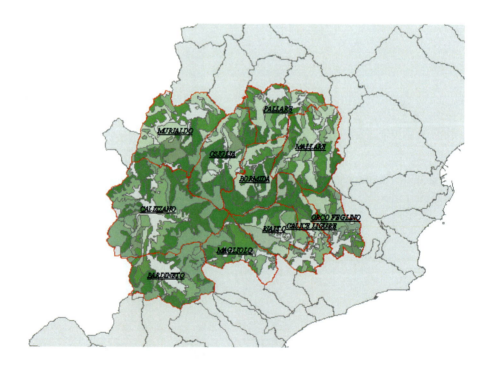

Figure 2.2a. Forest parcels present on the territory.

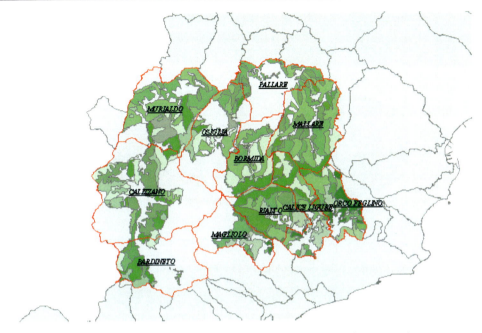

Figure 2.2b. Forest parcels present on the territory without normative constraints.

Besides, in the present work, attention wants to be given to legislation, as well as to a particular care of the natural resources and to the real possibility of biomass collection. As a consequence, the following parcels, among those represented in figure 2.2b, have been discarded:

- private properties that could not be included because of the lack of permission from the owner;
- slope greater than 50%;
- distance from major roads greater than 100 m;
- areas of bionaturalistic importance;
- areas characterized by forest fires and hydro geological risk.

The total remaining area available for biomass exploitation is equal to $5.6 \cdot 10^3$ *hectares*; this corresponds to and a biomass quantity equal to $8.6 \cdot 10^5$ m^3. The typologies of biomasses present on the territory under study are reported in figure 2.3. Other kinds of biomasses should be added as inputs of the model because of the agreements among the local industries:

- $4.4 \cdot 10^4$ $m^3 \cdot year^{-1}$ of scraps from wood industries;
- $17.6 \cdot 10^3$ $m^3 \cdot year^{-1}$ of scraps from pruning of public urban vegetation, pallets, and cratings;
- $7 \cdot 10^3$ $m^3 \cdot year^{-1}$ of chestnut from a wood trader in Val Bormida.

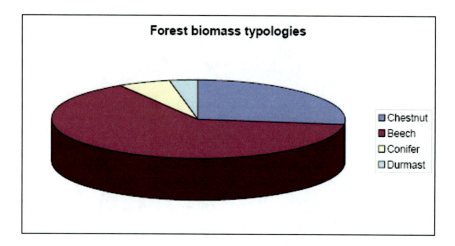

Figure 2.3. Main typologies of biomass present on the territory.

The biomass coming from the industrial scraps is used entirely (100%), while forest biomasses are used according to the legislative constraints previously discussed.

Results

The considered optimization problems have been modelled through the use of an optimization software (*Lingo 8.0*, Lindo Systems Inc.). First of all, the static optimization problem has been solved for different plant capacities. That is to say, that the optimization problem has been solved for different values of parameter *CAP*. For each run, the optimal values of the decision variables and of the objective function has been found. Incrementing *CAP*, if the available biomass is sufficient, the optimization problem is feasible, otherwise the Lingo software finds a "not feasible solution".

A sensitivity analysis of the cost function respect to the parameter that represents the plant capacity (*CAP*) is carried out. This analysis has been carried out till the solution is feasible. The solution is not feasible when there is not enough biomass to satisfy the capacity under study, i.e. the sensitivity analysis is performed till a capacity that is not sustainable is found. In the following, results for possible plant capacities (*CAP* = 1 *MW*, 2 *MW*, 3 *MW*, etc.) are reported. The DSS has furnished feasible solutions till a plant capacity of 6 *MW*, without reaching the 7 *MW* capacity. With the only aim of testing the decision model, the sensitivity analysis has been performed for values of *CAP* that do not correspond to real installed plants. It can be seen that the optimization problem is feasible till *CAP*=6.42 *MW*. Of course, CAP=6.42 MW cannot be a real plant capacity. This value should be observed only as a result of the sensitivity analysis of the proposed decision model.

In table 2.1, the costs for the plant owner (regarding biomass collection and transportation) are reported. They correspond to the optimal values of equation (4) for different values of *CAP*. It is important to note that the value of the objective function is zero for *CAP* = 1 *MW* and *CAP* = 2 *MW* because, in this case, the DSS selects only the scraps coming from the private trader, whose unit costs are set equal to zero because.

Table 2.1. Objective function values for different plant capacities

Objective function value	
MW	€
1	- 0
2	0
3	65804,8
4	401101,4
5	841893,4
6	1480792
6,42	1958196

As it is evident in table 2.1, the costs trend is increasing: a maximum value of the cost function of about $2 \cdot 10^6$ €, for the alimentation of a plant of 6,42 *MW* is reached. The optimal biomass use that corresponds to the scenarios reported in table 2.1 is reported in table 2.2. Note that, for *CAP*=1 and 2, the only biomasses used are industrial scraps, while for higher plant capacities there are different type of biomasses. Different heating values (*PCI*) corresponds to different biomass kinds. As a consequence, depending on the typology that is burnt, the same quantity of material does not involve the same energy production. The heating value for the forest biomasses is assumed to be equal to 11.3 MJ/kg, while scraps from pruning and from industries are considered to have a value of 12.5 MJ/kg and 16.7 MJ/kg, respectively. In table 2.2b, the results for the case of maximum sustainable plant capacity (*CAP*=6,42 MW) are reported in detail. Specifically, the use of forest parcels is the maximum admissible and the used percentage respect to the total forest biomass present in the territory is equal to 3.4%.

Table 2.2a. Used biomass for each plant capacities (in MW)

Used biomass	
MW	m^3
1	20530
2	41061
3	60451
4	74030
5	86418
6	98976
6,42	104105

Table 2.2b. Typologies of used biomass for CAP =6.42 MW

Used biomass	
MW	6,42
m^3	104105
% Industrial scraps	100%
% Used forest biomass	3,4%
Parcels number	373

The objective function values for different values of *CAP* are reported in figure 2.4. In particular, in the same figure, together with costs, are reported the benefits (calculated taking into account the produced energy) for the different plant capacities. From a cost-benefit analysis, comparing the costs curve with the benefits line, coming from the sales of produced energy, it comes out that a reasonable benefit is obtained for a plant with capacity between 4 and 5 *MW*. Further analysis have then been performed on a plant with a capacity of 5 *MW*. In fact, from the analysis of previous results, it can be said that the availability of forest biomass and other typologies of biomass (discards of primary workmanship of the wood and prunings of the public green) allows the installation of a plant with power equal to 5 MW.

In table 2.3 is reported the optimal solution for a plant capacity equal to 5 MW. The used forest parcels are the 50% (in number of 192) of the total forest parcels (374).

Then, in order to better qualify the obtained results and the sustainability of the plant, an evaluation of the produced emissions has been performed, and in particular of those emissions produced by the vehicles used for the biomass transport. In order to move the biomass from the single forest parcel to the plant, camions able to take 20 tons of material, with diesel engine, have been considered. Once the number of necessary vehicles has been calculated, on the basis of the distance among each parcel and the plant, it has been possible to calculate the pollutant emissions in the atmosphere due to transport operations.

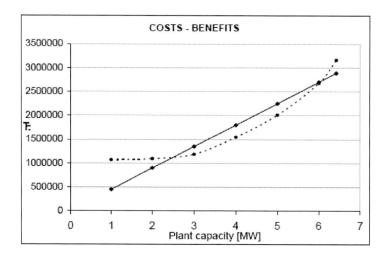

Figure. 2.4. Cost-benefits analysis for every plant capacity .

Table 2.3. Biomass exploitation scenario for CAP = 5 MW

Used biomass	
MW	5
m^3	86418
%Industrial scraps	100%
% Used forest biomass	1,7%
Parcels number	192

Finally, the dynamic decision model has been applied. From the previous results, $CAP = 5$ MW has been considered. The primary objective of this decision problem is to define the collection activity in the first five years of the plant life, highlighting the yearly costs. In Figure 2.5, the quantities of biomass to be collected in the Bardineto commune have been reported as an example. The number of parcels is here equal to six with different quantities of collected every year

The communes from which is collected most are Mallare and Calice Ligure. In figure 2.6, the collected biomass in the these communes is shown over the first five years.

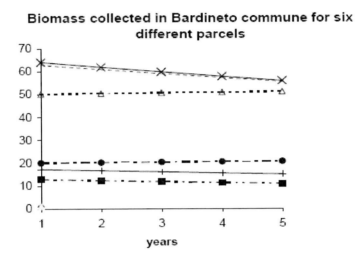

Figure 2.5. Biomass collected in the Bardineto commune over five years for a plant capacity of 5MW.

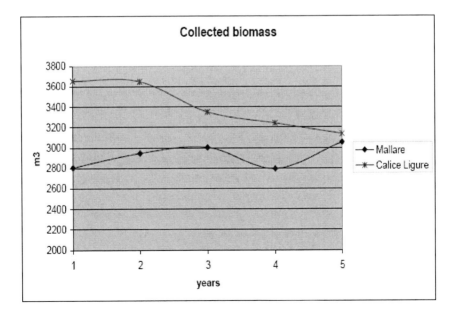

Figure 2.6. Biomass collected in the Mallare and Calice Ligure communes over five years for a plant capacity of 5MW.

2.3. System Implementation

The developed EDSS allows experts to plan the biomass use in a given region, in order to minimize costs and guarantee the environmental sustainability. In particular, the EDSS is based on three modules: the GIS-based interface, the database, and the optimization module, subdivided in strategic planning, tactical planning and operational management decisions. The users can view the territory via a GIS oriented interface. The territory is divided in parcels, characterized by a biomass type. As a first step, the users can customize the problem and create scenarios. By default, the system appoints as eligible all the parcels. However, it is possible to eliminate those parcels that can not be considered for harvesting because, for example, they are hardly reachable, environmentally protected, or the owner does not allow their use. Moreover, it is possible to add other biomass collection sites, such as for example biomass deriving from agriculture/industrial production, by inserting location, biomass quantity, and purchase cost. As regards plant typologies, it is possible to pick the plant directly on the map through the interface. Moreover, one can choose the plant typology and change parameters. After that, a procedure calculates the distance of the selected parcels from the first available road and from the plant location. All data and calculations are stored in a database. Then, the optimization procedure is called. When the optimization procedure ends, the optimal results are shown on the map. For a suitable management of the information, the data planned in the GIS module and the results deriving from the optimization module are stored in a relational database. Communication with the database is managed by a proper ODBC (Open Database Connectivity) interface, while the optimization module is called within the MS Visual Basic 6.0 program by a specific Lingo 8.0 component.

3. Logistics Aspects of Solid Waste Management in Urban Areas: Formalization of Multi-Objective Optimization Problems

The collection and proper treatment and disposal of the increasing amounts of solid waste represents a daily challenge to almost all municipalities. On the other hand, solid wastes can be seen as a resource when their treatment and management lead to material recovery and energy production.

There are two main possibilities in order to recover the produced waste: material recovery and energy recovery. Material recovery can be performed through separate collection, recycling of different materials, or through plants that separate specific materials or that transform the refuse in something that can be used. Energy production can be performed, through several kinds of plants, on different kinds of materials present in the refuse: organic material, untreated waste, treated waste, Refuse Derived Fuel, etc.

Material recovery influences energy recovery for two main reasons:

-   the quantity of material that is recovered is not dedicated to energy production;
-   every material composing the refuse has a specific heating value: the kinds of material that are recovered strongly influence the energetic efficiency of the material transformed into energy.

The main objective is to define suitable strategies that can allow proper material recovery and energy production programs. The first step for the formalization of Decision Support Systems for solid waste management is the choice of the decision variables. In the following, the decision variables necessary for MSW are reported (for further details, see Fiorucci et al. (2003)). After that, in the following sections, a discussion is reported concerning the formalization of objectives and constraints for multi-objective decision models.

3.1. Introduction

The complexity of planning a Municipal Solid Waste (MSW) management system depends on the necessity of taking simultaneously into account conflicting objectives. It is really difficult for planners to develop a sustainable approach to waste management and to integrate strategies aiming at producing the best practicable and environmentally sustainable option. To formalize these strategies, in the last two decades, considerable research efforts have been directed towards the development of optimization models for MSW flow allocation. Several examples of mathematical programming models have been developed for MSW management planning, such as, for example, in Chang and Chang [1998], Fiorucci et al. [2003], Costi et al. [2003]. The necessity of taking into account economic, technical, and normative aspects, paying particular attention to environmental problems (which usually cannot be dealt with by economic quantifications only) is more and more felt. Such a reason has led several authors to propose multi-criteria decision approaches.

The nature of the MSW management problem is a multi-criteria one, and should be treated by a suitable technique, depending on the specific case study. Multi-criteria problems (MCDM-Multi-Criteria Decision Making) are classified as follows (J. Malczewski, GIS and Multicriteria Decision Analysis):

- Multiobjective decision problems
- Multi-attribute decision problems.

The attributes are proprieties of the entities of the real world; measurable quantity or quality of a certain entity (decision objects). The objective is an indication regarding the system state that wants to be achieved (it indicates the directions of improvements of the attributes). Often the decisions are taken not by a single individual but from many decision makers. When there are many decision makers or decision groups it is necessary to distinguish between a team and a coalition. The first one has consistent preferences, while, for the second, this may not be true. Besides, there might be two types of decisions: competitive and independent. Decision problems can be referred or to easily predictable situations (deterministic, certainty situation: all information are known and there is a deterministic connection between decision and effect) or situations that are predictable in a very difficult way (uncertainty decision problem). There are two causes of uncertainty for a decision process: the validity of information, and future events that can affect the preference of different alternatives. When uncertainty is related to information only for the decisional situation, they are probabilistic or stochastic decision problems. They are fuzzy decision problems when uncertainty is related to the description of the meaning of the events, phenomena.

In this chapter, a multi-objective decision making (MODM) approach to sustainable MSW management is presented, with the aim to support the decision on the optimal flows of solid waste to be sent to landfill, recycling and to the different treatment plants. To achieve this goal, in the proposed approach the decision makers (DM) are interactively involved in the decision process, following the reference point methodology.

3.2. The MSW Decision Problem

Consider a decision framework in which a DM needs support in facing a MSW planning problem. Specifically, given a MSW configuration (that is that the number and type of plants in the MSW system is fixed a priori), the DM aims at establishing the optimal waste flows, and the plants size. The model of such a system is similar to the one in Costi et al., [2001 and 2004], where the decisional variables also include the sizes of the plants and the flows among them, but only a single objective, the economic cost, is taken into account. In the municipality, the total daily MSW production can be partitioned into eleven typologies of materials, namely, paper, plastic, plastic bags, plastic bottles, glass, organic, wood, metals, textiles, scraps, and inert matter. The structure of the overall MSW system is depicted in figure 3.1, where five types of plants are represented and the flows among them are indicated.

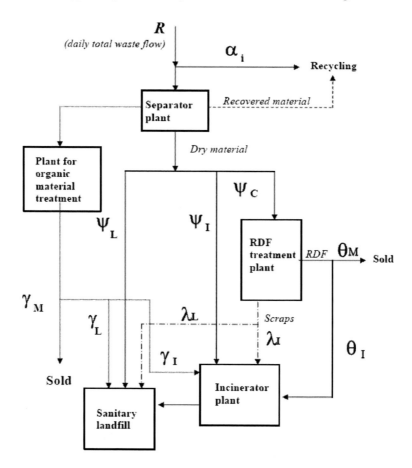

Figure 3.1. The MSW management system.

Apart from R, which represents the total daily MSW production, all symbols represent flow percentages. More specifically, for every branching point, the following convention is adopted: the symbol associated with an outgoing link represents the percentage of the flow corresponding to the unique incoming link.

The total waste flow is partly gathered (percentage α_i) by a separate collection and then sent to recycling. Note that recycling is not possible for three kinds of the above materials, that is, heavy plastics, scraps and inert matter, whereas the other eight materials can be separately collected by different methods. Besides to separate collection, material recovery is also possible by dividing the various materials in a separator plant. From such a plant, three flows may come out:

- the metals that can be sent to recycling;
- the organic material that must be sent to a treatment plant (humid material);

a fraction of material, with low humidity and high heating value (dry material), that can be burnt (percentage ψ_I), or sent to the plant for Refuse Derived Fuel (RDF) production (percentage ψ_C), or disposed in the sanitary landfill (percentage ψ_L).

The RDF plant produces fuel, which can be sold to industries (percentage θ_M) or burnt in the incinerator (percentage θ_I), and scraps, which can be sent either to the incinerator (percentage λ_I) or to the landfill (percentage λ_L).

The organic material collected for recycling can be directly sent to a composting plant because it is pure enough to produce compost for agricultural use. The humid material is treated in the organic material treatment plant, which produces Stabilized Organic Material (SOM). SOM can be sold (percentage γ_M), burnt in the incinerator (percentage γ_I), or sent to the landfill (percentage γ_L). Clearly, material recovery takes place not only through recycling but also through the various treatment plants which provide SOM, RDF and metals. Energy recovery by MSW combustion has to be taken into account as well. As recycling modifies the composition of the refuse sent to incineration, it influences the heating value of the refuse that has to be burnt, and hence energy recovery. The purpose of the DM is to determine the various flows of the different materials in the whole MSW management system in order to satisfy a number of technological and normative constraints and minimizing four main objectives: the economic cost of material treatment, the quantity of unrecycled waste, the quantity of waste sent to the landfill, and the emissions of the incinerator. In the following sections the details of the mathematical formulation and of the approach of the multi-objective decision problem (MODM) are illustrated.

3.3. The MODM Approach

In the considered context, the multi-objective problem can be in general expressed as a vector optimization problem (VOP):

$$\min_{\underline{x}\in X} \underline{F}(\underline{x})$$

(3.1)

where $F(x)=[f_j(x), j \in J=\{1,...,m\}]^T$ and X represents the feasible decision space.

The MODM approach used in this work follows the Reference Point Analysis [Wierzbicki et al., 2002], adapted to the case study, and an iterative solution for the interaction with the DMs, following the experience of the Satisficing Trade-Off Method (STOM) developed by Nakayama [Wierzbicki et al., 2002].

The kind of information required by the procedure proposed in this paper is different from the one used in STOM: the reason of this choice is the necessity of making the meaning of the evaluation quite clear for the DMs involved in the specific application context considered, so that they can easily provide the information needed.

The first step is to define the utopia solution $q^u \in Q$ for all the objectives. This can be found solving the following problems:

$$q_j^u = f_j(\underline{x}_{(j)}^u) = \min_{x \in X \cap X^a} f_j(\underline{x}), j \in J \tag{3.2}$$

where $X_a=\{f_j(x) \leq q_{ja} \ \forall j \in A\}$, being A the subset of objectives for which a level has been provided by the DM. Instead, the nadir solution can be appropriately fixed by selecting the maximum values assumed by the objectives:

$$q_j^n = \max_{h=1,...,m} f_j(\underline{x}_{(h)}^u), j \in J \tag{3.3}$$

Then, the objective functions are normalized by means of the following substitution

$$q_j \leftarrow \frac{q_j - q_j^u}{q_j^n - q_j^u}, j \in J \tag{3.4}$$

The achievement function to be maximized is:

$$\sigma(\underline{q},\underline{\overline{q}}) = \min_{j=1,...,m}(\overline{q}_j - q_j) + \varepsilon \sum_{j=1}^m (\overline{q}_j - q_j) \tag{3.5}$$

subject to the set of constraint X_{ea}. As theoretically justified in [Wierzbicki, 2002], the parameter ε can be computed as $\varepsilon=1/(M-1)$, being M a suitable upper bound on the trade-offs among the objectives. The initial efficient solution is identified by maximizing the achievement function $\sigma(\underline{q},\overline{q}^0)$ defined in (3.5), being $\overline{q}^0$ the initial aspiration levels either fixed at the utopia solution or directly provided by the DM, and q_j. The results of the maximization regard the optimal value of each objective function, to which corresponds the normalized solution that is comparable with the pre-defined reference point. Then, the DM evaluates if the levels of the objectives associated with the current solution are satisfying, and, in the affirmative case, the procedure is terminated. If none of the objective levels is satisfying the DM, the procedure can be either terminated, not being able to provide any support, or re-initialised by setting a different set of aspiration levels. Actually, in such a case it could seem appropriate to revise some of the constraints that specify X_{ea}, in particular relaxing some of the acceptability conditions. Finally, let U_k the set of indexes of the objectives whose level is considered not satisfying and Sk the complementary set of indexes of the objectives considered satisfying at the k-th iteration. The procedure aims at identifying a trade-off in an implicit way asking the DM to indicate for at least one of the objective $j \in S_k$ an increase (recall that a minimization is considered) Δq_j^k that the DM is willing to accept in order to possibly improve the objectives in U_k.

The procedure then computes a new reference point from the objective levels of the current efficient solution as $\overline{q}_j^{k+1} = \overline{q}_j^k + \Delta q_j^k \quad \forall j \in S_{kTO} \subseteq S_k$, where S_{kTO} is the set of the objectives in Sk for which the DM is willing to accept an implicit trade-off, and $\overline{q}_j^{k+1} = \overline{q}_j^k$ $\forall j \in J \backslash S_{kTO}$.

Then, a new candidate efficient and acceptable solution (x_{k+1}, q_{k+1}) is found by maximizing the order consistent achievement function as follows

$$\max_{x \in X^k} \sigma(\underline{q}, \overline{\underline{q}}^{k+1}) \qquad (3.6)$$

being $X_k = X_{k-1} \cup \{q_j \leq \overline{q}_j^k + \Delta q_j^k, \forall j \in S_{kTO}\}$. In this way the new reference points are taken into account in (3.6) and a relevant set of new constraints are added which impose the maximum worsening level accepted by the DM. This interaction continues till the DM is satisfied for all the objectives. As recent approaches to MODM have pointed out [Wierzbicki et al., 2002], information provided during the decision making process (also called "progressive" information), generally lead to identify decisions that are easily recognized to be consistent with the DM's preference and then finally accepted. In addition, the use of progressive information does not require that the DM expresses definitive and accurate preference judgements only once, but lets the DM free to revise the preference at each step of the decision process, taking into account the current solution point at which the judgements previously provided have led to.

3.4. The Formalization of the MODM Decision Problem

The primary decision variables correspond to the flows of materials and represent the components of the decision vector x. The following decision variables are considered: α_i $(i = 1,...,11)$, ψ_C, ψ_I, ψ_L, λ_L, λ_I θ_M, θ_I $\gamma_L, \gamma_I, \gamma_M$ (see figure 3.1).

3.4.1.Objectives

Four objective functions are considered: minimizing economic costs, minimizing unrecycled waste, minimizing waste sent to landfill, and minimizing incinerator emissions. For brevity, the complete formalization of these functions is not reported. Further information can be found in [Fiorucci et al., 2003, Costi et al., 2004]. The first objective function fl(x) is related to economic costs. Three main components are assumed for f1(x), that are, recycling cost $C^r(\underline{x})$, maintenance costs $C^g(\underline{x})$, and benefits $B(\underline{x})$ related to either energy or RDF production, leading to the following expression:

$$f_1(\underline{x}) = C^r(\underline{x}) + C^g(\underline{x}) - B(\underline{x}) \tag{3.7}$$

All these costs are function of the previously defined decision variables.

Unrecycled material, in this model, is simply the total waste produced R minus the waste separately collected, namely:

$$f_2(\underline{x}) = R - \sum_{i=1}^{9} \alpha_i r_i \tag{3.8}$$

The quantity of waste per year sent to the landfill (called $\overline{Q}_5(\underline{x})$ in this work) is function of the decision variables (and of the different parameters that characterize each treatment plant efficiency). The third objective is:

$$f_3(\underline{x}) = \overline{Q}_5(\underline{x}) \tag{3.9}$$

Finally, emission concentrations and quantities depend on the chemical reactions, which take place among the various elements present in the entering refuse. Every material present in the wastehas a specific percentage of S, Cl, C, N, O, H, F, that can give the following compounds: CO_2, H_2O, HCl, O_2, N_2, SO_2, HF. The quantities produced depend on the mole numbers, on the flows entering the incinerator plant, and on the efficiency of exhaust gas treatment. In the proposed approach, only HCl emissions have been taken into account.

The fourth objective is:

$$f_4(\underline{x}) = M_{HCl}(\underline{x}) \tag{3.10}$$

where $M_{HCl}(\underline{x})$ is the overall amount of chlorine entering daily the incinerator plant.

3.4.2.Constraints

Different classes of constraints have been included in the formalization of the mathematical optimization problem: minimum recycling constraints, treatment plants' size constraints, flow conservation constraints, RDF and SOM composition constraints, incineration emissions constraints, landfill saturation constraints.

3.5. The Case Study

The model proposed in this paper has been applied to a case study concerning the municipality of Genova where refuse disposal is a very critical problem. With a daily waste production of 1355 t, the current solution is the disposal in a unique landfill, whose residual capacity is rapidly decreasing. For the sake of brevity the data relevant to the case under concern are not reported here, but they can be found in Fiorucci et al. [2003], where the MSW management problem was faced without introducing a multi-objective formulation. The preliminary step that must be performed in order to apply the MODM approach to the MSW case study is to identify for each objective function fj both the utopia qju, and the nadir qjn solutions and to normalise the function with respect to the interval [qju, qjn]. Table 3.1 reports the four objective functions considered together with their dimensions and the computed utopia and nadir solutions.

The case has been analyzed by two different decision makers, DM1 and DM2, showing different attitudes in selecting the initial reference solution and in interacting with the methodology. The first decision maker, DM1, is not able to initially identify a feasible satisfying reference point. So, DM1 simply accepts to start the method from the (unfeasible) utopia point. Then, the method computes the first solution from this reference and presents it to DM1.

Table 3.1. Utopia and Nadir computation

f_j	Dim.	q_{ju}	q_{jn}
f1	M€	45.732	64.027
f2	Tons	376.616	880.750
f3	Tons	0.020	0.100
f4	Mg/m3	3.392	10.000

Table 3.2. Iteration sequence for DM1

k	Reference Point				Objective Values			
	$\bar{q}_1$	$\bar{q}_2$	$\bar{q}_3$	$\bar{q}_4$	q_1	q_2	q_3	q_4
1	0	0	0	0	0.34	0.34	0.34	0.16
2	0.5	0	0	0	0.65	0.15	0.05	0.00
3	0.5	0.45	0	0	0.56	0.23	0.06	0.00
4	0.5	0.45	0.06	0	0.52	0.30	0.08	0.00

Table 3.2 reports the iteration sequence characterizing DM1. The first column of the table denotes the iteration, the other eight columns respectively reports first the references used and the objective values obtained by the method from such references. Note that the values reported for both the references and the objective have been normalized with respect to the interval [qju, qjn] for j=1,...,4.

After having analyzed the results obtained in the first iteration, DM1 is willing to accept a worsening for the satisfying objective 1, accepting costs that are in the middle between the normalized utopia and nadir ($\bar{q}_1 = 0.5$, see table 2), with the aim to achieve a possible improvement in at least one of the not satisfying objectives. Then, DM1 proceeds with the other iterations as summarized in table 2. The objective values obtained at the iteration 2 are quite good for q_j, j=2,3,4, but the projection on the efficient frontier provided by minimizing $\sigma(\underline{q},\bar{q})$ leads to a cost value that is considered too high. Then, instead of reducing the reference variation introduced for the objective 1, i.e., performing again the first iteration,

DM1 tries to exploit the improved consciousness about the objective levels that can be actually achieved, and fixes a new reference variation for the objective 2. This kind of behaviour continues until the iteration 4. As a matter of fact, this last iteration is considered a worsening by DM1 and, also in view of other steps, DM1 feels quite satisfied with the solution given by iteration 4. To transform the subjective satisfaction into an objective evaluation of the quality of the solution is quite a hard task in a multi-objective problem. In the proposed decision problem, where the objectives are normalized, a star coordinate system representation may help both the DM to view the solution with respect to the reference, the nadir and the utopia points, and the DSS specialist to have a more objective evaluation of the quality of results. Figure 2 shows the solution at the iteration 4 and the related reference point; the nadir coordinates are the end points of the star, whereas the utopia ones are in the centre of the star. The solution can be assessed as adequate also since it is almost included in the area delimited by the convex hull of the reference points and because its objectives are lower than the mean of the solutions obtained.

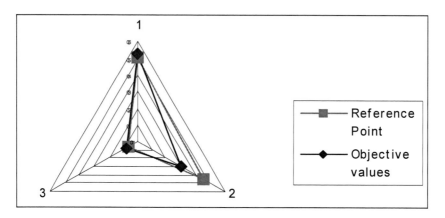

Figure 3.2. The star representation of the result of the iteration process for DM1.

A similar process characterizes the iterations performed by DM2. However in this case, the behavior of DM2 in the decision process has some important differences with respect to DM1. While DM1 has started from the unfeasible utopia as reference point, DM2, who shows some awareness about the possible outcomes that one can expect for the considered MSW management problem, starts from a reference point that, satisfying all the problem constraints, results to be feasible. The attitude by DM2 in the iterations of decision algorithm is different with respect to DM1. While DM1 relaxed the reference point according to ideas suggested from the solution obtained in the current iteration, DM2 needs to firm up on improving the amount of recycled waste (objective q_2), taking into account the effects on the other objectives.

The iteration of the decision process is summarized by the values in the table 3.3.

Moving from iteration 1 to 2, q_1 and q_3 remain the same while q_2 is lowered and q_4 is increased. From iteration 2 to 3, q_1 rises while the other objectives decrease. However, DM2 is not very satisfied by q_1 and wants to lower it. At iteration 4, q_1 is lowered, and q_2 and q_3 increase a bit. DM2 is satisfied by this iteration. Figure 3.3 reports the star representation for the result of the DM2 iteration process.

Table 3.3. Iteration sequence for DM2

k	Reference Point				Objective Values			
	$\bar{q}_1$	$\bar{q}_2$	$\bar{q}_3$	$\bar{q}_4$	q_1	q_2	q_3	q_4
1	0.50	0.50	0.03	0.10	0.54	0.26	0.07	0.00
2	0.50	0.20	0.03	0.10	0.54	0.24	0.07	0.04
3	0.50	0.10	0.03	0.10	0.59	0.19	0.06	0.00
4	0.45	0.10	0.03	0.10	0.56	0.21	0.08	0.00

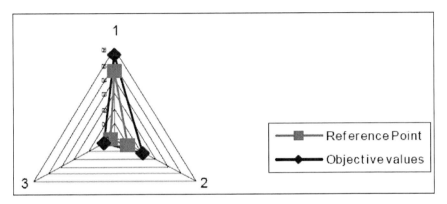

Figure 3.3. The star representation of the result of the iteration process for DM2.

To reach the final decision, DM1 and DM2 should discuss their two solutions in order to agree upon a compromise. To solve the decision the following questions have to be answered: is it worthwhile to spend additional 0.54M€ a year (moving costs from 55.18M€ to 55.56€, corresponding to the gap from q_1=0.52 for DM1 to q_1=0.56 for DM2), in order to reduce of 25.21 tons per year the amount of unrecycled waste (moving from 525.7€ to 501.8 tons per year corresponding to the gap from q_2=0.30 for DM1 to q_2=0.21 for DM2)? Which solution is most feasible and robust according to the decisional variables?

4. Conclusion

The complexity of environmental problems requires the development and application of new tools capable of processing not only numerical aspects, but also experience from experts and wide public participation, which are all needed in decision-making processes. Environmental decision support systems (EDSSs) are among the most promising approaches to confront this complexity (Poch, 2004). A wide range of EDSS have been developed in the last two decades, and there are many more currently under development. Rizzoli and Young (1999) seeks to provide a framework within which the different EDSS can be placed and compared according to the approaches used in their development.

The role of Decision Support Systems (DSS) for environmental systems planning and management, with specific attention to the use of forest biomasses for energy production, and energy and material recovery from the planning and management of urban solid waste is here investigated. The two mentioned issues are treated in two different sections to highlight the legislative and technological differences beyond the "treatment" and the "possibilities of use" of biomasses and solid wastes. The chapter is divided in two main sections: energy production from renewable resources, and material and energy recovery from solid urban waste. The first one is focused on forest biomass exploitation and includes a detailed mathematical description of the formalized decision model and its solution applied to a real case study (Savona, Italy). The formalized model is the generalization (resulted from new research activity and modelling efforts, with new results) of a model developed by the same authors (Freppaz et al., 2004). Attention is paid to the forest ecosystem, to the necessity of preserving satisfying hydrogeological conditions, and to social and economical issues. The second one regards a more qualitative discussion about the role of Decision Support Systems (mainly based on the DSS we have been developing, and, partly, a review of two previous works (Fiorucci et al., 2003; Costi et al. 2004)) for solid waste management in urban areas. In this case, the structure of DSSs able to help decision makers of a municipality in the development of incineration, disposal, treatment and recycling integrated programs, taking into account all possible economic costs, technical, normative, and environmental issues, is described. Future developments of the presented DSSs certainly regard the quantification of probabilistic and uncertainty aspects, and the application of multi-objective procedures and systematic sensitivity analysis. Then, regarding forest biomass exploitation, the issues that need a deeper study, with respect to the present state of the DSS implementation, are: a detailed study about biomass growth and productivity as a function of trees age; a deeper analysis of CO_2 adsorption, the optimization model regarding the logistic of biomass collection and transporting, the location of the "wood collection points", the wood inventory management, the management of energy and material flows when drying plants are used before combustion, the possibility to use short rotation techniques, and the evaluation of the material flows, produced emissions and costs in connection with the use of different types of plants (combustion, pyrolisis, gasification). Instead, as regards the decision model for solid waste management, possible improvements are mainly relevant to the evaluation of the emissions into atmosphere. Finally, dynamic features related to waste generation, logistics aspects, and impacts from landfill should be conveniently modelled..

References

Alhumod, J (2005). Municipal solid waste recycling in the Gulf Co-operation Council states. *Resources, Conservation and Recycling,* **45** (2), p. 142-158

G. D'Antonio, M. Fabbricino (1998) Soluzione ottimale della adozione di uno schema integrato di gestione dei rifiuti solidi urban. *RS Rifiuti Solidi*, vol. XII, No 4, July-August, p.230-240.

Badran, M:F., El-Haggar, S.M. (2005) .Optimization of municipal solid waste management in Port Said – Egypt *Waste Management in press*

Basosi, R., Maltagliati, S., Vannuccini, L (1999). Potentialities development of renewable energy sources in an integrated regional system: Tuscany. *Renewable Energy*, **16**(1-4), p. 1167-1173.

Begon, M., Mortimer M. (1981). *Population Ecology*. Blackwell, Oxford.

Benayoun, R., de Montgolfier, J., Tergny, J. ,Larichev, O.I. (1971). Linear programming with multiple objective functions: STEP method (STEM), *Mathematical Programming* **1** 366-375.

Bernetti, I. Il mercato delle biomasse per scopi energetici: un modello di offerta, *Rivista di Economia Agraria*, anno LIII n. 3.

Berryman, A. (1981). Population systems: a general introduction Y.H. Chang, Ni-Bin Chang (1998) Optimization analysis for the development of short team solid waste management strategies using presorting process prior to incinerators. *Resources, Conservation and Recycling* **24**, 7-32

Broman, PG. (2000). Environmental Management System (EMS) - A tool for the holistic improvement of environmental. *TAPPI Proceedings - Environmental Conference and Exhibit*, v 1, *Setting the Environmental Course of 21st Century*, p 75-79

Chang, Y.H., and N. Chang, Optimization analysis for the development of short team solid waste management strategies using presorting process prior to incinerators, *Resources, Conservation and Recycling* **24** 7-32, 1998.

Chang, N.B., Chen, Y.L., Wang, S.F., 1997. A fuzzy interval multi-objective mixed integer programming approach for the optimal planning of solid waste management systems. *Fuzzy sets and systems*, **89** 35-60.

N. Chang, E.Davila, B. Dyson, R. Brown (2005) Optimal design for sustainable development of a material recovery facility in a fast-growing urban setting. *Waste Management* **25**, p. 833-846

N. Chang, S.F.Wang (1997) A fuzzy goal programming approach for the optimal planning of metropolitan solid waste management systems. *European Journal of Operational Research* **99**, p. 303-321

Chang, N.B.,Wei,Y.L. (2000). Siting recycling drop-off stations in urban area by genetic algorithm-based fuzzy multi-objective nonlinear integer programming, *Fuzzy sets and systems*,**114** p. 133-149.

Clarke, M.J., Maantay, J.A. (2006). Optimizing recycling in all of New York City's neighborhoods: Using GIS to develop the REAP index for improved recycling education, awareness, and participation. *Resources Conservation and Recycling*, p.128-148

Costi, P., Fiorucci, P., Minciardi, R., Robba, M., Sacile, R. (2001). Optimizing solid waste management in metropolitan areas, IFAC Workshop on Modeling and Control in Environmental Issues, Yokohama (Japan), p.181-186

Costi, P., Minciardi, R., Robba, M., Rovatti, M., and R. Sacile, A comprehensive decisional model for the development of environmentally sustainable programs for urban solid waste management, *Waste Management*, **24**, pp. 277-295, 2004.

Daskalopoulos, E., Badr, O., Probert, S.D. (1998). An integrated approach to municipal solid waste management. *Resources, Conservation and Recycling*, **24** p. 33-50

Del Brio JA (2003). Influence of the perception of the external environmental pressures on obtaining the ISO 14001 standard in Spanish industrial companies. *International Journal of Production Research*, vol. 41, n 2, p 337-348

Denzer, R (2005). Generic integration ef environmental decision support systems-state of the art. *Environmental Modelling and Software*, **20** (10), p. 1217-1223

Ediger V.S, and Kentel E. (1999). Renewable energy potential as an alternative to fossil fuels in Turkey. *Energy Conversion and Management*, **40**(7), 743-755.

P.C. Faaij Andre Energy, Bio-energy in Europe: changing technology choices, *Policy* **34** (2006) 322–342

Faaij A., Meuleman B., Turkenburg W., Wijk A.V., Bauen A., Rosillo-Calle F. and Hall D. (1998). Externalities of biomass based electricity production compared with power generation from coal in the Netherlands, *Biomass and Bioenergy*, **14** (2), p. 125-147

Fiorucci, P., Minciardi R., Robba, M., Sacile, R. (2003). Solid waste management in urban areas: development and application of a decision support system. *Resources Conservation and Recycling* **37** (4) 301-328

D. Freppaz, R. Minciardi, M. Robba, M. Rovatti, R. Sacile, A.C. Taramasso (2004). Optimizing forest biomass exploitation for energy supply at a regional level. *Biomass and Bioenergy*, **26** (1), p. 15-25;

Regione Liguria (Assessorato all'energia), Enea, Comunità montane Alta Val Bormida e Alta Valle Scrivia (1997) – *Piano locale per lo sviluppo e la promozione dell'uso energetico delle fonti rinnovabili* – BIOMASSE;

Gardiner, L.R., Steuer, R.E. (1994). Unified interactive multiple objective programming. *European Journal of Operational Research*, **74** p. 391-406

Ghose, M.K., Dikshit A.K, Sharma, S.K (2000). A GIS based transportation model for solid waste disposal – A case study on Asansol municipality. *Waste Management, in press*

Graham RL, English BC, Noon CE (2000). A Geographic Information System-based modelling system for evaluating the cost of delivered energy crop feedstock. *Biomass and Bioenergy*, p. 309-329.

Hall DO, Scrase JI (1998). Will biomass be the environmentally friendly fuel of the future? *Biomass and Bioenergy*, **15**(6), p. 451-456.

Hillring B (2002). Rural development and bioenergy experiences from 20 years of development in Sweden. *Biomass and Bioenergy,* Volume **23** (6), *p.* 443-451

Hokkanen, J., Salminem, P. (1997). Choosing a solid waste management system using multicriteria decision analysis. *European Journal of Operational Research*, **98** p. 19-36.

K. Holmgren, D. Henning, (2004). Comparison between material and energy of municipal waste from an energy perspective. A study of two Swedish municipalities. *Resources, Conservation and Recycling* **43**, p. 51–73.

Hoogwijk, M., Faaij, A., Van den Broek, R., Berndes, G., Gielen, D., Turkenburg, W. (2003). Exploration of the ranges of the global potential of biomass for energy. *Biomass and Bioenergy*, **25** (2), p. 119–133.

Kuiper L.C., Sikkema R., Stolp J.A.N. (1998). Establishment needs for short rotation forestry in the EU to meet the goals of the Commission's White Paper on renewable energy. *Biomass and Bioenergy*, **15**(4-5), p. 375-367.

Marakas, G.M. (1999). Decision Support Systems in the Twenty-First Century, Prentice-Hall, Inc. NJ.

Martinot E. (1998). Energy efficiency and renewable energy in Russia-Transaction barriers, market intermediation, and capacity building. *Energy Policy*, **26**(11), p. 905-915.

Matthies, M., Giupponi, C., Ostendorf, B. (2005). Environmental decision support systems: current issues, methods and tools. *Environmental Modelling and Software, in press*

Miettinen, K.M. (1999). Nonlinear multiobjective optimization, Kluwer Academic Publisher.

Mitchell C.P. (1995) "New cultural treatments and yield optimisation", *Biomass and Bioenergy*,**9**, p.11-34.

Mitchell CP. (2000) Development of decision support system for bioenergy applications. *Biomass and Bioenergy*, p. 265-278.

Nagel J. (2000) Determination of an economic energy supply structure based on biomass using a mixed-integer linear optimisation model. *Ecological Engineering*, p. S91-S102.

Nagel J. (2000) Biomass in energy supply, especially in the state of Brandeburg, Germany. *Ecological Engineering*, **16** p. S103-S110.

Ville Niutanen, Jouni Korhonen (2003). Towards a regional management system - waste management scenarios in the Satakunta Region, Finland, *International Journal of Environmental Technology and Management*, Vol. 3 (2), p. 131 - 156

Noon C.E. and Daly M.J. (1996). GIS-based resource assessment with BRAVO. *Biomass and Bioenergy*, vol.10 (2-3), p. 101-109.

Pari L (2001). Energy production from biomass: the case of Italy. *Renewable Energy*, **22**, p. 21-30.

Poch, M., Comas, J., Rodriguez-Roda, Ignasi; Sanchez-Marre, Miquel; Cortes, Ulises (2004). Designing and building real environmental decision support systems. *Environmental Modelling and Software*, **19** (9), p 857-873

Quesada, I., Grossmann, I.E. (1995). Global optimisation of bilinear process networks with multicomponent flows. *Computers and Chemical Engineering*, **19** (12), p. 1219-1242.

Reeves, G.R., Hedin, S.R. (1993). A generalized interactive goal programming procedure. *Computers and Operations Research*, 20, p. 747-753.

Rizzoli A.E. and Young W.J. (1997). Delivering environmental decision support systems: software tools and techniques. *Environmental Modelling and Software*, vol.12, n2-3, , pp.237-249.

Salvia, M., Cosmi, C., Macchiato, M., Mangiamele, L. (2002). Waste management system optimization for Southern Italy with MARKAL model. *Resources, Conservation and Recycling,* Volume 34 (2), p. 91-106

Sarkis, J. (2000). A comparative analysis of DEA as a discrete alternative multiple criteria decision tool. *European Journal of Operational Research*,123 p. 543-557.

Shekdar, A.V., Mistry, P.B. (2001). Evaluation of multifarious solid waste management systems-A goal programming approach, *Waste Management and Research*,19, p. 391-402

Tchobanoglous G., Theisen H., Samuel A.V (1993). Integrated solid waste management: *Engineering principles and management issues*, Mc Graw Hill, p. 978

Ushiyma I. (1999). Renewable energy in Japan. *Renewables Energy*, **16**(1-4), 1174-1179.

Van Kooten, G. Cornelis (2005). Certification of sustainable forest management practices: A global perspective on why countries, *Policy and Economics*, v 7 (6), p 857-867

Varela M., Lechón Y., Sáez R. (1999). Environmental and socioeconomic aspects in the strategic analysis of a biomass power plant integration. *Biomass and Bioenergy*, **17** p. 405-413.

Voivontas D, Assimacopoulos D, Koukios EG (2001). Assessment of biomass potential for power production: a GIS based method. *Biomass and Bioenergy*, p. 101-112.

Wenk, M.S. (2004). The European Union's Eco-Management and Audit Scheme (EMAS): An analysis of the development process *Proceedings of the A and WMA's 97th Annual Conference and Exhibition*; Sustainable Development: Gearing up for the Challenge, p. 4419-4426 ISSN: 1052-6102 , Publisher: Air and Waste Management Association, Pittsburgh, PA 15222, United States

Wierzbicki, A.P., Makoski, M.,Wessels, J. (2002). *Model-based decision support methodology with environmental applications*, Kluwer Academic Publishers

In: Energy Research Developments
Editors: K.F. Johnson and T.R. Veliotti

ISBN: 978-1-60692-680-2
© 2009 Nova Science Publishers, Inc.

Chapter 10

NORTH AMERICAN OIL SANDS: HISTORY OF DEVELOPMENT, PROSPECTS FOR THE FUTURE[*]

Marc Humphries

Energy Policy
Resources, Science, and Industry Division

Abstract

When it comes to future reliable oil supplies, Canada's oil sands will likely account for a greater share of U.S. oil imports. Oil sands account for about 46% of Canada's total oil production and oil sands production is increasing as conventional oil production declines. Since 2004, when a substantial portion of Canada's oil sands were deemed economic, Canada, with about 175 billion barrels of proved oil sands reserves, has ranked second behind Saudi Arabia in oil reserves. Canadian crude oil exports were about 1.82 million barrels per day (mbd) in 2006, of which 1.8 mbd or 99% went to the United States. Canadian crude oil accounts for about 18% of U.S. net imports and about 12% of all U.S. crude oil supply.

Oil sands, a mixture of sand, bitumen (a heavy crude that does not flow naturally), and water, can be mined or the oil can be extracted in-situ using thermal recovery techniques. Typically, oil sands contain about 75% inorganic matter, 10% bitumen, 10% silt and clay, and 5% water. Oil sand is sold in two forms: (1) as a raw bitumen that must be blended with a diluent for transport and (2) as a synthetic crude oil (SCO) after being upgraded to constitute a light crude. Bitumen is a thick tar-like substance that must be upgraded by adding hydrogen or removing some of the carbon.

Exploitation of oil sands in Canada began in 1967, after decades of research and development that began in the early 1900s. The Alberta Research Council (ARC), established by the provincial government in 1921, supported early research on separating bitumen from the sand and other materials. Demonstration projects continued through the 1940s and 1950s. The Great Canadian Oil Sands company (GCOS), established by U.S.-based Sunoco, later renamed Suncor, began commercial production in 1967 at 12,000 barrels per day.

The U.S. experience with oil sands has been much different. The U.S. government collaborated with several major oil companies as early as the 1930s to demonstrate mining of and in-situ production from U.S. oil sand deposits. However, a number of obstacles, including the remote and difficult topography, scattered deposits, and lack of water, have resulted in an uneconomic oil resource base. Only modest amounts are being produced in Utah and

---

[*] Excerpted from GAO Report RL34258, dated December 11, 2007.

California. U.S. oil sands would likely require significant R&D and capital investment over many years to be commercially viable. An issue for Congress might be the level of R&D investment in oil sands over the long-term.

As oil sands production in Canada is predicted to increase to 2.8 million barrels per day by 2015, environmental issues are a cause for concern. Air quality, land use, and water availability are all impacted. Socio-economic issues such as housing, skilled labor, traffic, and aboriginal concerns may also become a constraint on growth. Additionally, a royalty regime favorable to the industry has recently been modified to increase revenue to the Alberta government. However, despite these issues and potential constraints, investment in Canadian oil sands will likely continue to be an energy supply strategy for the major oil companies.

Acronyms and Abbreviations

AEUB	Alberta Energy and Utility Board
API	American Petroleum Institute
ARC	Alberta Research Council
ARCO	Atlantic Richfield Company
CCA	Capital Cost Allowance
CONRAD	Canadian Oil Sands Network for Research and Development
COS	Canadian Oil Sands
CSS	Cyclic Steam Stimulator
EIA	Energy Information Administration
GCOS	Great Canadian Oil Sands Company
GHG	greenhouse gases
IEA	International Energy Agency
mbd	million barrels per day
NEB	National Energy Board
OPEC	Organization of Petroleum Exporting Countries
PADD	Petroleum Administration for Defense District
R&D	research and development
ROI	return on investment
SAGD	steam-assisted gravity drainage
SCO	synthetic crude oil
SIRCA	Scientific and Industrial Research Council of Alberta
USGS	United States Geological Survey
VAPEX	Vapor Extraction Process

Introduction

Current world oil reserves are estimated at 1.292 trillion barrels. The Middle East accounts for 58% of world oil reserves, and the Organization of Petroleum Exporting

Countries (OPEC) accounts for 70%. The Middle East also leads in reserve growth and undiscovered potential, according to the Energy Information Administration (EIA).[1]

The United States' total oil reserves are estimated at 22.7 billion barrels, a scant 1.8% of the world's total (see Appendix A). U.S. crude oil production is expected to fall from 5.4 million barrels per day (mbd) in 2004 to 4.6 mbd in 2030, while demand edges up at just over 1% annually. Net imports of petroleum are estimated by the EIA to increase from 12.1 mbd (58% of U.S. consumption) to 17.2 mbd (62% of U.S. consumption) over the same time period.[2]

When it comes to future reliable oil supplies, Canadian oil sands will likely account for a larger share of U.S. oil imports. Oil sands account for about 46% of Canada's total oil production, and oil sand production is increasing as conventional oil production declines. Since 2004, when a substantial portion of Canada's oil sands were deemed economic, Canada has been ranked second behind Saudi Arabia in oil reserves. Canadian crude oil exports were about 1.82 million barrels per day in 2006, of which 1.8 mbd or 99% went to the United States. Canadian crude oil accounts for about 18% of U.S. net imports and about 12% of all U.S. crude oil supply.

An infrastructure to produce oil, upgrade, refine, and transport it from Canadian oil sand reserves to the United States is already in place. Oil sands production is expected to rise from its current level of 1.2 (mbd) to 2.8 mbd by 2015. However, infrastructure expansions and skilled labor are necessary to significantly increase the flow of oil from Canada. For example, many refineries are optimized to refine only specific types of crude oil and may not process bitumen from oil sands. One issue likely to be contentious is the regulatory permitting of any new refinery capacity because of environmental concerns such as water pollution and emissions of greenhouse gases.

Challenges such as higher energy costs, infrastructure requirements, and the environment, may slow the growth of the industry. For example, high capital and energy input costs have made some projects less economically viable despite recent high oil prices. Canada ratified the Kyoto Protocol in 2002, which bound Canada to reducing its greenhouse gas (GHG) emissions significantly by 2012 but according to the government of Canada they will not meet their Kyoto air emission goals by 2012. The Pembina Institute reports that the oil sands industry accounts for the largest share of GHG emissions growth in Canada.[3]

Major U.S. oil companies (Sunoco, Exxon/Mobil, Conoco Phillips, and Chevron) continue to make significant financial commitments to develop Canada's oil sand resources. Taken together, these companies have already committed several billion dollars for oil sands, with some projects already operating, and others still in the planning stages. Many of these same firms, with the U.S. government, did a considerable amount of exploration and development on "tar sands" in the United States, conducting several pilot projects. These U.S. pilot projects did not prove to be commercially viable for oil production and have since been abandoned. Because of the disappointing results in the United States and the expansive reserves in Canada, the technical expertise and financial resources for oil sands development has shifted almost exclusively to Canada and are likely to stay in Canada for the foreseeable future. However, with current oil prices above $60 per barrel and the possibility of sustained high prices, some oil sand experts want to re-evaluate the commercial prospects of U.S. oil sands, particularly in Utah.

This CRS report examines the oil sands resource base in the world, the history of oil sands development in the United States and Canada, oil sand production, technology, development, and production costs, and the environmental and social impacts. The role of government — including direct financial support, and tax and royalty incentives — is also assessed.

World Oil Sands Reserves and Resources[4]

Over 80% of the earth's technically recoverable natural bitumen (oil sands) lies in North America, according to the U.S. Geological Survey (USGS) (see Appendix B). Canadian oil sands account for about 14% of world oil reserves and about 11% of the world's technically recoverable oil resources.

What Are Oil Sands?

Oil sands (also called tar sands) are mixtures of organic matter, quartz sand, bitumen, and water that can either be mined or extracted in-situ[5] using thermal recovery techniques. Typically, oil sands contain about 75% inorganic matter, 10% bitumen, 10% silt and clay, and 5% water.[6] Bitumen is a heavy crude that does not flow naturally because of its low API[7] (less than 10 degrees) and high sulfur content. The bitumen has high density, high viscosity, and high metal concentration. There is also a high carbon-to-hydrogen molecule count (i.e. oil sands are low in hydrogen). This thick, black, tar-like substance must be upgraded with an injection of hydrogen or by the removal of some of the carbon before it can be processed.

Oil sand products are sold in two forms: (1) as a raw bitumen that must be blended with a diluent[8] (becoming a bit-blend) for transport and (2) as a synthetic crude oil (SCO) after being upgraded to constitute a light crude. The diluent used for blending is less viscous and often a by-product of natural gas, e.g., a natural gas condensate. The specifications for the bit blend (heavy oil) are 21.5 API and a 3.3% sulfur content and for the SCO (light oil) are 36 API and a 0.015% sulfur content.[9]

U.S. Oil Sand Resources

The USGS, in collaboration with the U.S. Bureau of Mines, concluded in a 1984 study that 53.7 billion barrels (21.6 billion measured plus 32.1 billion speculative) of oil sands could be identified in the United States. An estimated 11 billion barrels of those oil sands could be recoverable. Thirty-three major deposits each contain an estimated 100 million barrels or more. Fifteen percent were considered mineable and 85% would require in-situ production. Some of the largest measured U.S. oil sand deposits exist in Utah and Texas. There are smaller deposits located in Kentucky, Alabama, and California. Most of the deposits are scattered throughout the various states listed above. As of the 1980s, none of these deposits were economically recoverable for oil supply. They are still not classified as reserves (see figure 1).

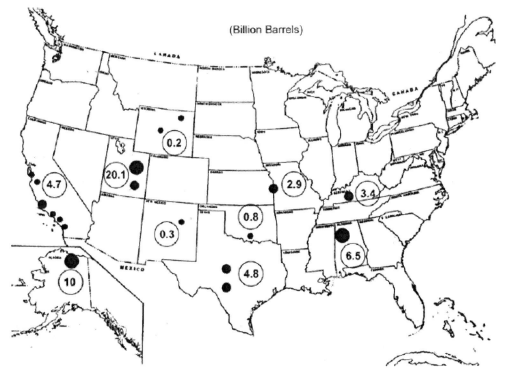

Source: Major Tar Sand and Heavy Oil Deposits of the United States, Interstate Oil Compact Commission, 1984, p. 2.

Figure 1. Tar (Oil) Sand Deposits of the United States.

Canadian Oil Sand Resources

Canadian oil sand resources are located almost entirely in the province of Alberta. The Alberta Energy and Utility Board (AEUB) estimates that there are 1.6 trillion barrels of oil sands in place, of which 11% are recoverable (175 billion barrels) under current economic conditions (see table 1). Mineable reserves at the surface account for 35 billion barrels (20%) and in-situ reserves at 141 billion barrels (80%). The AEUB estimates that the ultimate amount to be discovered (ultimate volume-in place) is 2.5 trillion barrels: about 2.4 trillion in-situ and 140 billion surface-mineable. Of this ultimate discovered amount, about 314 billion barrels are expected to be recovered (175 billion barrels in reserves now and another 143 billion barrels anticipated. See table 1).[10] However, EIA estimates only 45.1 billion barrels (reserve growth and undiscovered potential) to be added to Canada's reserve base by 2025.[11]

Oil sands occur primarily in three areas of Alberta: Peace River, Athabasca, and Cold Lake (see figure 2 below). Current production is 1.1 million barrels per day and is expected to reach 2.0 mbd by 2010 and 3.0 mbd by 2015.[12] According to the International Energy Agency (IEA), Canada's oil sands production could exceed 5.0 mbd by 2033 but would require at least $90 billion in investment.[13]

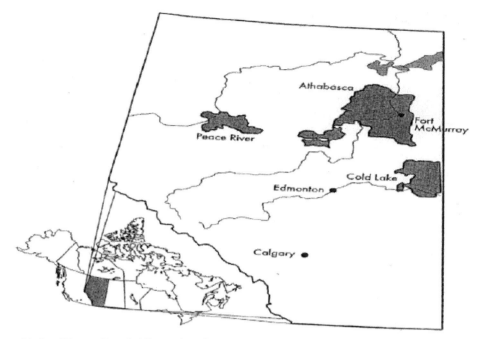

Source: National Energy Board, Alberta, Canada.

Figure 2. Oil Sands Areas in Alberta, Canada.

Table 1. Canada's Bitumen Resources

Billion Barrels	Ultimate Volume in Place	Initial Volume in Place	Ultimate Recoverable Volume	Initial Established Reserves	Cumulative Production	Remaining Established Reserves
Mineable						
Athabasca	138.0	113.0	69.0	35.0	2.5	32.7
In Situ						
Athabasca	N/A	1,188.0	N/A	N/A	N/A	N/A
Cold Lake	N/A	201.0	N/A	N/A	N/A	N/A
Peace River	N/A	129.0	N/A	N/A	N/A	N/A
Subtotal	2,378.0	1,518.0	245.0	142.8	1.26	141.5
Total	2,516.0	1,631.0	314.0	177.8	3.76	174.2

Source: Alberta Energy Utility Board.

As a result of recent high oil prices, 44 new oil sands projects are planned for Alberta between 2004 and 2012, 26 in-situ and 18 surface-mining.[14] If all projects were to go forward, an estimated C$60 billion would be required for construction. Several of the projects are expansions of current operations. The National Energy Board (NEB) projects as much as C$81.6 billion being spent between 2006 and 2016.[15] Eighty-two percent of the projected investment — expected to peak in 2008 — is directed towards the Fort McMurray/Woods Buffalo Region of Alberta. A total of C$29 billion was spent on oil sands development between 1996 and 2004.[16]

History of Development

Role of Industry and Government

U.S. Oil Sands

Interest in U.S. oil sand deposits dates back to the 1930s. Throughout the 1960s and 1970s, 52 pilot projects involving mining and in-situ techniques were supported by the U.S. government in collaboration with major oil companies such as Conoco, Phillips Petroleum, Gulf Oil, Mobil, Exxon, Chevron, and Shell. Several steam-assisted technologies were being explored for in-situ production. These sources have had little economic potential as oil supply. The Energy Policy Act of 2005 (P.L. 109-58), however, established a public lands leasing program for oil sands and oil shale[17] R&D.

Based on the Canadian experience with oil sands production, it was established that commercial success in mining oil sands is a function of the ratio of overburden to oil sand thickness.[18] This ratio should not exceed one. In other words, the thickness of the overlying rock should not be greater than the thickness of the sand deposit. It was estimated by the USGS that only about 15% of the U.S. resource base has a ratio of one or less.

Major development obstacles to the U.S. oil sands resource base include remote and difficult topography, scattered deposits, and the lack of water for in-situ production (steam recovery and hot water separation) or undeveloped technology to extract oil from U.S. "hydrocarbon-wetted" deposits.[19] The Canadian technology may not be suited for many U.S. deposits. In Texas, deposits were considered by Conoco Oil to be too viscous to produce in-situ. A Bureau of Mines experiment with oil sands production in Kentucky proved to be commercially infeasible. In Utah, there were attempts at commercial production over the past three decades by several oil companies but projects were considered uneconomic and abandoned. As of 2004, some oil sands were being quarried on Utah state lands for asphalt used in road construction, and a small amount of production is taking place in California.[20] "Since the 1980s there has been little production for road material and no government funding of oil sands R&D," according to an official at the Department of the Interior.[21]

A 2006 conference on oil sands held at the University of Utah indicated renewed interest in U.S. oil sands but reiterated the development challenges mentioned above. Speakers also pointed out new technologies on the horizon that are being tested in Utah.[22] Conference organizers concurred that long-term research and development funding and huge capital development costs would be needed to demonstrate any commercial potential of U.S. oil sand deposits. A recent report[23] on U.S. unconventional fuels (an interagency and multistate collaboration) makes a number of general recommendations (for the development of oil sands and other unconventional fuels), which include economic incentives, establishing a regulatory framework, technology R&D, and an infrastructure plan. A recommendation specific to oil sands calls for closer U.S. collaboration with the government of Alberta to better understand Canadian oil sands development over the last 100 years. The report's task force estimates that based on a "measured" or "accelerated" development pace scenario,[24] U.S. oil sand production could reach 340,000-352,000 barrels per day by 2025.[25]

Canadian Oil Sands

Canada began producing its oil sands in 1967 after decades of research and development that began in the early 1900s. Wells were drilled between 1906 and 1917 in anticipation of finding major conventional oil deposits. The area around Fort McMurray, Alberta, was mapped for bituminous sand exposures in 1913 by Canada's Federal Department of Mines. By 1919, the Scientific and Industrial Research Council of Alberta (SIRCA), predecessor to the Alberta Research Council (ARC),[26] became interested in oil sands development. One of its newly recruited scientists, Dr. Karl Clark, began his pioneering work on a hot-water flotation process for separating the bitumen from the sand. In this separation process, the mined oil sand is mixed with water and a sodium hydroxide base and rotated[27] in a horizontal drum at 80 degrees centigrade. Dr. Clark's efforts led to a pilot plant in 1923 and a patented process by 1929. He continued to improve the process through several experimental extraction facilities through the 1940s.

The technical feasibility was demonstrated in 1949 and 1950 at a facility in Bitumont, Alberta, located on the Athabasca River near Fort McMurray. The technology being tested was largely adopted by the early producers of oil sands — Great Canadian Oil Sands (GCOS), Ltd., and Syncrude. Sunoco established GCOS, Ltd., in 1952 and then invested $250 million in its oil sands project. Another major player in the oil sands business in Canada was Cities Services, based in Louisiana. Cities Services purchased a controlling interest in the Bitumont plant in 1958, then in 1964, along with Imperial Oil, Atlantic Richfield (ARCO), and Royalite Oil, formed the Syncrude consortium.[28]

The ARC continued its involvement with oil sands R&D throughout the 1950s and 1960s. Several pilot projects were established during that period. Suncor [29] began construction of the first commercial oil sands production/separation facility in 1964 and began production in 1967, using the hot water extraction method developed and tested by ARC. In 1967, Suncor began to produce oil sands at a rate of 12,000 barrels per day.

Just a year later, in 1968, the government of Alberta deferred an application by Syncrude Canada for a $200 million, 80,000 barrel oil sands facility. Eventually, in 1978, the Energy Resources Conservation Board of Alberta approved Syncrude's proposal to build a $1 billion plant that would produce up to 129,000 barrels per day.

However, ARCO, which represented 30% of the project, pulled out of the consortium as costs of the plant climbed toward $2 billion. At that point (1978) the federal and provincial governments joined in. The federal government purchased a 15% share, Alberta a 10% share, and Ontario 5%, making up the 30% deficit. At the time, the Canadian government was promoting the goal of energy self-sufficiency, and the Alberta government agreed to a 50/50 profit-sharing arrangement instead of normal royalties for Syncrude.[30]

The Alberta Energy Company[31] purchased 20% of Syncrude and then sold 10% of its share to Petrofina Canada, Ltd., and Hudson Bay Oil and Gas, Ltd.[32] The consortium grew from four to nine owners. From 1983 to 1988 Syncrude spent $1.6 billion to boost production to 50 million barrels per year. In 1984, the government of Alberta agreed to a new royalty structure for oil sands producers coinciding with Syncrude's capital expansion plans. In 1985, the Alberta government announced that existing oil sands operations and new plants would not be taxed on revenues, and the petroleum gas revenue tax would be phased out. During the same time-frame, Syncrude's cash operating costs were just under $18 per barrel with total costs over $20 per barrel,[33] while the market price of oil fluctuated under $20 per barrel.

Because of huge capital requirements, oil sands producers lobbied for continued royalty relief and thought the government should "defer tax and royalty revenues until project expansions were completed."[34] In 1994, the National Oil Sands Task Force (an industry/government group) was created, and the Canadian Oil Sands Network for R&D (CONRAD) agreed to spend $105 million annually to boost production and trim costs. Costs continued to fall ($15.39/bbl in 1992[35] to under $14/ bbl in 1994[36]) as Syncrude ownership continued to change. In 1996, the National Oil Sands Task Force recommended a package of royalty and tax terms to ensure consistent and equal treatment of projects, because oil sand projects previously were treated on a project-by-project basis. The implementation of favorable royalty treatment is discussed below.

The ARC has had a successful partnership with the private sector in oil sands research and development. As a result of favorable royalty and tax terms and Alberta's $700 million R&D investment in oil sands extraction (from 1976-2001), the private sector has invested billions of dollars of development capital in oil sand projects.[37] Syncrude has said that "partnering with ARC gave us the ability to explore a potentially valuable technology."[38]

Oil Sands Production Process

Oil sands production measured only 1.3% of total world crude oil production in 2005. By 2025 it may reach 4.1% of total world production. But more importantly, it may mean U.S. access to extensive North American oil reserves and increased energy security.

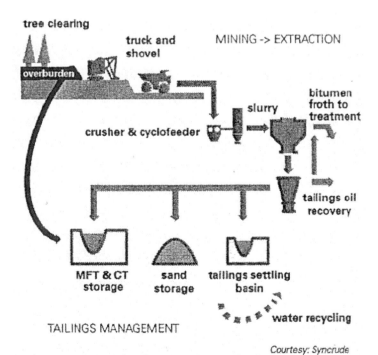

Courtesy: Syncrude

Source: *Oil Sands Technology Roadmap*, Alberta Chamber of Resources, January 2004, p. 21.

Figure 3. Major Mining Process Steps.

Oil sands are either surface-mined or produced in-situ. Mining works best for deposits with overburden less than 75 meters thick. Mining requires a hydraulic or electric shovel that loads the sand into 400-ton trucks, which carry the material to a crusher to be mixed into a slurry. Using pumps and pipelines, the slurry is "hydro transported" to an extraction facility to extract bitumen (see figure 3). This process recovers about 90% of the bitumen.[39]

In 2005, mining accounted for about 52% of Alberta's oil sand production (572,000 b/d); in-situ accounted for about 48% (528,000 b/d), one-third of which was produced using the Cold Production method in which oil sands are light enough to flow without heat. The in-situ approach, which was put into commercial production in 1985,[40] is estimated to grow to 926,000 barrels per day by 2012. Currently, the largest production projects are in the Fort McMurray area operated by Syncrude and Suncor (see table 4 for leading producers of oil sands).

Extraction Process

The extraction process separates the bitumen from oil sands using warm water (75 degrees Fahrenheit) and chemicals. Extracting the oil from the sand after it is slurried consists of two main steps. First is the separation of bitumen in a primary separation vessel. Second, the material is sent to the froth tank for diluted froth treatment to recover the bitumen and reject the residual water and solids. The bitumen is treated either with a naphtha solvent or a paraffinic solvent to cause the solids to easily settle. The newer paraffinic treatment results in a cleaner product.[41] This cleaner bitumen is pipeline quality and more easily blended with refinery feedstock. After processing, the oil is sold as raw bitumen or upgraded and sold as SCO.

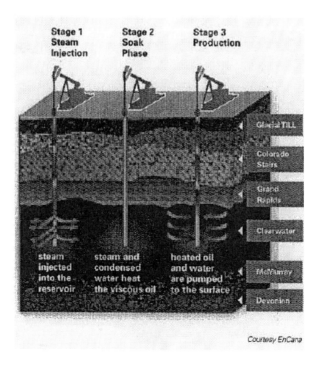

Figure 4. Continued on next page.

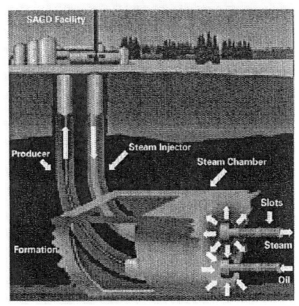

SAGD Oil Production Technology

Source: Oil Sands Technology Roadmap, p. 28.

Figure 4. In-SITU Recovery.

Table 2. Leading Oil Sands Producers
(barrels per day)

Project Owner	Type of Project	2002	2003	2006 1st Quarter	Planned Production Targets
Suncor	Mining	206,000	217,000	264,400	410,000
Syncrude	Mining	230,000	212,000	205,000	560,00
Athabasca Oil Sands (Shell, Chevron, and Marathon Oil)a	Mining	N/A	130,000	77,400	525,000
Imperial Oil	In-situ	112,000	130,000	150,000	180,000
CNRL	In-situ	N/A	35,000	122,000	500,000
Petro Canada	In-situ	4,500	16,000	21,000 (2005)	100,000
EnCana	In-situ	N/A	5,300	36,000	250,000

Source: *Oil Sands Industry Update*, Alberta Economic Development, 2004 and 2006.
[a]. Marathon Oil Corp. acquired Western Oil Sands, Inc. on October 18, 2007.

Production Technology

For in-situ thermal recovery, wells are drilled, then steam is injected to heat the bitumen so it flows like conventional oil. In-situ production involves using various techniques.

One technique is the Cyclic Steam Stimulator (CSS), also known as "huff and puff." CSS is the most widely used in-situ technology. In this process, steam is added to the oil sands via vertical wells, and the liquefied bitumen is pumped to the surface using the same well.

But a relatively new technology — steam-assisted gravity drainage (SAGD) — has demonstrated that its operations can recover as much as 70% of the bitumen inplace. Using SAGD, steam is added to the oil sands using a horizontal well, then the liquefied bitumen is pumped simultaneously using another horizontal well located below the steam injection well (see figure 4). The SAGD process has a recovery advantage over the CSS process, which only recovers 25%-30% of the natural bitumen. Also, the lower steam to oil ratio (the measurement of the volume of steam required to extract the bitumen) of SAGD results in a more efficient process that uses less natural gas.[42] SAGD operations are limited to thick, clean sand reservoirs, but it is reported by the industry that most of the new in-situ projects will use SAGD technology.[43] A number of enhanced SAGD methods are being tested by the Alberta Research Council. They could lead to increased recovery rates, greater efficiency, and reduced water requirements.

The emerging Vapor Extraction Process (VAPEX) technology operates similarly to SAGD. But instead of steam, ethane, butane, or propane is injected into the reservoir to mobilize the hydrocarbons towards the production well. This process eliminates the cost of steam generators and natural gas. This method requires no water and processing or recycling and is 25% lower in capital costs than the SAGD process. Operating costs are half that of the SAGD process.[44]

A fourth technique is cold production, suitable for oil sands lighter than those recovered using thermal assisted methods or mining. This process involves the coproduction of sand with the bitumen and allows the oil sands to flow to the well bore without heat. Imperial Oil uses this process at its Cold Lake site. Oil sand produced using in-situ techniques is sold as natural bitumen blended with a diluent for pipeline transport.

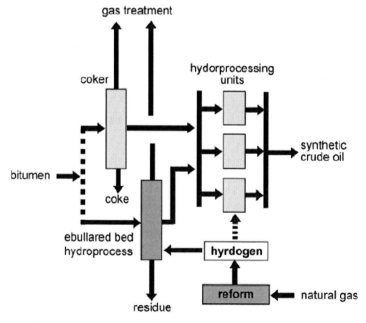

Source: Oi l Sands Technology Roadmap, p. 41.

Figure 5. Upgrading to SCO.

The overall result of technology R&D has been to reduce operating costs from over $20/barrel in the early 1970s to $8-12/barrel in 2000. While technology improvements helped reduce some costs since 2000, total costs have risen significantly as discussed below, because of rising capital and energy costs.[45]

Upgrading[46]

Upgrading the bitumen uses the process of coking for carbon removal or hydro- cracking for hydrogen addition (see figure 5). Coking is a common carbon removal technique that "cracks" the bitumen using heat and catalysts, producing light oils, natural gas, and coke (a solid carbon byproduct). The coking process is highly aromatic and produces a low quality product. The product must be converted in a refinery to a lighter gas and distillate. Hydrocracking also cracks the oil into light oils but produces no coke byproduct. Hydrocracking requires natural gas for conversion to hydrogen. Hydrocracking, used often in Canada, better handles the aromatics. The resulting SCO has zero residues which help keep its market value high, equivalent to light crude.

Partial upgrading raises the API of the bitumen to 20-25 degrees for pipeline quality crude. A full upgrade would raise the API to between 30-43 degrees — closer to conventional crude. An integrated mining operation includes mining and upgrading. Many of the mining operations have an on-site upgrading facility, including those of Suncor and Syncrude. Suncor uses the coking process for upgrading, while Syncrude uses both coking and hydrocracking and Shell uses hydrocracking. (For the complete oil sands processing chain, see figure 6.)

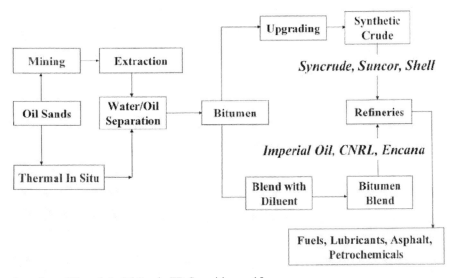

Source: *Overview of Canada's Oil Sands*, TD Securities, p. 15.

Figure 6. Oil Sands Processing Chain.

A major trend among both mining and in situ producers is to integrate the upgrading with the refinery to cut costs; e.g., linking SAGD production with current refinery capabilities. Long-term processing success of oil sands will depend on how well this integration takes place and how well the industry addresses the following issues:

- cost overruns,
- cost effective upgrading, reducing highly aromatic, high-sulfur SCO,
- and
- dependence on and price of natural gas for hydrogen production (originally used because of its low price but now considered by some to be too expensive).

The wide heavy-oil/light-oil price differential has been an incentive to increase upgrading. The price for heavy crude was as low as $12 per barrel in early 2006 and its market is limited by refineries that can process it and by its end use as asphalt. In its June 2006 report, the NEB describes numerous proposals for building upgraders.[47]

Cost overruns for the integrated mining projects or expansions, sometimes as much as 50% or more of the original estimates, have been a huge problem for the industry. The main reasons cited by the COS report are poor management, lack of skilled workers, project size, and engineering issues.

Cost of Development and Production

Operating and total supply costs have come down significantly since the 1970s. Early supply costs were near C$35 per barrel (in 1970s dollars). Reductions came as a result of two major innovations in the production process. First, power shovels and energy efficient trucks replaced draglines and bucketwheel reclaimers, and second, hydrotransport replaced conveyor belts to transport oil sands to the processing plant.[48]

Operating costs include removal of overburden, mining and hydro transport, primary extraction, treatment, and tailings removal. The recovery rate, overburden volumes, cost of energy, transport distances, and infrastructure maintenance all have an impact on operating costs.

Supply costs (total costs) include the operating costs, capital costs, taxes and royalties, plus a 10% return on investment (ROI). When compared to conventional new oil production starts, an oil sands project may have operating costs over 30% higher than the world average for conventional new starts. However, its nearly nonexistent royalty and tax charge makes the total cost per barrel of energy significantly less than the conventional oil project. The NEB in its Energy Market Assessment estimated that between US$30-$35 per barrel oil is required to achieve a 10% ROI.[49]

Operating costs for mining bitumen were estimated at around C$9-$12 per barrel (C$2005) — an increase of up to C$4 per barrel since the 2004 NEB estimates. Supply cost of an integrated mining/upgrading operation is between C$36 and $40/barrel for SCO — a dramatic increase over the C$22-$28 estimate made in 2004. These supply costs for an integrated mining/upgrading operation were expected to decline with improvements in technologies (see table 3). However, natural gas prices rose 88% and capital costs rose 45% over the past two years.

Operating costs for SAGD in-situ production in 2005 were about C$10-$14 per barrel of bitumin, up from C$7.40 per barrel in 2004. Recovery rates are lower than with mining, at 40%-70%, and the price of energy needed for production is a much larger factor. The SAGD operations are typically phased-in over time, thus are less risky, make less of a "footprint" on the landscape than a mining operation, and require a smaller workforce. SAGD supply cost for Athabasca oil sand rose from between C$11-$17/barrel (bitumen) to C$18-$22/barrel;

using the CSS recovery technique, supply costs are estimated higher at between C\$20-\$24/barrel, an increase from C\$13-\$19/barrel. Cost increases/decreases for in-situ operations are largely dependent on the quality of the reservoir and natural gas prices, but as SAGD and other new technologies (e.g. VAPEX) become more efficient, industry is expecting some cost declines. SAGD (in-situ) supply costs are less sensitive to capital costs than mining projects because the capital investment is far less.

Natural gas is a major input and cost for mining, upgrading, and in situ recovery: Mining requires natural gas to generate heat for the hot water extraction process, upgraders need it for heat and steam, and in situ producers use natural gas to produce steam which is injected underground to induce the flow of bitumen. Natural gas accounts for 15% of the operating costs in mining operations compared to 60% of operating costs in SAGD in-situ production. The major cost for thermal in-situ projects (SAGD, CSS) is for the natural gas that powers the steam-producing generators. For SAGD projects, 1 thousand cubic feet is needed per barrel of bitumen. Reducing the steam-to-oil ratio (SOR) — the quantity of steam needed per barrel of oil produced — is critical for lowering natural gas use and costs.[50] SAGD has a lower SOR than CSS projects but cannot be used for all oil sand in-situ production. However, most new in-situ projects will use SAGD.

Canadian oil sand producers continue to evaluate energy options that could reduce or replace the need for natural gas. Those options include, among other things, the use of gasification technology, cogeneration, coal, and nuclear power.

Table 3. Estimated Operating and Supply Cost by Recovery Type (C\$2005 Per Barrel at the Plant Gate)

	Crude Type	Operating Cost	Supply Cost
Cold Production - Wabasca, Seal	Bitumen	6-9	14-18
Cold Heavy Oil Production and Sand (CHOPS) - Cold Lake	Bitumen	8-10	16-19
Cyclic Stream Stimulation (CSS)	Bitumen	10-14	20-24
Steam Assisted Gravity Drainage (SAGD)	Bitumen	10-14	18-22
Mining/Extraction	Bitumen	9-12	18-20
Integrated Mining/Upgrading	SCO	18-22	36-40

Source: Canada's Oil Sands, Opportunities and Challenges to 2015, National Energy Board, Canada, June 2006.
Note: Supply costs for the first five technologies do not include the coat of upgrading bitumen to SCO.

Tax and Royalty on Oil Sands

In 1997 the Alberta government implemented a "Generic Oil Sands Royalty Regime"[51] specific to oil sands for all new investments or expansions of current projects. Since then, oil sand producers have had to pay a 1% minimum royalty based on gross revenue until all capital costs including a rate of return are recovered. After that, the royalty is either 25% of net project revenues or 1% of the gross revenues, whichever is greater.[52] The 1% prepayout royalty rate is in stark contrast to conventional world royalties. Net project revenues (essentially net profits before tax) include revenues after project cash costs, such as operating costs, capital, and R&D are deducted. Royalty payments may be based on the value of bitumen or SCO if the project includes an upgrader. Currently, 51% of oil sand projects (or

75% of production volume) under the Generic Royalty regime are paying the 25% royalty rate. Two major oil sands producers, Suncor and Syncrude (accounting for 49% of bitumen production) have "Crown Agreements" in place with the province that have allowed the firms to pay royalties based on the value of synthetic crude oil (SCO) production with the option to switch to paying royalties on the value of bitumen beginning as early as 2009. Royalties paid on bitumen, which is valued much lower than SCO, would result in less revenue for the government. The agreements expire in 2016.

Royalty revenues from oil sands fluctuated widely between 1997 and 2005. For example, royalties from oil sands were less than $100 million in 1999, then rose to $700 million in 2000/2001, but fell in 2002/2003 to about $200 million as production continued to rise. Royalties from oil sands rose dramatically in 2005/2006 to $1 billion, and the Government of Alberta forecasts royalties of $2.5 billion in 2006/2007 and $1.8 billion in 2007/2008.[53] Oil price fluctuations are the primary cause for such swings in royalty revenues.

The Albertan provincial government established a Royalty Review Panel in February 2007 to examine whether Alberta was receiving its fair share of royalty revenues from the energy sector and to make recommendations if changes are needed. In its September 2007 report, the panel concluded that "Albertans do not receive their fair share from energy development."[54] When the oil sands industry was ranked against other heavy oil and offshore producers such as Norway, Venezuela, Angola, United Kingdom, and the U.S. Gulf of Mexico, Alberta received the smallest government share.[55] This is, however, a difficult comparison to make because it is not among oil sand producers only and the fiscal regimes of the various producing countries is dynamic. However, based on a general analysis by T.D. Securities, typically, on average, world royalty rates could add as much as 45% to operating costs while the 1% rate may add only 3% to operating costs.[56]

The Panel recommended keeping the "pre-payout, post-payout" framework intact (see footnote 52), which would retain the 1% pre-payout royalty rate, but in the post-payout phase, firms would be required to pay a higher net revenue royalty rate of 33% plus continue to pay the 1% base royalty.

On October 25, 2007, the Alberta Government announced and published its response to the Royalty Review Panel's report.[57] It retained the "pre-payout," "postpayout" royalty framework but concluded that a sliding-scale rate structure would best achieve increasing the government's share of revenues from oil sands production. The pre-payout base rate would start at 1%, then increase for every dollar above US$55 per barrel (using the West Texas Intermediate or WTI price) reaching a maximum increase of 9% when prices are at or above $120 per barrel. In the post-payout phase, the net revenue rate will start at 25%, then rise for every dollar oil in priced above US$55 per barrel, reaching a maximum of 40% of net revenues when oil is $120 per barrel or higher. The new rate structure will take effect in 2009. The Government of Alberta has initiated negotiations with Suncor and Syncrude in an attempt to include them under the new oil sands royalty framework by 2009.

Oil sand firms pay federal and provincial income taxes and some differences exist in the tax treatment of the oil sands and conventional oil industries. Since the Provincial 1996 Income Tax Act, both mineable and in-situ oil sand deposits are classified as a mineral resource for Capital Cost Allowance (CCA) purposes which means mineral deposits receive higher cost deductions than conventional oil and gas operations (i.e. acquisition costs and intangible drilling costs).[58] The provincial government of Alberta has agreed to the 2007 federal budget proposal to eliminate the CCA deduction for oil sands. The Royalty Review

Panel also supported this change in its report. The federal government of Canada, however, provided some balance by reducing the general federal corporate income tax rate from 22.1% to 15% beginning in 2012.[59]

U.S. Markets

Oil sand producers continue to look to the United States for the majority of their exports. Seventy-five percent of Canadian nonconventional oil exported to the United States is delivered to the Petroleum Administration for Defense District (PADD)[60] II in the Midwest. This region is well positioned to receive larger volumes of nonconventional oil from Canada because of its refinery capabilities. Several U.S.- based refinery expansions have been announced that would come online between 2007-2015. If Canada were to reach its optimistic forecasted oil sands output level of 5 mbd in 2030, and maintained its export level to the United States at around 90%, it would be exporting about 4.5 mbd to the United States. This would mean that imports from Canada would reach nearly 30% of all U.S. crude oil imports. U.S. refinery capacity is forecast to increase from 16.9 mbd in 2004 to nearly 19.3 mbd in 2030,[61] a 2.4 mbd increase — significant but perhaps not enough to accommodate larger volumes of oil from Canada, even if refinery expansions would have the technology to process heavier oil blends. Canada is pursuing additional refinery capacity for its heavier oil.

Pipelines

Oil sands are currently moved by two major pipelines (the Athabasca and the Corridor, not shown in figure 7) as diluted bitumen to processing facilities in Edmonton. After reaching

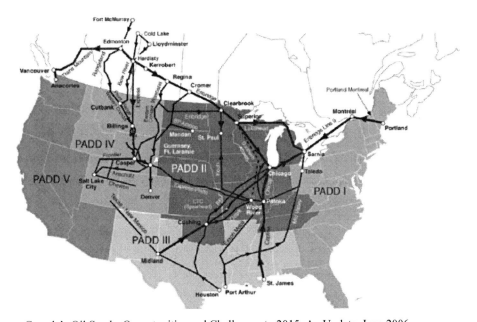

Source: Canada's Oil Sands, Opportunities and Challenges to 2015: An Update, June 2006.

Figure 7. Major Canadian and U.S. (Lower 48) Crude Oil Pipelines and Markets.

refineries in Edmonton, the synthetic crude or bitumen is moved by one of several pipelines to the United States (see figure 7). The Athabasca pipeline has capacity of 570,000 barrels per day (b/d) while the Corridor has capacity of less than 200,000 b/d. Current pipeline capacity has nearly reached its limit. However, there are plans to increase Corridor's capacity to 610,000 b/d by 2010.

A number of new pipeline projects have been proposed or initiated that would increase the flow of oil from Canada to the United State's PADDs II, III, and V. Most of the new projects are scheduled to come online between 2008 and 2012. In addition, a couple of U.S. pipelines reversed their flow of crude oil (from south to north) to now carry Canadian heavy crude, originating from oil sands, to Cushing Oklahoma and Southeast Texas. Pipeline capacity could be a constraint to growth in the near term but the NEB predicts some excess pipeline capacity by 2009. An estimated $31.7 billion has been invested in pipeline projects for oil sands in western Canada.[62]

Environmental and Social Issues

The Federal Government of Canada classified the oil sands industry as a large industrial air pollution emitter (i.e., emitting over 8,000 tons $CO2$/year) and expects it to produce half of Canada's growth in greenhouse gas (GHG) emissions[63] (about 8% total GHG emissions) by 2010. The oil sands industry has reduced its "emission intensity" by 29% between 1995-2004 while production was rising. $CO2$ emissions have declined from 0.14 tons/bbl to about 0.08 tons/bbl or about 88 megatons since 1990.[64] Alberta's GHG goals of 238 megatons of $CO2$ in 2010, and 218 megatons $CO2$ in 2020 are not expected to be met.[65] Reducing air emissions is one of the most serious challenges facing the oil sands industry. However, according to the Pembina Institute, a sustainable energy advocate, greenhouse gas emissions intensity ($CO2$/barrel) from oil sands is three times as high as that from conventional oil production.[66] The industry believes if it can reduce energy use it can reduce its emissions. As emissions per barrel of oil from oil sands decline overall, the Canadian government projects that total GHG emissions will continue to rise through 2020, attributing much of the increase to increased oil sands production.[67]

Water supply and waste water disposal are among the most serious concerns because of heavy use of water to extract bitumen from the sands. For an oil sands mining operation, about 2-3 barrels of water are used from the Athabasca river for each barrel of bitumen produced; but when recycled produced water is included, 0.5 barrels of "make-up" water is required, according to the Alberta Department of Energy. Oil sands projects currently divert 150 million cubic meters of water annually from the Athabasca River but are approved to use up to 350 million cubic meters.[68] Concerns, however, arise over the inadequate flow of the river to maintain a healthy ecosystem and meet future needs of the oil sands industry. Additionally, mining operations impact freshwater aquifers by drawing down water to prevent pit flooding.

The freshwater used for in-situ operations is needed to generate steam, separate bitumen from the sand, hydrotransport the bitumen slurry, and upgrade the bitumen to a light crude. For SAGD operations, 90-95% of all the water used is recycled. Since some water is lost in the treatment process, additional freshwater is needed. To minimize the use of new freshwater supplies, SAGD operators use saline water from deeper underground aquifers. The use of saline water, however, generates huge volumes of solid waste which has posed serious disposal problems.

Wastewater tailings (a bitumen, sand, silt, and fine clay particles slurry) also known as "fluid fine tailings" are disposed in large ponds until the residue is used to fill mined-out pits. Seepage from the disposal ponds can result from erosion, breaching, and foundation creep.[69] The principal environmental threat is the migration of tails to a groundwater system and leaks that might contaminate the soil and surface water.[70] The tailings are expected to reach 1 billion cubic meters by 2020. Impounding the tailings will continue to be an issue even after efforts are made to use alternative extraction technology that minimizes the amount of tails. Tailings management criteria were established by the Alberta Energy and Utilities Board/Canadian Environmental Assessment Agency in June 2005. Ongoing extensive research by the Canadian Oil Sands Network for Research and Development (CONRAD) is focused on the consolidation of wastewater tailings, detoxifying tailings water ponds, and reprocessing tailings. Some R&D progress is being made in the areas of the cleanup and reclamation of tailings using bioremediation and electrocoagulation.[71]

The National Research Council of Canada (NRC) is conducting research to treat wastewater tailings and recover their byproduct residual bitumen, heavy metals, and amorphous solids (fertilizers). A pilot project is underway to clean and sort tailings, and recover metals such as aluminum and titanium.[72]

Surface disturbance is another major issue. The oil sands industry practice leaves land in its disturbed state and left to revegetate naturally. Operators, however, are responsible over the long term to restore the land to its previous potential.[73] Under an Alberta Energy Utility Board directive (AEUB), Alberta's Upstream Oil and Gas Reclamation and Remediation Program has expanded industry liability for reclaiming sites. The directive requires a "site-specific liability assessment" that would estimate the costs to abandon or reclaim a site.[74]

The government of Alberta's Department of the Environment established a "Regional Sustainable Development Strategy" whose purpose is, among other things, to "ensure" implementation of management strategies that address regional cumulative environmental impacts.[75] The oil sands industry is regulated under the Environmental Protection and Enhancement Act, Water Act, and Public Lands Act. Oil sands development proposals are reviewed by AEUB, Alberta Environment, and the Alberta Sustainable Resource Development at the provincial level. Review at the federal level may also occur.

Issues for Congress

The Energy Policy Act of 2005 (P.L. 109-58) describes U.S. oil sands (along with oil shale and other unconventional fuels) as a strategically important domestic resource "that should be developed to reduce the growing dependence of the United States on politically and economically unstable sources of foreign oil imports."[76] The provision also requires that a leasing program for oil sands R&D be established. Given U.S. oil sands' strategic importance, but limited commercial success as discussed above, what level of federal investment is appropriate to reach U.S. energy policy goals? While an estimated 11 billion barrels of U.S. oil sands may be significant if it were economic, it represents a small share of the potentially recoverable resource base of unconventional fuels (e.g., 800 billion barrels of potentially recoverable oil from oil shale and another 20 billion barrels of recoverable heavy oil). Where is the best return on the R&D dollar invested for increased domestic energy supply and what are the long-term prospects for commercial application of unconventional fuels technology? Another important consideration to look at is where the oil industry is investing its capital and R&D for oil sands projects.

In light of the environmental and social problems associated with oil sands development, e.g., water requirements, toxic tailings, carbon dioxide emissions, and skilled labor shortages, and given the fact that Canada has 175 billion barrels of reserves and a total of over 300 billion barrels of potentially recoverable oil sands (an attractive investment under current conditions demonstrated by the billions of dollars already committed to Canadian development), the smaller U.S. oil sands base may not be a very attractive investment in the near-term.

U.S. refinery and pipeline expansions are needed to accommodate Canadian oil sands developments. Those expansions will have environmental impacts, but the new infrastructure could strengthen the flow of oil from Canadian oil sands. This expanded capacity will likely lead to even greater investment in Canada.

Whether U.S. oil sands are developed, Congress will continue to be faced with regulatory matters. Oil imports from oil sands are likely to increase from Canada and the permitting of new or expanded oil refineries will continue to be an issue because of the need to balance concerns over the environment on one hand and energy security on the other.

Prospects for the Future

Because capital requirements for oil sands development has been enormous and risky, government involvement was seen as being essential in Canada, particularly during sustained periods of low oil prices. This private sector/government partnership in R&D, equity ownership, and public policy initiatives over the last 100 years has opened the way for the current expansion of the oil sands industry in Alberta.

Ongoing R&D efforts by the public and private sectors, sustained high oil prices, and favorable tax and royalty treatment are likely to continue to attract the increasing capital expenditures needed for growth in Canada's oil sands industry. Planned pipeline and refinery expansions and new upgrading capacity are underway to accommodate the increased volumes of oil sands production in Canada. U.S. markets will continue to be a major growth area for oil production from Canadian oil sands. Currently, about 5% of the total oil refined in the United States is from Canada's oil sands.

Even though prospects for Canadian oil sands appear favorable, factors such as water availability, waste water disposal, air emissions, high natural gas costs, insufficient skilled labor, and infrastructure demands may slow the pace of expansion.

Prospects for commercial development of U.S. oil sands are uncertain at best because of the huge capital investment required and the relatively small and fragmented resource base. The Task Force on Strategic Unconventional Fuels reported that oil sands comprise only about 0.6% of U.S. solid and liquid fuel resources, while oil shale accounts for nearly 25% of the total resource base.[77]

Appendix A

Table A1. Estimated World Oil Resources
(in billions of barrels)

Region and Country	Proved Reserves	Reserve Growth	Undiscovered	Total
OECD				
United States	22.4	76.0	83.0	180.4
Canada[78]	178.8	12.5	32.6	223.8
Mexico	12.9	25.6	45.8	84.3
Japan United States	0.1	0.1	0.3	0.5
Australia/ New Zealand	1.5	2.7	5.9	10.1
OECD Europe	15.1	20.0	35.9	71.0
Non-OECD				
Russia	60.0	106.2	115.3	281.5
Other Non-OECD Europe/Eurasia	19.1	32.3	55.6	107.0
China	18.3	19.6	14.6	52.5
India	5.8	3.8	6.8	16.4
Other Non-OECD Asia	10.3	14.6	23.9	48.8
Middle East	743.4	252.5	269.2	1,265.1
Africa	102.6	73.5	124.7	300.8
Central and South America	103.4	90.8	125.3	319.5
Total	1,292.5	730.2	938.9	2,961.6
OPEC	901.7	395.6	400.5	1,697.8
Non-OPEC	390.9	334.6	538.4	1,263.9

Sources: Proved Reserves as of January 1, 2006: *Oil & Gas Journal*, vol. 103, no. 47 (December 19, 2005), p. 46-47. Reserve Growth Total and Undiscovered, 1995-2025; U.S. Geological Survey, *World Petroleum Assessment 2000*, website [http://greenwood.cr.usgs.gov/WorldEnergy/DDS-60]]. Estimates of Regional Reserve Growth: Energy Information Administration, *International Energy Outlook 2006*, DOE/EIA-0484(2006) (Washington, DC, June 2006), p. 29.

Note: Resources Include crude oil (including lease condensates) and natural gas plant liquids.

Appendix B

**Table B1. Regional Distribution of Estimated Technically
Recoverable Heavy Oil and Natural Bitumen
(in billions of barrels)**

Region	Heavy Oil		Natural Bitumen (oil sands)	
	Recovery Factor[a]	Technically Recoverable	Recovery Factor[a]	Technically Recoverable
North America	0.19	35.3	0.32	530.9
South America (Venezuela)	0.13	265.7	0.09	0.1
W. Hemisphere	0.13	301.0	0.32	531.0
Africa	0.18	7.2	0.10	43.0
Europe	0.15	4.9	0.14	0.2
Middle East	0.12	78.2	0.10	0.0
Asia	0.14	29.6	0.16	42.8
Russia	0.13	13.4	0.13	33.7[b]
E. Hemisphere	0.13	133.3	0.13	119.7
World		434.3		650.7

Source: U.S. Department of the Interior. U.S. Geological Survey Fact Sheet, FS 070-03 August 2003.
Note: Heavy oil and natural bitumen are resources in known accumulations.
[a] Recovery factors were based on published estimates of technically recoverable and in-place[79] oil or bitumen by accumulation. Where unavailable, recovery factors of 10% and 5% of heavy oil or bitumen in-place were assumed for sandstone and carbonate accumulations, respectively.
[b] In addition, 212.4 billion barrels of natural bitumen in-place is located in Russia but is either in small deposits or in remote areas in eastern Siberia.

References

[1] DOE, EIA, International Energy Outlook, 2006, p. 29.
[2] U.S. Department of Energy, EIA, Annual Energy Outlook, 2006.
[3] Oil Sands Fever, The Environmental Implications of Canada's Oil Sand Rush, by Dan Woynillowicz, et. al, The Pembina Institute, November 2005.
[4] Reserves are defined by the EIA as estimated quantities that geological and engineering data demonstrate with reasonable certainty to be recoverable in future years from known reservoirs under existing economic and operating conditions. Resources are defined typically as undiscovered hydrocarbons estimated on the basis of geologic knowledge and theory to exist outside of known accumulations. Technically recoverable resources are those resources producible with current technology without consideration of economic viability.
[5] In-situ mining extracts minerals from an orebody that is left in place.

[6]   Canada's Oil Sands: Opportunities and Challenges to 2015, An Energy Market Assessment, National Energy Board, Canada, May 2004, p. 5.

[7]   API represents the American Petroleum Institute method for specifying the density of crude petroleum. Also called API gravity.

[8]   Diluents are usually any lighter hydrocarbon; e.g., pentane is added to heavy crude or bitumen in order to facilitate pipeline transport.

[9]   Canada's Oil Sands, May 2004, p. 10.

[10] Canada's Oil Sands, May 2004, p. 4

[11] DOE, EIA, International Energy Outlook, 2006, p. 29

[12] Canada's Oil Sands, NEB, June 2006.

[13] World Energy Investment Outlook, 2003 Insights, International Energy Agency (IEA), 2003.

[14] Canadian Oil Sands, May 2004, p. 25.

[15] The U.S.-Canadian dollar exchange rate fluctuates daily. As of early October 2007 the exchange rate is U.S.$1 = C$0.9969. In December 2006 the exchange rate was U.S.$1 = C$1.15.

[16] Oil Industry Update, Alberta Economic Development, Spring 2005.

[17] Oil shale is a compact rock (shale) containing organic matter capable of yielding oil.

[18] U.S. Tar-Sand Oil Recovery Projects — 1984, L.C. Marchant, Western Research Institute, Laramie, WY, p. 625.

[19] Hydrocarbon-wetted oil sand deposits require different technology for bitumen extraction than that used for Alberta's water-wetted deposits. Oil sands are characterized as having a wet interface between the sand grain and the oil coating; this allows for the separation of oil from the grain. U.S. oil sands do not have a wet interface making the separation difficult.

[20] Phone communication with B. Tripp, Geologist, Utah Geological Survey, May 2004.

[21] Phone communication with Richard Meyers, Department of the Interior specialist in oil sands, September 2004.

[22] Presentation by Earth Energy Resources, Inc., at the Western U.S. Oil Sands Conference, University of Utah, September 21, 2006.

[23] Development of America's Strategic Unconventional Fuels Resources, Initial Report to the President and the Congress of the United States, Task Force on Strategic Unconventional Fuels, September 2006.

[24] The measured pace is based on sufficient private investment capital as a result of government policies but little direct government investment. An accelerated pace would imply a global oil supply shortage and rely more on significant government investment.

[25] Development of America's Strategic Unconventional Fuels Resources, Initial Report to the President and the Congress of the United States, Reference no. 17.

[26] The ARC was established in 1921, housed at the University of Alberta in Edmonton, and funded by the provincial government of Alberta. Its mandate was to document Alberta's mineral and natural resources. Today, the ARC is a wholly-owned subsidiary of the Alberta Science and Research Authority (ASRA) within Alberta's Ministry of Innovation and Science. The ARC has an annual budget of $85 million.

[27] The Influence of Interfacial Tension in the Hot-Water Process for Recovering Bitumen From the Athabasca Oil Sands, by L.L. Schramm, E.N. Stasiuk, and D. Turner, presented at the Canadian International Petroleum Conference, paper 2001-136, June 2001.

[28] Syncrude Canada Ltd. when first organized as a consortium of major oil companies comprised: Imperial Oil (an affiliate of Exxon), Atlantic Richfield (ARCO), Royalite Oil (later combined with Gulf Canada), and Cities Services R&D (See The Syncrude Story, p. 5). Its ownership has changed over the years as indicated in the text. Its current ownership structure is as follows: Canadian Oil Sands Ltd. (31.74%), Imperial Oil (25%), Petro- Canada Oil and Gas (12%), Conoco Phillips Oil Sands Partnership II (9.03%), Nexen Inc. (7.23%), Murphy Oil Co. Ltd. (5%), Mocal Energy Ltd. (5%) and the Canadian Oil Sands Limited Partnership (5%).

[29] GCOS, Ltd., was later renamed Suncor.

[30] A Billion Barrels for Canada, The Syncrude Story, pp. 44-45.

[31] The Alberta Energy Company (AEC) was created by the government of Alberta in 1975. Fifty percent was publicly owned. The government phased out is equity interest and in 1993 sold its remaining interest. The AEC and PanCanadian Energy Corporation merged in 2002 and became EnCana. EnCana sold its interest in Syncrude in 2003. For more details see Alexander's Oil and Gas Connection, "Company News North America," January 15, 2004.

[32] The Syncrude Story, pp. 72-73.

[33] Ibid, p. 98-99.

[34] Ibid, p.104

[35] Ibid, p. 122.

[36] Ibid, p. 136.

[37] The Alberta Energy Research Institute: Strategic Research Plan, 2003.

[38] ARC, Guide to the ARC, 2001-02. The ARC's more recent focus on developing in-situ technologies is beginning to shift back to surface mining R&D. They believe that their role is to help many of the newcomers to the industry develop "best practices" technology. The ARC sees itself as an ongoing player in the R&D business because of the huge challenges related to environmental quality, cost reductions, and the need for new upgrading technologies and refinery expansions.

[39] Oil Sands Technology Roadmap, p. 20.

[40] Oil Sands Industry Update, AED, June 2006, p. 7.

[41] Oil Sands Technology Roadmap: Unlocking the Potential, Alberta Chamber of Resources, January 2004 p. 23.

[42] According to the National Energy Board Report, one thousand cubic feet of natural gas is required per barrel of bitumin for SAGD operations. Canada's Oil Sands, May 2004.

[43] Canada's Oil Sands, June 2006 p. 4.

[44] Canada's Oil Sands, Opportunities and Challenges to 2015, An Energy Market Assessment, May 2004, National Energy Board, Canada, p. 108.

[45] Canada's Oil Sands, June 2006.

[46] Overview of Canada's Oil Sands, TD Securities, January 2004, p. 19.

[47] NEB, June 2006, pp. 20-21.

[48] COS, 2004, p. 9.

[49] COS, 2006, p. 5.

[50] COS, 2004, p. 18.
[51] The generic oil sands royalty regime consists of three parts: the lease sale, a minimum 1% pre-payout gross revenue royalty, and a 25% post-payout net revenue royalty. The payout period is the time it takes a firm to recover all allowable capital costs including a rater of return.
[52] Oil and Gas Fiscal Regime, Alberta Resource Development of Western Canadian Provinces and Territories, p. 39, 1999.
[53] Oil Sands, Benefits to Alberta and Canada, Today and Tomorrow, Through a Fair, Stable and Competitive Fiscal Regime, Canadian Association of Petroleum Producers, May 2007, Appendix B.
[54] Our Fair Share, Report of the Alberta Royalty Review Panel, September 18, 2007, p. 7.
[55] Ibid., p. 27.
[56] Overview of Canada's Oil Sands, T.D. Securities, January 2004, p. 7.
[57] The New Royalty Framework, October 25, 2007.
[58] Oil and Gas Taxation in Canada, January 2000, PriceWaterhouseCoopers.
[59] Canadian Department of Finance, Economic Statement, October 30, 2007.
[60] There are 5 PADD's in the United States. PADDs were created during World War II as a way to organize the distribution of fuel in the United States.
[61] DOE/EIA, Annual Energy Outlook 2006 with Projections to 2030, February 2006.
[62] "Oil Sands Producers Facing Pipeline Capacity Constraints," The Energy Daily, August 7, 2007.
[63] Greenhouse gas emissions include carbon dioxide, methane, nitrous oxide, hydrofluorocarbons, perfluorocarbons, and sulfur hexafluoride.
[64] COS, 2004, p. 62.
[65] Ibid., p. 63.
[66] Oil Sands Fever, by Dan Woynillowicz, et al., The Pembina Institute, November 2005.
[67] COS, June 2006, p. 39.
[68] Oil Sands Fever, op. cit.
[69] Canada's Oil Sands (water conservation initiatives), pp. 66-68.
[70] Canada's Oil Sands, p. 68.
[71] Ibid., p. 69.
[72] For more on byproducts, see Canada's Oil Sands, p. 70.
[73] Ibid., p. 71.
[74] Ibid.
[75] Oil Sands Industry Update, AED, June 2006, p. 29.
[76] Section 369 of Energy Policy Act of 2005.
[77] Development of America's Strategic Unconventional Fuels Resources, September 2006, p. 5.
[78] Oil sands account for 174 billion barrels of Canada's total 179 billion barrel oil reserves. Further, the Alberta Energy and Utilities Board estimates that Alberta's oil sands contain 315 billion barrels of ultimately recoverable oil. Canada's Oil Sands: Opportunities and Challenges to 2015: An Update, June 2006, National Energy Board.
[79] In-place oil is a continuous ore body that has maintained its original characteristics.

In: Energy Research Developments
Editors: K.F. Johnson and T.R. Veliotti, pp. 313-333 ISBN 978-1-60692-680-2
ⓒ 2009 Nova Science Publishers, Inc.

Chapter 11

TIDAL ENERGY EFFECTS OF DARK MATTER HALOS ON EARLY-TYPE GALAXIES

T. Valentinuzzi, R. Caimmi and M. D'Onofrio
Astronomy Dept., Padua Univ., Padova, Italy

Abstract

Tidal interactions between neighboring objects span across the whole admissible range of lengths in nature: from, say, atoms to clusters of galaxies i.e. from micro to macrocosms. According to current cosmological theories, galaxies are embedded within massive non-baryonic dark matter (DM) halos, which affects their formation and evolution. It is therefore highly rewarding to understand the role of tidal interaction between the dark and luminous matter in galaxies. The current investigation is devoted to Early-Type Galaxies (ETGs), looking in particular at the possibility of establishing whether the tidal interaction of the DM halo with the luminous baryonic component may be at the origin of the so-called "tilt" of the Fundamental Plane (FP). The extension of the tensor virial theorem to two-component matter distributions implies the calculation of the self potential energy due to a selected subsystem, and the tidal potential energy induced by the other one. The additional assumption of homeoidally striated density profiles allows analytical expressions of the results for some cases of astrophysical interest. The current investigation raises from the fact that the profile of the (self + tidal) potential energy of the inner component shows maxima and minima, suggesting the possible existence of preferential scales for the virialized structure, i.e. a viable explanation of the so called "tilt" of the FP. It is found that configurations related to the maxima do not suffice, by themselves, to interpret the FP tilt, and some other relation has to be looked for.

1. Introduction

According to current cosmological theories, about 85% of existing mass in the universe is in the form of (non baryonic) dark matter (hereafter quoted as DM), whose tidal energy effects on the embedded (baryonic) matter could be large. Since the first evidence of DM presence in galaxy clusters (e.g., Zwicky, 1933), the existence of massive, non baryonic halos is consistent with present-day CMB surveys, large scale galaxy clusters studies, and necessary, e.g., for a viable explanation of flat rotation curves well outside visible disks of spiral galaxies (for an exhaustive review on DM refer to, Freeman and McNamara, 2006). The current investigation aims to provide further insight on the tidal action induced by massive halos on hosted galaxies, taking into consideration special sequences of two-component systems, intended to model early-type galaxies (hereafter quoted as ETGs) and their hosting halos.

The idea of exploring two-component systems, a stellar spheroid completely embedded in a DM halo, moves from the fact that the virial potential energy (hereafter quoted as VPE) of the stellar component, shows a non-monotonic trend as a function of the radius, as opposed to one-component systems. This behavior is induced by the DM halo tidal potential, and is more effective for shallower DM halo density profiles. The occurrence of extremum points in potential energy could be highly rewarding, as in mechanics they correspond to stationary points and may be special configurations for the system. These extremum points could be a key to the explanation of the so called "tilt" of the Fundamental Plane (FP) (see, Bender et al., 1992).

The current investigation is based on two ETGs density profiles of astrophysical interest, using the formalism of the two-component virial theorem for an explicit expression of the VPE of the stellar subsystem embedded in the DM halo. The models and related special sequences of two-component systems, intended to represent ETGs, are defined in Section 2.. An analysis of VPE extremum points, with the further restriction of energy conservation, is performed and discussed in Section 3.. Comparisons between model predictions and both data from observations and results from computer simulations, are made in Section 4.. Conclusions are drawn in Section 5..

2. Models

Hernquist cuspy mass density profiles shall be used for the bright stellar component as best compromise between a reliable description of ETG luminosity profiles, and the need of dynamical and photometric analytical quantities when exploring models. The explicit formulation is:

$$\rho_*(r) = \frac{M_*}{2\pi} \frac{r_*}{r(r_*+r)^3} = \frac{\rho_N}{2\pi} \frac{1}{s(1+s)^3} \tag{1}$$

$$s \equiv \frac{r}{r_*} \quad ; \quad \rho_N \equiv \frac{M_*}{r_*{}^3}$$

where M_* is the total mass, r_* is a scaling radius, and the asterisk denotes the stellar subsystem. Aiming to explore the effect of either a cored or a cuspy DM density profile on the VPE, Hernquist stellar profiles shall be coupled to either Hernquist cuspy DM profiles:

$$\rho_h(r) = \frac{M_h}{2\pi} \frac{r_h}{r(r_h+r)^3} = \frac{\rho_N}{2\pi} \frac{\mathcal{R}\beta}{s(\beta+s)^3} \tag{2}$$

or Plummer cored DM profiles:

$$\rho_h(r) = \frac{3M_*}{4\pi} \frac{\mathcal{R}r_h{}^2}{(r_h{}^2+r^2)^{5/2}} = \frac{3\rho_N}{4\pi} \frac{\mathcal{R}\beta^2}{(\beta^2+s^2)^{5/2}} \tag{3}$$

where M_h is the total mass, r_h is a scaling radius, $\beta \equiv r_h/r_*$, $\mathcal{R} \equiv M_h/M_*$, and the index, h, denotes the DM halo. Accordingly, HH and HP models result by use of Eqs. (1), (2), and (1), (3), respectively. For a selected stellar profile, HH and HP models represent limiting situations of the largest and the lowest tidal effect (for assigned DM halo mass), inside the effective radius of the stellar subsystem, R_e, respectively.

As described in earlier attempts (Limber, 1959; Brosche et al., 1983), the fundamental quantities involved in the virial theorem for two-component systems, are the self potential energy of each component:

$$\Omega_u = 2\pi \int_0^\infty \rho_u(r)\varphi_u(r)r^2 \, dr, \qquad u = *, h \tag{4}$$

the interaction potential energy:

$$W_{uv} = 2\pi \int_0^\infty \rho_u(r)\varphi_v(r)r^2 \, dr, \qquad u = *, h, \qquad v = h, * \tag{5}$$

and the virial of external forces, or tidal potential energy:

$$V_{uv} = -4\pi \int_0^\infty \rho_u(r)\frac{d\varphi_v(r)}{dr}r^3 \, dr, \qquad u = *, h, \qquad v = h, * \tag{6}$$

where φ is the gravitational potential, and spherical symmetry has been assumed. The interaction potential energy, W_{uv}, and the tidal potential energy, V_{uv}, are not, in general, equal[1].

The virial theorem for each component can be written as:

$$V_u = \Omega_u + V_{uv} = 2T_u, \qquad u = *, h, \qquad v = h, * \tag{7}$$

where V represents the VPE. In one-component systems, it is usually named "the virial of the system" (Clausius, 1870). In two-component models, the "system" is the component under investigation subjected to the tidal action of the other one, in other words Eq.(7) represents the VPE of the u-subsystem affected by the tidal potential of the v-subsystem.

Comparing model results with observations requires the particularization of Eq.(7) to the real case of interest (i.e., $u = *$, $v = h$) for HH and HP models:

$$V_*^{(HH)} = \Omega_* + V_{*h} = -\Omega_N \left[\frac{1}{6} + \mathcal{R}\,V_{*h}^{(HH)}(\beta)\right] \tag{8}$$

$$V_*^{(HP)} = \Omega_* + V_{*h} = -\Omega_N \left[\frac{1}{6} + \mathcal{R}\,V_{*h}^{(HP)}(\beta)\right] \tag{9}$$

$$\Omega_N = GM_*^2/r_* \tag{10}$$

where the explicit expressions of $V_{*h}^{(HH)}$ and $V_{*h}^{(HP)}$ can be found in the Appendix.

In general, the VPE is a function of 4 independent parameters, meaning that it is defined in a 5 dimensional hyperspace. The dependence on M_* can be neglected in that it only acts as a scaling parameter, which means:

$$V_* = f_4(M_*, r_*, \beta, \mathcal{R}) = M_*\, f_3(r_*, \beta, \mathcal{R}) \tag{11}$$

---

[1]It can be seen that in HH and HP models these quantities are equal in the limit of $\beta \to 1$, even if the mass density profiles of the two components are different.

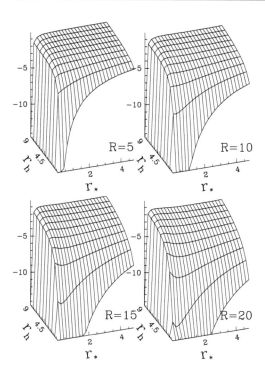

Figure 1. The plot of the bright component VPE, V_*, in units of GM_*^2, for HP models with different $\mathcal{R}$; r_* and r_h are expressed in kpc. For $\mathcal{R} > 10$ the VPE appears to be non-monotonic along r_* axis; this threshold in $\mathcal{R}$ is model dependent (HH models yield $\mathcal{R} > 20$).

where the numerical suffix of the general function indicates the dimensionality of the hyper-surface. It is worth noticing that via Eq. (7) V_*/M_* is σ_*^2, the stellar rms peculiar velocity averaged by mass (provided systematic rotation is negligible). For a fixed value of $\mathcal{R}$ the VPE function is defined in a 3 dimensional space; in other words it becomes a surface, as:

$$V_* = f_2(r_*, \beta) = f_2(r_*, r_h) \tag{12}$$

and either r_* or $\beta = r_h/r_*$ can be chosen as independent parameters. From here on, when the VPE surface is mentioned, it is intended that $\mathcal{R}$ is therein fixed.

Fig. 1 is a 3D-plot of the bright component VPE in units of GM_*^2 for HP models (HH models exhibit a similar trend, and are not presented here). In the case under discussion the following system of equations:

$$\begin{cases} \dfrac{\partial V_*}{\partial M_*} = 0 \\[6pt] \dfrac{\partial V_*}{\partial r_*} = 0 \\[6pt] \dfrac{\partial V_*}{\partial r_h} = 0 \\[6pt] \dfrac{\partial V_*}{\partial \mathcal{R}} = 0 \end{cases} \tag{13}$$

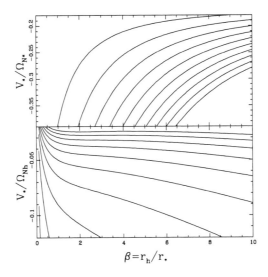

Figure 2. The plot of the VPE for HP models normalized to the stellar component (top panel, $\Omega_{N*} \equiv GM_*^2/r_*$, $\mathcal{R}$ increases from left to right) and to the dark component (bottom panel, $\Omega_{Nh} \equiv GM_h^2/r_h$, $\mathcal{R}$ increases from bottom to top) for $1 \le \mathcal{R} \le 10$, and steps of 1. In the former case no extremum points are present, while in the latter they appear for $\mathcal{R} > 4.3$. This dichotomy is a consequence of the path dependence of the extremum points in the VPE 3D hypersurface.

has no solution, meaning that the 3D VPE hypersurface exhibits no extremum point. On the other hand, extremum points can take place along selected peculiar paths on the VPE hypersurface. Their occurrence is strongly depending on the definition of the path; for example, a path along the r_* axis of Fig. 1 at fixed r_h (i.e. the case of a frozen halo) implies extremum points above a threshold, $\mathcal{R} \sim 10$; on the contrary a path along the r_h axis at fixed r_* yields no extremum point regardless of $\mathcal{R}$ value, as sketched in Fig. 2. The occurrence of extremum points, for shallow density profiles in the limit of a frozen DM halo, has been widely investigated in earlier attempts (see, Secco, 2000, 2001).

3. The *Energy-Conservation Paradigm*

The importance of focusing on a precise path along the VPE surface, to univocally determine the presence of extremum points has been stressed. One of the most interesting physically meaningful paths to be chosen is dictated by energy conservation. Accordingly, a fixed relation between $\mathcal{R}$, β and r_* must be taken into consideration. The path definition related to energy conservation, together with the properties exhibited by the corresponding family of curves, in particular the occurrence of extremum points, shall be quoted as the *energy-conservation paradigm*.

Consider the evolution of a forming galaxy from an initial total energy E_o state (i.e. the maximum expansion state) till the final virialization E_{vir} state. The collapsing spheroid will go through complex processes of gas dissipation, supernovae and black hole feedback, major and minor merging, etc., which will finally lead to a (not necessarily virialized) state where gas has been exhausted by star formation and energy dissipation may safely be neglected.

In this view, energy conservation holds from an initial dissipationless configuration to the virialized state[2], which implies, according to the virial theorem, the total energy E is half the potential energy at the end of the evolution. More specifically, the following relation holds:

$$E_{vir} = \frac{GM_*{}^2}{r_{*(vir)}} f\left[\mathcal{R}_{(vir)}, \beta_{(vir)}\right] = E \qquad (14)$$

where $f[...]$ represents a function of the variables inside brackets which are, in general, different for different models. In this way, it is always possible to express the stellar scaling radius, r_* in terms of the total energy, E, and a function of only β and $\mathcal{R}$ via Eq.(14), as:

$$r_{*(vir)} = \frac{GM_*^2}{E} f\left[\mathcal{R}_{(vir)}, \beta_{(vir)}\right] \qquad (15)$$

Substituting r_* in Eqs.(8) and (9), the stellar VPE in the *energy-conservation paradigm*, takes the expression:

$$V_*^{(HH)} = E f_*^{(HH)}\left[\mathcal{R}_{(vir)}, \beta_{(vir)}\right] \qquad (16)$$

$$V_*^{(HP)} = E f_*^{(HP)}\left[\mathcal{R}_{(vir)}, \beta_{(vir)}\right] \qquad (17)$$

for HH and HP models, respectively, and both the functions f_*, are presented in the Appendix. The *energy-conservation paradigm* reduces the VPE physical space to two dimensions only. Topologically, the VPE surface is intersected by an iso-energy plane, yielding a curve with varying r_* and r_h, but constant total energy. The dependence on E, exhibited by Eqs. (16) and (17), can be neglected in that it only acts as a scaling parameter, which means extremum points depend only on $\mathcal{R}_{(vir)}$ and $\beta_{(vir)}$, for a selected model.

This holds true also for complicated evolution of the initial assembly phases of the proto-galaxy; in fact, even if the mass density profiles are going through many changes, the presence or the absence of extremum points depend only on the final relaxed state of the galaxy, owing to the onset of energy conservation due to dissipationless evolution.

3.1. Analysis of Extremum Points

As shown in Fig. 3, the *energy-conservation paradigm* yields isoenergetic VPE curves where extremum points occur above a threshold in $\mathcal{R}$. Accordingly, for a fixed value of $\mathcal{R}$, there is only a single configuration of HH or HP models maximizing the VPE. On the other hand, extremum points occur only above a relatively high threshold, $\mathcal{R} \sim 10$ for HP models, and $\mathcal{R} \sim 20$ for HH models. Even if values above $\mathcal{R} \sim 10$ seem to be in contradiction with data from observations (e.g., Cappellari et al., 2006), analyzing the

---

[2]This process is perfectly described by N-Body simulations of pure gravitational collapse of two component dark and visible matter subsystems. The total initial energy is approximately conserved (dissipation is caused only by blown off particles) and the final virialized state is obviously strictly connected to the initial energy of the system which, in turn, depends on the dissipative history formation of the spheroid.

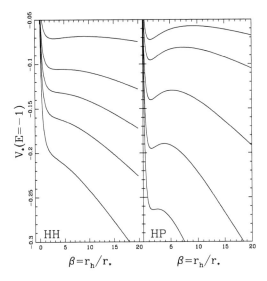

Figure 3. The VPE as a function of the fractional scaling radius, $\beta = r_h/r_*$, for HH models (left panel) and HP models (right panel), normalized to a total energy value, $E = -1$, in the *energy-conservation paradigm*. Curves refers to (from bottom to top) $\mathcal{R} = 10, 13, 16, 20, 30$, for HH models, and $\mathcal{R} = 5, 7, 10, 15, 20$ for HP models, respectively.

properties of VPE extremum points is still justified, in that maxima and minima in potential energy might represent special configurations for the system under consideration.

Extremum points of VPE isoenergetic curves described by Eqs.(16) and (17), are selected in the following way:

$$\frac{\partial}{\partial \beta} f_*^{(HH)}[\mathcal{R}_{(vir)}, \beta_{(vir)}] = 0 \tag{18}$$

$$\frac{\partial}{\partial \beta} f_*^{(HP)}[\mathcal{R}_{(vir)}, \beta_{(vir)}] = 0 \tag{19}$$

which provides a relation between the fractional mass, $\mathcal{R}$, and the scaling fractional radius, β, as:

$$\widetilde{\mathcal{R}}^{(HH)} = f^{(HH)}(\widetilde{\beta}) \tag{20}$$

$$\widetilde{\mathcal{R}}^{(HP)} = f^{(HP)}(\widetilde{\beta}) \tag{21}$$

For the sake of simplicity, in the following sections, only the maxima will be analyzed, as all considerations and conclusions apply equivalently to the minima; accordingly, from here on, the tilde shall denote a quantity calculated only on the VPE maximum, and configurations corresponding to a VPE maximum shall be quoted as MVPE configurations (i.e., Maximum Virial Potential Energy configurations). The plot of Eqs.(20) and (21) is presented in Fig. 4 (top panel).

The DM to stellar mass ratio inside the stellar scaling radius, $\widetilde{r}_*$, is:

$$\widetilde{\mathcal{M}}^{(HH)}(\widetilde{\beta}) = \frac{4\,\widetilde{\mathcal{R}}^{(HH)}}{(1+\widetilde{\beta})^2} \tag{22}$$

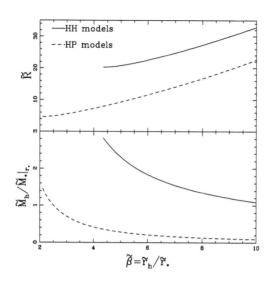

Figure 4. Global DM to stellar mass ratio, $\widetilde{\mathcal{R}}$ (top panel), and its counterpart within the stellar scaling radius, $\widetilde{\mathcal{M}}$ (bottom panel), as a function of the fractional scaling radius, $\tilde{\beta}$, for MVPE configurations related to HH models (full lines) and HP models (dashed lines). A similar trend shown in both cases is probably related to a coinciding stellar density profile. A lower threshold for the occurrence of maxima takes place for cored halos, with respect to cuspy halos.

Table 1. Values of the fractional scaling radius, $\tilde{\beta}$, the global fractional mass, $\widetilde{\mathcal{R}}$, and the fractional mass within the stellar scaling radius, $\widetilde{\mathcal{M}}$, for a number of representative MVPE configurations related to HH and HP models, respectively, in the *energy-conservation paradigm*.

$\tilde{\beta}$	$\widetilde{\mathcal{R}}^{(HH)}$	$\widetilde{\mathcal{R}}^{(HP)}$	$\widetilde{\mathcal{M}}^{(HH)}$	$\widetilde{\mathcal{M}}^{(HP)}$
03	-	05.55	-	0.70
04	-	07.29	-	0.41
05	20.76	09.32	2.30	0.28
06	22.49	11.59	1.83	0.20
07	24.74	14.06	1.54	0.15
08	27.27	16.72	1.34	0.12
09	30.00	19.57	1.20	0.10
10	32.88	22.61	1.08	0.09
15	48.90	40.59	0.76	0.05
18	59.58	53.55	0.66	0.04
20	67.09	63.10	0.60	0.03

$$\widetilde{\mathcal{M}}^{(HP)}(\widetilde{\beta}) = \frac{4\widetilde{\mathcal{R}}^{(HP)}}{(1+\widetilde{\beta}^2)^{3/2}} \tag{23}$$

for HH and HP models, respectively. The plot of Eqs.(22) and (23) is presented in Fig.4 (bottom panel). It can be seen that that fractional masses $\widetilde{\mathcal{M}}$ of MVPE configurations with $\widetilde{r}_* < \widetilde{r}_h/5$ (i.e., $\widetilde{\beta} > 5$), and the related trend, are consistent with observations: in fact larger $\widetilde{\mathcal{M}}$ values occur for larger $\widetilde{r}_*$, or lower $\widetilde{\beta}$ (see, e.g., Cappellari et al., 2006).

The energy conservation paradigm makes a simple but very powerful tool to univocally define MVPE configurations for two-component models of ETGs and their hosting halos. The whole set of MVPE configurations, $(\widetilde{\mathcal{R}}, \widetilde{\beta})$, can be determined for any assigned model with specified density profile for each subsystem. As soon as the $\widetilde{\mathcal{R}}$ value is fixed, a MVPE configuration is defined and model predictions can be compared with results deduced from observations.

3.2. On the Restriction of Energy Conservation

Though dissipative processes have a central role in galaxy formation, the *energy-conservation paradigm* is a useful tool in dealing with the late evolution, where energy conservation (in absence of merger events) holds to a first extent.

Imagine a cold gas collapsing into the potential well of a massive halo. The proto-galaxy undergoes all sorts of dissipative processes till the main bulk of stars is produced, and the system becomes globally dissipationless with total energy, E. Assuming energy conservation, implies:

$$E = E_{vir} \tag{24}$$

Let E_o, ΔE, be the total energy at the beginning of evolution and the change up to the (fiducial) end of the dissipative phase, respectively. In this view, Eq. (24) takes the more general form:

$$E_o - \Delta E = E = E_{vir} \tag{25}$$

on the other hand, according to Eqs. (20) and (21), MVPE configurations are defined by a selected model via $\widetilde{\mathcal{R}}_{(vir)}$, regardless of the total energy, E, which acts as a mere scaling parameter.

In conclusion, energy conservation may be considered in a more general scenario, where energy dissipation occurs but the energy change, ΔE, is supposed to be known, and the special case, $\Delta E = 0$, corresponds to energy conservation during the whole evolution.

4.   Linking Tidal Potential Energy with Observations

It can be shown that the application of the virial theorem to two-component systems yields (see, Ciotti et al., 1996):

$$M_* = c_2^* R_e \sigma_o^2 \tag{26}$$

where c_2^* is a coefficient which depends on the DM halo to stellar mass ratio $\mathcal{R}$, fractional scaling radius β, and radial orbital anisotropy.

Following the same procedure adopted in an earlier attempt (Ciotti et al., 1996), by solving Jeans equation for a two-component system, the aperture projected velocity dispersion for the stellar component is obtained as:

$$\sigma_{ap}^2(R_{ap}) = \frac{2\pi}{M_p^*(R_{ap})} \int_0^{R_{ap}} \Upsilon_* I(R)\sigma_p^2(R)R\mathrm{d}R \qquad (27)$$

where $M_p^*(R_{ap})$ is the projected stellar mass inside the aperture radius R_{ap}, $I(R)$ is the projected luminosity profile, $\sigma_p^2(R)$ is the projected velocity dispersion profile and Υ_* is the stellar mass-to-light ratio, assumed constant throughout the galaxy. For MVPE configurations, with the position, $\sigma_{ap}^2 = \sigma_o^2$, the coefficient c_2^*, which comprises the effects of the DM halo (leaving aside non-homology and radial orbital anisotropy from our analysis), can be expressed as a function of either $\widetilde{\mathcal{R}}$ or $\widetilde{\beta}$, by substituting Eq.(20) and Eq.(21) into Eq.(26). Accordingly, the following relations hold:

$$c_2^{*(HH)} = f_\beta^{(HH)}(\widetilde{\beta}) = f_{\mathcal{R}}^{(HH)}(\widetilde{\mathcal{R}}) \qquad (28)$$

$$c_2^{*(HP)} = f_\beta^{(HP)}(\widetilde{\beta}) = f_{\mathcal{R}}^{(HP)}(\widetilde{\mathcal{R}}) \qquad (29)$$

where the functions are calculated for spherical, isotropic models. The result is plotted in Fig.5: being $\widetilde{\mathcal{R}}$ monotonically increasing with $\widetilde{\beta}$, c_2^* also does with both $\widetilde{\mathcal{R}}$ and $\widetilde{\beta}$. This is going on the right way, because larger c_2^* means, as deduced from the FP edge on view, less massive galaxies, and this is compatible with wide amounts of DM (i.e., larger $\mathcal{R} = M_h/M_*$), and relatively smaller structures (i.e., larger $\beta = r_h/r_*$, which means smaller r_*); the vice versa holds true also.

The $\widetilde{\mathcal{R}}$-$\widetilde{\beta}$ relation plotted in Fig.4 has been splitted in Fig.5 by replacing c_2^* with either $\widetilde{\beta}$ (left panel) or $\widetilde{\mathcal{R}}$ (right panel), respectively. Values of c_2^* within the shaded region, correspond to MVPE configurations with unreasonably high $\widetilde{\mathcal{R}}$ values ($\widetilde{\mathcal{R}} > 25$ for HH models). Other models with halo inner logarithmic slopes ranging between 0 (P profiles) and 1 (H profiles), would correspond to MVPE configurations with more reasonable $\widetilde{\mathcal{R}}$ values.

4.1. MVPE Configurations and the FP

The most diffuse representation of the FP for ETGs writes:

$$\log R_e = A \log \sigma_0 + B\langle I \rangle_e + C \qquad (30)$$

where A, B and C, are constant coefficients (for each wavelength filter bandpass) derived by means of a multiple regression fit of the effective radius R_e (the radius of the circle encircling half the total galaxy luminosity), the central projected velocity dispersion, σ_o, and the average effective surface brightness within the effective area in flux units, $\langle I \rangle_e$.

The deviation of the A coefficient from the value of 2 predicted by the Virial Theorem is commonly known as the "tilt problem" of the FP. Several attempts have been made to understand the origin of the observed tilt. Among them we recall here: 1) the radial orbital anisotropy (see, e.g., Nipoti et al., 2002) in the velocity distribution of ETGs; 2) the weak

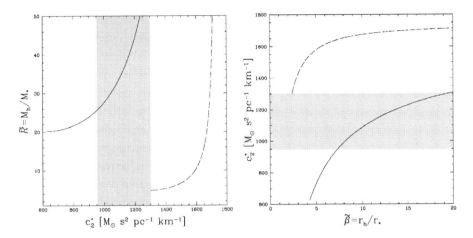

Figure 5. Plot of $\widetilde{\mathcal{R}}$ as a function of c_2^* (left panel), and as a function of $\widetilde{\beta} = \widetilde{r}_h/\widetilde{r}_*$ (right panel), for HH (full lines) and HP (dashed lines) models. Values of c_2^* within the shaded region, correspond to MVPE configurations with unreasonably high $\widetilde{\mathcal{R}}$ values ($\widetilde{\mathcal{R}} > 25$ for HH models). Models related to DM density profiles with inner slope between their HP and HH counterparts should, in principle, lie within the shaded region and exhibit lower $\widetilde{\mathcal{R}}$ values.

homology (see, e.g., Bertin et al., 2002; Trujillo et al., 2004) of the structure of ETGs, that is the deviation of the light profiles of the galaxies from the simple de Vaucouleurs $r^{1/4}$ law; 3) the Initial Mass Function and Star Formation (see, e.g., Renzini and Ciotti, 1993; Chiosi et al., 1998), that is the variation of the M/L ratio with the ETG mass/luminosity; 4) the Dark Matter (see, e.g., Ciotti et al., 1996; Borriello and Salucci, 2001), that is a trend of the tilt with the dark-to-bright mass ratio; 5) the existence of a maximum in the Clausius potential energy of ETGs (see, Secco, 2000, 2001, 2005). Up to nowadays none of these explanations for the FP tilt have been definitely confirmed. In particular, the last attempts (Secco, 2000, 2001, 2005) deal with MVPE configurations but under the restriction of a frozen halo i.e. fixed r_h (see Fig. 1) regardless of energy conservation, and using different scaling laws with respect to the current investigation.

In what follows, the FP tilt of ETGs shall be considered in the light of the *energy-conservation paradigm*. More specifically, MVPE configurations shall be placed on the 3D observational parameter space, (R_e, σ_o, I_e). To get a physical and immediate overview of the problem, the κ-plane of Bender et al. (1992) shall be used:

$$\kappa_1 = (\log \sigma_o^2 + \log R_e)/\sqrt{2} \sim M_{dyn} \tag{31}$$

$$\kappa_2 = (\log \sigma_o^2 + 2\log I_e - \log R_e)/\sqrt{6} \tag{32}$$

$$\kappa_3 = (\log \sigma_o^2 - \log I_e - \log R_e)/\sqrt{3} \sim (M/L)_{dyn} \tag{33}$$

$$I_e \equiv M_*/(2\pi \Upsilon_* R_e^2) \tag{34}$$

where $M_{dyn} = M_* + M_h$ is the total mass. The combination of Eq.(26) and Eqs.(31)-(34) yields:

$$\kappa_1 = (\log M_* - \log c_2^*)/\sqrt{2} \tag{35}$$

$$\kappa_2 = (3 \log M_* - \log c_2^* - 6 \log R_e - 2 \log \Upsilon_* - 13.6)/\sqrt{6} \tag{36}$$

$$\kappa_3 = (\log \Upsilon_* - \log c_2^* + 6.8)/\sqrt{3} \tag{37}$$

To close the system of Eqs.(31)-(33), an additional equation is needed, provided by the empirical Faber-Jackson law:

$$\sigma_o = 220 \left(\frac{L}{L_o}\right)^{0.25} [\text{kms}^{-1}] \quad ; \quad L_o \equiv 10^{10} h^{-2} L_\odot \tag{38}$$

which combined with Eq.(26) and (36), assuming a dimensionless Hubble parameter, $h = 0.7$, yields, after some algebra:

$$\sigma_o^2 = 0.340 L^{0.5} = \frac{M_*}{c_2^* R_e} \rightarrow R_e = 3 \frac{M_*^{0.5} \Upsilon_*^{0.5}}{c_2^*} \tag{39}$$

$$\kappa_2 = (5 \log c_2^* - 5 \log \Upsilon_* - 16.5)/\sqrt{6} \tag{40}$$

In conclusion, κ_2 and κ_3 depend only on Υ_* and c_2^*, while κ_1 on M_* and c_2^*.

The comparison of model predictions with observational data has been made with a sample of 9000 ETGs, in the redshift range $0.01 \leq z \leq 0.3$, selected from the Sloan Digital Sky Survey first data release (SDSS DR1) using morphological and spectral criteria (see, Bernardi et al., 2003, for further details). The sample is highly representative of nearby ETGs and has been found to produce a relatively thin FP. The SDSS DR1 data have been plotted together with model predictions in Figs. 6, 7, 8, for fixed values of Υ_*. It is apparent that MVPE configurations with assumed constant Υ_* cannot reproduce the whole set of data within a fiducial range, $4 \leq \Upsilon_* \leq 12$. In addition, mapping some region in the κ_1-κ_3 plane does not necessarily imply a corresponding mapping in the κ_1-κ_2 plane. Finally, the scatter along κ_1 axis shown by MVPE configurations exceeds on both sides the scatter shown by the data, for reasonable values of M_*. At this stage, a number of preliminary considerations can be drawn, namely:

1. There is no apparent correlation between MVPE configurations and the stellar mass to light ratio, Υ_*.

2. Dealing with MVPE configurations only, within the *energy-conservation paradigm*, yields a restricted range in c_2^* (i.e., in κ_3), for both HP and HH models, no matter the value of Υ_*, while data from observations span a wider range.

3. There is no apparent correlation between MVPE configurations and total or stellar mass.

The first point comes from the fact that, for an assigned DM to stellar mass ratio, $\mathcal{R}$, MVPE configurations select a single effective radius, R_e. On the other hand, Υ_* is strictly connected with age and metallicity of the galaxian population instead of the peculiar structure and mass/light distribution of the system.

The second point is highly model dependent: though different models can be related to different ranges in c_2^*, still the range width shows little model dependence provided only MVPE configurations are dealt with.

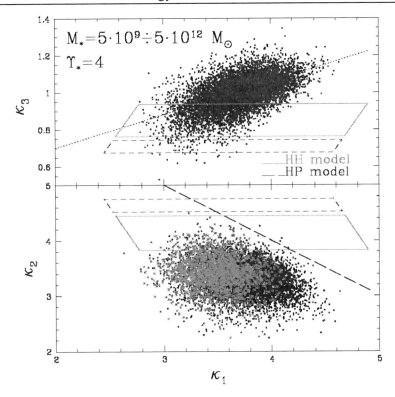

Figure 6. Comparison between observations and model predictions in the κ-space FP for ETGs within the mass range, $5\,10^9 \leq M_*/M_\odot \leq 5\,10^{12}$, assuming $\Upsilon_* = 4$. The DR1-SLOAN survey in r^* band is represented by dots, and the best fitting relation by a dotted line (top panel). The dashed line (bottom panel) is an empirical border line which seems not to be violated by any gravitationally bound astrophysical object. Full and dashed line parallelograms show the coverage of HH and HP models, respectively. ETGs matched by HH and HP models in the κ_1-κ_3 plane (top panel) are represented by starred points in the κ_1-κ_2 plane (bottom panel). Data from observations cannot be entirely matched by MVPE configurations in both planes.

The third point means that the MVPE configurations, according to Eq. (**??**), are defined by the pair $(\widetilde{\mathcal{R}}, \widetilde{\beta})$, regardless of the total mass, $M = M_* + M_h$. To describe the FP tilt, a relation between total mass ($\propto \kappa_1$) and total mass-to-light ratio ($\propto \kappa_3$) is needed. In conclusion the FP tilt cannot be explained, in the light of the *energy-conservation paradigm*, by MVPE configurations, assuming a constant Υ_*.

4.2. Testing with N–Body Simulations

The aim of the current subsection is to see to what extent MVPE configurations, within the *energy-conservation paradigm*, can be preferred final states of the transition from earlier non-virialized states, e.g. at turnaround. Keeping in mind that star formation processes, energy feedback, implementation and hierarchical merging are of extreme interest in the analysis of galaxy assembly processes, further attention shall be devoted to the evolution

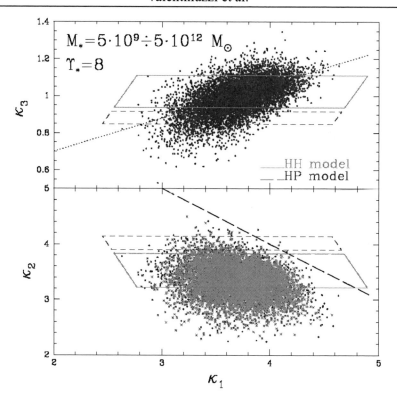

Figure 7. Same as Fig.6 but assuming $\Upsilon_* = 8$.

of the stellar subsystem induced by gravitational action only. If MVPE configurations are preferred final states of gravitationally bound two-component systems, they must be attained regardless of the formation process, in particular through a monolithic gravitational collapse of star particles in the potential well of a massive DM halo. In the following, the monolithic gravitational collapse shall be considered due to its intrinsic simplicity, but the basic ideas and the results hold true also for more complex galaxy formation processes.

4.2.1. The VPE for a Two-Component γ-model

For the sake of exploring numerical simulations performed in a recent investigation (Nipoti et al., 2006), an additional model must be considered, where each density profile (for fixed mass and scaling radius) depends on a single parameter, γ, for that reason called the γ-model (see, e.g., Dehnen, 1993; Tremaine et al., 1994). Two-component γ models shall be hereafter quoted as GG models. In the special case, $\gamma = 1$, γ density profiles reduce to Hernquist density profiles, and GG models coincide with HH models. Different values of γ parameter permit the exploration of different inner slopes for the two components. Following a similar procedure with respect to HH and HP models, within the *energy-conservation paradigm*, the counterpart of Eqs. (16)-(17) and (20)-(21) for GG models reads:

$$V_*^{(GG)} \;=\; E\, f_*^{(GG)}[\mathcal{R}_{(vir)}, \beta_{(vir)}] \tag{41}$$

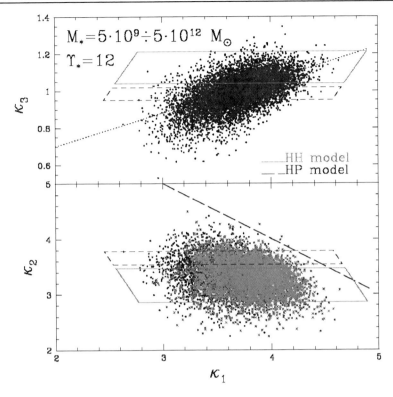

Figure 8. Same as Fig.6 but assuming $\Upsilon_* = 12$.

$$\widetilde{\mathcal{R}}^{(GG)} = f^{(GG)}(\widetilde{\beta}) \tag{42}$$

where integrations must be performed numerically. Approximate threshold values, $(\widetilde{\beta}, \widetilde{\mathcal{R}})$, defining MVPE configurations, are listed in Tab. 2 for different GG models. If two-component virialized density profiles yielded by computer simulations can be fitted by model parameters, $(\gamma_*, \gamma_h, \mathcal{R})$, then the pair, $(\widetilde{\beta}, \widetilde{\mathcal{R}})$, defining the related MVPE configurations, can be determined.

Values of parameters related to two-component, collisionless, virialized systems at the end of computer runs performed in a recent investigation (Nipoti et al., 2006), are listed in Tab.3. Cases (i)-(iv) and (vii) therein appear to be largely aspherical, as shown by the ratio of the principal axes of inertia, a, b, c, of the stellar subsystem.

In the special case of homeoidally striated ellipsoids (Roberts, 1962) where the boundaries are also similar and similarly placed (Caimmi, 1993; Caimmi & Marmo, 2003), the asphericity effect is expressed by a single shape factor which, for the larger elongation among cases of Tab.3, $c/a = b/a = 0.3$, equals about 1.6 instead of 2 related to spherical shapes (e.g. Caimmi, 1991). Accordingly, values of parameters listed in Tab.3 hold, to an acceptable extent, also for spherical configurations. The general situation could yield a larger discrepancy, but the shape of the DM halo at the end of computer runs has not been mentioned in the parent paper (Nipoti et al., 2006), even if small asphericities are expected for sufficiently high DM to stellar mass ratios, $\mathcal{R} \stackrel{>}{\sim} 2$. With this caveat in mind, the effect of oblateness or elongation shall be neglected in the following.

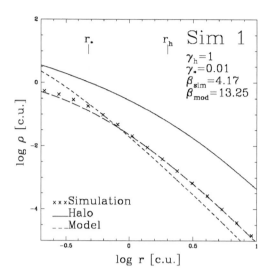

Figure 9. The best fitting (at the end of computer run Sim1 listed in Tab.3) GG model and related MVPE configuration. Captions: crosses - values of stellar density from the output of simulations; long-dashed line - best fitting γ_* model; full line - fixed γ_h halo model; short-dashed line - MVPE configuration with same $\mathcal{R}$ and γ_*. The stellar and DM scaling radius, r_* and r_h, related to the best fitting GG model at the end of computer run, are also represented. Both radius and density are in code units.

An inspection of Tab.3 shows that virialized configurations are obtained below the threshold, $\tilde{\mathcal{R}}_{min} \sim 10$, where MVPE configurations cannot take place (see Tab.2). Aiming to see if the same holds also above the threshold, a new set of 5 computer runs whit $\mathcal{R} = 25$, have been performed.

In order to maximize the resolution in the stellar subsystem and reduce computer time, the DM halo is modeled as a fixed gravitational potential induced by a spherical γ_h density profile, while stars are represented as particles ($N = 102,400$) of equal mass. At the end of the computation, the stellar subsystem is fitted by a γ_* density profile, and the related β is determined together with its counterpart, β_{MVPE}, related to the MVPE configuration with same $\mathcal{R}$. Values of input and output parameters are listed in Tab.4.

The presence of a massive DM halo, $\mathcal{R}=25$, implies a negligible tidal effect of the stellar subsystem, with the possible exception of the central region, where baryon concentration may be sufficiently large. This ensures that the technique used is admissible.

The best fitting (at the end of the computer run) GG model is compared to the MVPE configuration with same $\mathcal{R}$ in Figs. 9-12, for all cases listed in Tab. 4 except Sim3, which cannot be reproduced by γ density profiles. An inspection of Figs. 9-12 shows a difference between the best fitting γ_* model and its MVPE configuration counterpart with same $\mathcal{R}$, the latter being systematically steeper than the former.

The above results provide additional support to the idea that MVPE configurations are not preferred final states of gravitationally bound systems, within the *energy-conservation paradigm*. One of the caveats of the current analysis is hidden in the choice of the best fitting profile for the stellar component. In principle, the method used here can be adapted to all

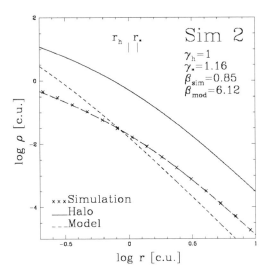

Figure 10. The best fitting (at the end of computer run Sim2 listed in Tab.3) GG model and related MVPE configuration. Captions as in Fig.9.

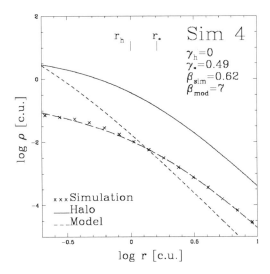

Figure 11. The best fitting (at the end of computer run Sim4 listed in Tab.3) GG model and related MVPE configuration. Captions as in Fig.9.

the possible final virialized mass density profiles, to select the related MVPE configuration. Simulated mass distributions at the end of the computer run have been fitted by γ density profiles, but better choices could in principle exist (e.g. Caimmi, 2006) which, on the other hand, are expected to yield similar results.

In conclusion, MVPE configurations can neither explain the tilt of the FP of ETGs, nor make preferential final states at the end of the transition from non equilibrium to virialized configurations, in the light of the current assumptions. To this aim, some other additional

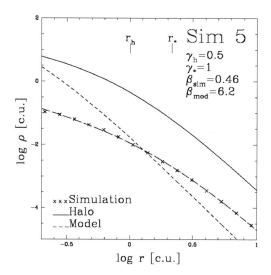

Figure 12. The best fitting (at the end of computer run Sim5 listed in Tab.3) GG model and related MVPE configuration. Captions as in Fig.9.

Table 2. Approximate threshold values, $(\widetilde{\beta}_{min}, \widetilde{\mathcal{R}}_{min})$, related to MVPE configurations for different GG models in the *energy-conservation paradigm*. MVPE configurations below $\widetilde{\mathcal{R}}_{min} \sim 10$ cannot occur in GG models.

GG models		
(γ_*, γ_h)	$\widetilde{\beta}_{min}$	$\widetilde{\mathcal{R}}_{min}$
$(0.0, 0.0)$	3.0	10
$(0.0, 1.0)$	7.5	18
$(1.0, 0.0)$	2.0	12
$(1.0, 1.0)$	4.5	20
$(1.5, 1.5)$	9.0	53
$(1.5, 1.0)$	3.8	25
$(1.5, 0.0)$	1.8	16

relation has to be looked for. On the other hand, the tidal potential energy induced by hosting DM halos on ETGs plays an essential role, even if not sufficient, for the interpretation of the FP of ETGs.

5.   Conclusions

ETGs embedded in DM halos have been modeled as two-component systems, where the stellar VPE is the sum of the self potential energy of the stellar subsystem, and the tidal potential energy of the DM halo. In general, the VPE can be represented as an hypersurface,

Table 3. Values of parameters related to two-component, collisionless, virialized systems at the end of computer runs performed in a recent investigation (Nipoti et al., 2006). The principal axes of inertia related to the stellar subsystem, are denoted as a, b, c, respectively. Cases (i)-(iv) and (vii) are largely aspherical.

ID	Name	c/a	b/a	γ_*	γ_h	β	$\mathcal{R}$
i	hpl1r2m002	0.34	0.35	0.12	1.45	10.70	2
ii	plpl05r2m002	0.30	0.34	1.37	0.57	01.78	2
iii	plpl05r5m002	0.47	0.72	1.48	0.58	01.59	5
iv	g0hq05r2m002	0.42	0.85	1.98	0.65	02.48	2
v	hpl4r2m005	0.99	0.99	0.00	0.18	00.56	2
vi	hpl8r2m005	0.99	1.00	0.00	0.34	00.32	2
vii	npl4r2m005	0.33	0.33	0.00	1.37	05.14	2
viii	npl4r4m005	0.97	0.98	0.02	1.98	02.78	4
xi	g15pl4r2m005	0.97	0.99	0.00	1.38	00.86	2

Table 4. Input and output parameters related to 5 computer runs performed in the current investigation, starting from a spherical-symmetric γ_* density profile $(N = 102, 400)$ within a fixed gravitational potential induced by a spherical-symmetric γ_h density profile. The fractional scaling radii, β_{sim} and β_{MVPE}, are related to the best fitting (at the end of the computer run) GG model and MVPE configuration with equal $\mathcal{R}$, respectively. The stellar subsystem at the end of Sim3 computer run cannot be reproduced by γ density profiles.

ID	Input parameters				Output parameters			
	γ_h	γ_*	β	$\mathcal{R}$	γ_*	β_{sim}	$\mathcal{R}$	β_{MVPE}
Sim1	1.00	1.00	2.00	25	0.01	4.17	25	13.25
Sim2	1.00	0.00	1.00	25	1.16	0.85	25	6.12
Sim3	0.00	1.00	2.00	25	-	-	25	-
Sim4	0.00	0.00	0.50	25	0.49	0.62	25	7.00
Sim5	0.50	0.00	0.50	25	1.00	0.46	25	6.20

according to Eqs. (11) and (12), where no absolute extremum point is present. In a reduced representation, called *energy-conservation paradigm*, extremum points take place above a threshold in the DM halo to stellar mass ratio, $\mathcal{R}$, and MVPE configurations are defined with that specific fractional scaling radius, $\tilde{\beta} = \tilde{r}_h/\tilde{r}_*$, which univocally represents the maximum of the VPE.

With the aim of investigating the role of MVPE configurations in large-scale objects, ETGs and their hosting DM halos have been modeled as two-component systems described by density profiles of astrophysical interest. Using the FP of ETGs, it has been shown that MVPE configurations cannot entirely fit data from observations, and that the scatter exhibited along κ_1 axis largely exceeds the scatter exhibited by data from observations, for all different values of (assumed constant) stellar mass to luminosity ratio, Υ_*. Furthermore, assuming the Faber-Jackson relation, it has been shown that regions where the data are matched by model predictions in the κ_1-κ_3 plane, have no counterpart in the κ_1-κ_2 plane, which furthermore rules out some reasonable Υ_* values for ETGs (see, e.g., Fig.6). In summary, MVPE configurations alone cannot explain the FP tilt of ETGs, at constant Υ_*, and an additional constraint is needed, which study is left to future work.

For the sake of completeness, it has been tested if MVPE configurations are preferred final states at the end of the transition from assigned initial conditions, using dynamical N-Body simulations. To this aim, 14 computer runs have been analyzed, and the related (with a possible exception) virialized configurations have been found to either lie below the minimum threshold in $\mathcal{R}$ for MVPE configurations, or exhibit systematically shallower density profiles. This has been interpreted as a further support to the idea, that MVPE configurations (as defined in the *energy-conservation paradigm*) make no preferred final state for two-component systems where only gravity is at work. A larger set of computer runs might provide further evidence for, or reason against, the last conclusion. On the other hand, the tidal potential energy induced by hosting DM halos on ETGs plays an essential role, even if not sufficient, for the interpretation of the FP of ETGs.

A. Analytical Values of Stellar VPE Related Quantities

We present here a number of analytical quantities found through the paper. First of all the values of the general functions describing the interaction term of the VPE for both our models:

$$\widetilde{V}_{*h}^{(HH)}(\beta) = \frac{1 + 4\beta - 5\beta^2 + 2\beta(\beta + 2)\ln\beta}{(\beta - 1)^4}$$

$$\widetilde{V}_{*h}^{(HP)}(\beta) = \frac{\mathcal{Q}(2\beta^4 + 11\beta^3 - 12\beta^2 - 4\beta + 1) + (6\beta^4 - 9\beta^2)\mathcal{W}}{\mathcal{Q}^7}$$

$$\mathcal{Q} \equiv \sqrt{\beta^2 + 1} \qquad \mathcal{W} \equiv \ln[(\sqrt{1 - \beta^{-2}} - 1)(\sqrt{1 + \beta^2} - 1)]$$

and the explicit expression of the functions describing the VPE in the *energy conservation paradigm*:

$$f_*^{(HH)} = \frac{2\beta[(\beta - 1)^4 - 6\mathcal{R}(\beta - 1)(5\beta + 1) + 12\mathcal{R}\beta(\beta + 2)\ln\beta]}{\mathcal{R}^2(\beta - 1)^4 + \beta(\beta - 1)^4 + 6\mathcal{R}\beta(\beta - 1)^2(\beta + 1) - 12\mathcal{R}\beta^2(\beta - 1)\ln\beta}$$

$$f_*^{(HP)} = \frac{16\beta\mathcal{X}^7 + 96\mathcal{R}\beta[\mathcal{X}(2\beta^4 + 11\beta^3 - 12\beta^2 - 4\beta + 1) + (6\beta^4 - 9\beta^2)\ln\mathcal{Y}]}{\mathcal{X}^3[9\pi\mathcal{R}^2\mathcal{X}^4 + 16\beta\mathcal{Z}] - 288\mathcal{R}\beta^3\mathcal{X}^2\ln Y}$$

$$\mathcal{X} \equiv \sqrt{\beta^2 + 1}$$

$$\mathcal{Y} \equiv \frac{(\sqrt{\beta^2 + 1} - 1)(\sqrt{\beta^2 + 1} - \beta)}{\beta}$$

$$\mathcal{Z} \equiv \beta^4 + 6\mathcal{R}\beta^3 - 12\mathcal{R}\beta^2 + 2\beta^2 - 12\mathcal{R}\beta + 6\mathcal{R} + 1$$

for HH and HP models, respectively.

References

Bender, R., Burstein D., Faber, S.M. 1992, *ApJ*, **399**, 462.

Bernardi, M., Sheth, R.K., Annis, J. and 28 coauthors 2003, *AJ*, **125**, 1817.

Bertin, G., Ciotti L., Del Principe, M. 2002, *A&A*, **386**, 149.

Borriello A., Salucci, P. 2001, *MNRAS*, **323**, 285.

Brosche, P., Caimmi, R., Secco, L. 1983, *A&A*, **125**, 338.

Caimmi, R. 1991, *Ap&SS*, **180**, 211.

Caimmi, R. 1993, *ApJ*, **419**, 615.

Caimmi, R., and Marmo C. 2003, *NewA*, **8**, 119.

Caimmi, R. 2006, *App. Math. Comp.*, **174**, 447.

Cappellari, M., Bacon, R., Bureau, M., and 12 cohautors 2006, *MNRAS*, **366**, 1126.

Chiosi, C., Bressan, A., Portinari, L., Tantalo R. 1998, *A&A*, **339**, 355.

Ciotti, L., Lanzoni, B., Renzini, A. 1996, *MNRAS*, **282**, 1.

Clausius, R. 1870, *Sitz. Niedewheinischen Gesellschaft*, Bonn, p.114.

Dehnen, W. 1993, *MNRAS*, **265**, 250.

D'Onofrio, M., Valentinuzzi, T., Secco, L., Caimmi, R., Bindoni, D. 2006, *NewA*, **50**, 447.

Freeman, K., McNamara, G. 2006, "In Search of Dark Matter", Springer Praxis Books, XVI.

Limber, D.N. 1959, *ApJ*, **130**, 414.

Nipoti, C., Londrillo, P., Ciotti, L. 2002. *MNRAS*, **332**, 901.

Nipoti, C., Londrillo, P., Ciotti, L. 2006, *MNRAS*, **370**, 681.

Renzini, A. and Ciotti, L. 1993. *ApJ*, **416**, L49.

Roberts, P.H. 1962, *ApJ* **136**, 1108.

Secco, L. 2000, *NewA*, **5**, 403.

Secco, L. 2001, *NewA*, **6**, 339.

Secco, L. 2005, *NewA*, **10**, 349.

Tremaine, S., Richstone, D.O., Byun, Y., and 6 cohautors 1994, *AJ*, **107**, 634.

Trujillo, I., Burkert, A., Bell, E.F. 2004, *ApJ*, **600**, L39.

Zwicky, F. 1933, *Helv. Phys. Acta* **6**, 110.

In: Energy Research Developments
Editors: K.F. Johnson and T.R. Veliotti

ISBN 978-1-60692-680-2
© 2009 Nova Science Publishers, Inc.

Chapter 12

APPLICATION OF COST FUNCTIONS FOR LARGE SCALE INTEGRATION OF WIND POWER USING A MULTI-SCHEME ENSEMBLE PREDICTION TECHNIQUE

Markus Pahlow[1,*] *Corinna Möhrlen*[2,†] *and Jess U. Jørgensen*[2,‡]
[1]Department of Civil and Environmental Engineering, Institute of Hydrology, Water Resources Management and Environmental Engineering, Ruhr-University Bochum, 44780 Bochum, Germany
[2]WEPROG, Eschenweg 8, 71155 Altdorf, Germany

Abstract

The use of ensemble techniques for wind power forecasting aids in the integration of large scale wind energy into the future energy mix and offers various possibilities for optimisation of reserve allocation and operating costs. In this chapter we will describe and discuss recent advances in the optimisation of wind power forecasts to minimise operating costs by using a multi-scheme ensemble prediction technique to demonstrate our theoretical investigations. In recent years a number of optimisation schemes to balance wind power with pumped hydro power have been investigated. Hereby the focus of the optimisation was on compensating the fluctuations of wind power generation. These studies assumed that the hydro plant was dedicated to the wind plant, which would be both expensive and energy inefficient in most of today's and expected future electricity markets, unless the wind generation is correlated and has a very strong variability. Instead a pooling strategy is introduced that also includes other sources of energy suitable to balance and remove the peaks of wind energy, such as biogas or a combined heat and power (CHP) plant. The importance of such pools of energy is that power plants with storage capacity are included to enable the pool to diminish speculations on the market against wind power in windy periods, when the price is below the marginal cost and when the competitiveness of wind power as well as the incentives to investments in wind power become inefficient and unattractive.

---

[*]E-mail address: markus.pahlow@rub.de
[†]E-mail address: com@weprog.com
[‡]E-mail address: juj@weprog.com

It will also be shown that the correlation of the produced wind power diminishes and the predictability of wind power increases as the wind generation capacity grows. Then it becomes beneficial to optimise a system by defining and applying cost functions rather than optimising forecasts on the mean absolute error (MAE) or the root mean square error. This is because the marginal costs of up and down regulation are asymmetric and dependent on the competition level of the reserve market. The advantages of optimising wind power forecasts using cost functions rather than minimum absolute error increase with extended interconnectivity, because this serves as an important buffer not only from a security point of view, but also for energy pricing.

1. Introduction

Wind power is considered one of the most important renewable energy sources in the near future. In the past years the number of wind farms world wide increased strongly, with a total installed capacity of more than 94 GW by the end of 2007 [1], with about 50% distributed in Germany, Spain and the USA. This high concentration suggests that from a global perspective there is space for ample additional wind power. The issue is therefore not whether it is feasible, but rather what does it cost? The main focus in wind power integration in the past has been on producing the most accurate forecast with minimal average error, but experience has shown that this is not necessarily optimal from a cost perspective. Balancing costs will increase with increasing volumes of wind power. The larger the forecasted error (in MW), the higher the balancing costs.

Another aspect is the concentration level of wind power, which has a side effect from a forecasting perspective. This is the correlated generation and forecast error and is mainly relevant for large amounts of offshore wind power, as it is planned e.g. in the North Sea to in order to reduce transmission costs. Such scenarios make it unfeasible to run an electricity network with a forecast optimisation target of minimal mean absolute error, because it implies short periods during which GW of fossil fuel based power plants will have to be started with short notice. Germany is such an example, where it is the large errors that dominate the balancing costs of wind power. Skewness of the balancing costs of negative and positive reserve gives advantage to conservative forecasts in such areas and requires curtailment or other scheduled plants to stop generation with short notice. Such scenarios are also regularly experienced in Spain. An EU 6th-Framework project is assisting the Transmission System Operators (TSO's) in developing tools to handle such cases with a so-called cluster management [2].

From these experiences, it appears that cost functions will not only benefit the market prices and balancing costs, but also act as a means to increase system security. For those countries, where the installed capacity has gone beyond the 10 GW level, it has become an important consideration in the daily operation.

Whether intermittency poses technical limits on renewables in the future is certainly also of concern for other forms of renewable energy sources [3], since OECD (Organisation for Economic Cooperation and Development) Europe and IEA's (International Energy Agency) World Energy Outlook [4] project up to 23% market share of non-hydro renewable energy by 2030. Natural variations of resource availability do not necessarily correspond with the (also varying) need of the consumers. Balancing supply and demand is therefore a critical issue and potentially requires backup by other means of energy supply. The variations can occur at any time scale: hourly changes in output require balancing of short-term fluctuations by the so-called operational reserve, while days with low output require balancing of longer-term output fluctuations by so-called capacity reserves. Conversely, exceptionally windy days or rainy seasons can produce a surplus of supply and there might arise issues of handling excess capacity, where grids are not sufficiently interconnected. TSO's buy balance or reserve capacity in advance to ascertain secure grid operation. However, in addition to the anticipated cost savings, reliable forecasting aids in reducing the aforementioned problems, enabling high wind energy penetration and at the same time ensuring power system security and stability.

The development of wind energy is country dependent and the development structure is a function of the wind resource, the political environment, the electrical grid and the market. This means that different strategies are required to solve forecasting and, in a broader perspective, optimisation of energy systems in different countries.

Therefore, it seems to be natural to focus efforts on optimisation targets that can reduce the required reserve of the growing intermittent energy generation.

2.   The Optimisation Problem

Large amounts of wind power can not be integrated seamlessly into the electrical grid. There is a need for a combination of wind energy forecasting, interconnectivity and storage capacity to ascertain smooth operation. Today's market participants are not delivering this mixture of services and there are costs associated with each of these services. Forecasting is the cheapest solution and may in fact be sufficient to get a mix of power implemented in an optimal way.

The application of wind power forecasting can generally be divided into two different kinds. The first kind is to reduce the need for balancing energy and reserve power, i.e., to optimise the power plant scheduling. The second kind is to provide forecasts of wind power feed-in for grid operation and grid security evaluation. Wind power forecasting may well be one of the most direct and valuable ways to reduce the uncertainty of the wind energy production schedule for the power system. Therefore, the objectives of a wind power forecast depend on the application [5]. It in fact has to be separated between the following targets:

- For optimised power plant scheduling and power balancing, an accurate forecast of the wind power generation for the whole control area is needed. The relevant time horizon depends on the technical and regulatory framework; e.g., the types of conventional power plants in the system and the trading gate closure times.

- For determining the reserve power that has to be held ready to provide balancing energy, a prediction of the accuracy of the forecast is needed. As the largest forecast errors determine the need for reserve power, these have to be minimised.

- For grid operation and congestion management, the current and forecast wind power generation in each grid area or grid connection point are needed. This requires a forecast for small regions or even single wind farms.

2.1. Energy Prices and Market Structures

Competitive energy prices in a globalised world require a mix of multiple sources of energy. This is in fact a prerequisite in order to keep the energy generation process cost efficient. Shortage of components or resources will increase the price of a given energy generation method, if there is major concurrent demand. Renewable energy systems are therefore likely to be more expensive to install, whereas they are cheaper to operate. This is also due to fear of increasing prices of fossil fuel, further destruction of our environment and climate change.

The different marginal costs for different generation is another argument that points towards highly mixed energy systems. The future of energy will therefore most likely contain a more complex mixture of energy generation and development, possibly with the exception of regions with excessive resources of one particular energy source. Consequently, we have to expect that the electricity systems are going to become more complex to manage.

There is a strong trend that the society is in favour of supporting new developments of renewable energy and in particular non-scheduled intermittent sources such as wind energy. We therefore focus on how to keep a sustainable price for wind power. The optimisation strategies that will be presented hereafter may result in higher energy prices in the short-term, but a relatively lower energy price in the long-term compared to non-cost optimisation scenarios. The optimisation target here is therefore to the benefit of the entire society, but first of all to the current wind farm owners in markets, where wind power is frequently the price maker.

There may be other valid optimisation targets such as minimal emissions. This would however favour nuclear energy and lead to the dependency on one natural resource, unknown issues regarding the nuclear waste and therefore risk of high energy prices. In addition, nuclear energy is not a flexible energy generation mechanism, which could effectively complement the intermittent energy units. Either the nuclear or the intermittent generator would have to generate less power than they potentially could, each one at their standard marginal cost.

3.   Optimisation Objectives

Production incentives naturally help to increase the amount of renewables. These incentives are typically defined on the national level. The incentives are in most cases not important for the optimisation problem, but they may encourage a phase shift of the generation by using a storage unit and thereby maximising the production while the market price is highest instead of delivering power when the intermittent power generation is natively highest.

3.1. Market Considerations

Differences in the market structures, politically defined support systems, demand profiles and the inertia level of the scheduled power generation does make the optimisation problem specific for each region, if not even for each end-user. However, there are some basic difficulties, which all markets will experience once the volume of intermittent energy resources reaches during certain time periods "price maker levels". The physical nature of the difficulties are in fact independent of how and by whom the energy is traded.

Our aim is therefore to present an optimisation strategy that works equally well in systems, where wind power is handled centralised or de-centralised, regardless whether it is managed by a TSO or not. The basic strategy is the same. The fundamental problem is how to trade the varying amount of wind power on the market and to determine the value. The value depends on how well the intermittent energy satisfies the demand, the predictability of the weather and also how eager the scheduled generators are to deliver power. The difficulty increases with the ratio between intermittent generation and demand, apart from the eager scheduled power generators such as CHP plants that also need to generate heat and therefore have very low marginal costs for electricity generation.

3.2.   Transition from Fixed Prices to the Liberalised Market

The incentive to operate wind optimal from an economic perspective might however not always be given, if there is a fixed price policy and the consumers are enforced to pay over tariffs. It is very difficult to clarify in such cases, whether the system can be optimised. The alternative approach is then to leave it up to the wind generation owners to optimise the system. Fixed price policies may also be limited to either a number of years or are valid only until a certain number of MW-hours has been produced. Wind farm owners are usually thereafter enforced to trade their energy on market terms. The transition to market terms can be difficult, if there is a public monopoly that dumps the price with wind energy and takes the loss back over tariffs. This is especially the case on markets that operate with the price-cross principle on the day ahead spot market (e.g. the Nordic states). This leaves almost no possibility to get a good price for the energy produced, if the monopoly has enough volume to meet the demand. Wind power that has to be traded on market terms could in such cases in the future also be traded outside the market directly to other participants, which may be capable to absorb the energy in their energy pool. There are in fact also initiatives in Germany to get the possibility to trade wind energy directly on the market, although wind energy is by law traded by the TSO's and paid by a fixed price tariff [6]. While it is not yet clear how such models will be implemented into the legislation, they are a signal that there is an interest by the society in optimising the cost of clean (wind) energy.

Even though price dumping, as described above, is neutral for the end-users if there is no or little export, it nevertheless significantly lessens the value of renewables and in particular wind energy for the wind turbines/farms that operate on the market without the possibility to get a fixed price. If price dumping leads to export, then the importing party is receiving energy not only as clean energy, but also for a very low price, partially because it is paid over governmental subsidy in the area where it was produced. It is expensive for a society in the producing area to practise such export in the long run. It also means that such a country either has to produce the bulk of energy from renewables, or otherwise it will create a negative imbalance in the country's allowance of carbon emissions when exporting renewables.

In other words, price dumping prevents wind energy from becoming competitive and to develop to a non-subsidised energy source, at least in Europe. This is in the longer term not in the society's interest.

To conclude, a single wind power plant on its own will not be able to compete with a pool of many plants, unless it is located in unusual predictable weather conditions and hence could be considered a semi-scheduled plant. However, there are also means to operate as an individual on the market by outsourcing the handling of the wind power trading to parties that are specialised in this and may pool the energy of their aggregated customer's power to a larger portfolio. Sufficiently large wind generators may in the future also consider to forecast and trade themselves in order to get the market price in addition to some incentives, but also the penalties. Nevertheless, this is expected to be the most profitable way forward for wind energy in the future.

3.3. The Skew Competition in the Trading of Wind

The wind generators are some of the weakest parties on the energy market seen from a trading perspective. Everybody with a good weather forecast and with a reasonable approximation and experience on the market can predict the wind energy generation in windy periods and therefore also the impact on the market price from wind. This is not so much a problem as long as the fraction of wind energy of the total generation is less than $\sim 10\%$. Above this level, this becomes a more serious problem, since the wind power traders are forced to bid on the energy to a relatively low price or the traders risk to be forced to sell the energy to a pool for a lower price after gate closure.

As previously mentioned, it may be the most beneficial way to trade wind energy with a pool that contains sufficient storage to create additional uncertainty on the market regarding the available energy than what the weather forecast provides. In such cases, the market can no longer predict the energy generation from wind power and speculate as aggressively against the wind power trader. The market can as an example not know, if the wind energy pool decides to store energy for some hours instead of delivering the power and thereby take the peak off the wind power generation. The scheduled generators can in time periods of strong wind speculate against the wind power trader by setting a price that lies below the marginal costs, because they know that the wind will be available for a marginal price, if the wind generation was bid in with a higher price than their own. If the scheduled generators know that wind generation may not be available for a near zero price, because it may be withheld from the market for storage, then the risk becomes too high and the scheduled generator will stop such speculations.

Pooling can hence be regarded as a security measure against unfair speculation. The pool will in that case be used in daily operation to phase shift wind generation such that the correlation between the demand and total output of the pool is highest.

3.4. Uncertainty Considerations

Another factor that has impact on the market price and competition level is the uncertainty of the weather forecast. Uncertain weather requires that more energy generation is active, either as primary energy or as reserve. This again increases the average energy price, because a larger fraction of the available generation will be required for balancing.

Steep ramping of wind power is an indicator of uncertainty in the weather and often even a cause of such. Nevertheless, steep ramping requires more scheduled capacity online, because the efficiency of the ramping generators is lower during the ramp.

On the other hand, there are also times of low uncertainty and high wind power generation. In these times, the traders of wind energy have to do something extraordinary to get rid of the energy. Although it may still be possible to bring the energy into the market by using dumping prices, it may not cover the marginal costs anymore. However, once a certain percentage of the total consumption is exceeded, other methods will have to be applied. These include energy pooling, export, methods to increase the demand and other trading strategies.

4.   Optimisation Schemes

In this section we describe a number of optimisation schemes that are either already in operation, or are likely to become part of the near future's energy trading at the markets.

4.1.   Pooling of Energy

The optimisation problem when looking at pools of energy is relatively complex. Therefore, an example will be used in the following to illustrate the optimisation targets and possibilities of this scheme.

Let us introduce the Balance Responsible Party (BRP) as the party that ascertains that the wind together with other generation units in a pool follows the planned schedule for the pool. The BRP may not be a generator, but some party that has the required tools to effectively predict and manage the different pool members in an optimal way seen from the pool members point of view.

The first milestone for the BRP is to bring together a sufficient amount of uncorrelated wind power in the pool. The lower the correlation on timescales greater or equal than an hour, the lower the forecasting error and the higher the energy price.

Low correlation means that the area aggregation over individual sites gives a smoother output signal, which is easier to forecast. The same smoothing is then implicitly applied to the forecast. The net result is that the phase on the shortest timescale of the individual wind farms is invisible on the aggregated power generation.

A large volume of wind power also helps to level out negative impact from problems due to restrictions on the grid or reduced turbine availability, although these may only appear as small errors in a large pool.

One of the key parameters for a BRP is to secure a permanently high level of uncertainty of when and how much energy the pool is delivering to the network. As discussed earlier, this is because the wind power trading becomes vulnerable to speculation from stronger market participants, if it can easily be derived how much wind energy needs to be traded on the market. This is not a trivial task for the BRP, because there are also other constraints, which limit the possible "confusion" level introduced to the market.

Hence, the major tools that the BRP needs apart from the wind farm capacity are:

- A storage unit

- An ensemble prediction system for the difference between demand and wind power

- A short-term prediction system for wind power and demand

- An optimisation tool

An optimisation tool hence should combine forecasts with grid constraints, market trends and the physical constraints of the storage unit. Congestion on the grid is a typical reason for higher prices, but mostly to the disadvantage of the energy pool, if it is dominated by wind power. The optimisation problem is in theory global, but in practise a limited area problem, because the global problem extends to political decisions that would have to be modelled by stochastic processes. Instead, optimisation should be applied on the local grid. For this, time dependent boundary conditions are required. These should in theory contain the large-scale global trends, but at some stage approach the average trends for the season. The optimisation tool will have to comprise a set of partial differential equations including the storage unit, a portion of the grid and the intermittent energy sources. The partial differential equations describe the total output and exchange between storage units and the intermittent source. The accuracy of the numerical solution is partially determined by the extent of the domain of dependence for the partial differential equations. The domain increases with forecast horizon and can well reach part of the boundary, if the boundary values are accurately predictable by a simple function or some other prediction tool.

An example of a predictable boundary condition could be the large-scale electricity demand, which could be approximated with a periodic function to simulate the diurnal cycle. The large scale demand will after some hours have an influence on how the storage system should be scheduled, but the time derivatives of our intermittent energy is then likely to be the dominant forcing term on the storage equation.

The domain of dependence for the solution increases with the forecast horizon, partly because the dependency domain of the weather forecast increases. However, such an energy system has as a good approximation no feedback on the weather, thus this system can be one way coupled. Also, trading of oil and the transmission of gas have both an influence on the pricing. Conflicts between employers and their labour (e.g. strikes), unavailability of multiple plants and extreme weather at offshore platforms can cause peaks in the spot market prices of gas and oil and consequently also electricity. Such peaks can have a dominant negative impact on the average market value of wind power, if there would be imbalance in the pool during an interval with high prices.

An objective optimisation process would hence require an algorithm that carries out simulations in a closed system, but with the possibility to control the time dependent boundary conditions. Typically, a strike would be known in advance and the time dependent boundary conditions could be manually adjusted by the user to take account for such effects. Although the optimisation problem should ideally encompass the entire globe, it appears that the impact from far distances on the next couple of days can be modelled equally well by subjective boundary conditions, as with attempts to objectively model such effects. The conditions that take place at far distances are rather consequences of unpredictable events that have spurious nature. Any objective algorithm would have to be tuned to discard observations that conflict with the present state of the system to prevent that the large scale numerical solution would become unstable. Large scale waves would propagate through the system and trigger new waves on the local scale and the final solution would destabilise the energy system and the result would be higher balancing costs.

It is important that objective systems only accept observations that are likely to be correct. They must lie within an a priori defined uncertainty interval or be discarded.

Spurious waves in an objective system are worse than no waves, even if they would be correct, because the user does often not know of the source of the observations and it may be difficult to trace back. It is therefore most important in the design of the optimisation tool that the end-user can work with and define a robust set of boundary conditions and can explore the sensitivity of the solution to a number of incidents that are each relatively unlikely events.

4.2.   The "Price Maker" Optimisation Problem

If the optimisation target is to predict an optimal price in an area between two market systems with a price difference due to a limited interconnection capacity between the two markets, it is considered a "price maker problem". This is because the energy would flow from the low price area to the high price area and the BRP would try to sell the energy in the direction of the high price area like other participants and there would be some likelihood of congestion on the line.

The "price maker" must therefore continuously predict the price of the neighbouring areas with more inertia and try to trade in the direction of the highest price. This involves usage of weather forecasting on larger scales and computation of demand and intermittent energy generation. In this case, it is no longer sufficient to just predict what may be correct for the BRP itself. Other market participants might have access to different forecast information and may conclude very different or similar scenarios. This is thus an application, where ensemble forecasting is helpful. Depending on the ensemble spread, the likelihood of high or low competition can be determined and therefore also the price level on the market.

The weather determines the upper limit for what the BRP can sell. Before gate closure a decision has to be taken on the basis of weather forecast information. However, the BRP may decide to sell less or more than he expects to be produced, depending on the likelihood of the weather and the expected balancing costs. The "price maker" is likely to also cause the bulk of the imbalance and therefore also the bulk of the balancing costs. The "price makers" sign of the error will correlate with the sign of the total imbalance. This means that the error in every settlement interval counts as a cost, while this is not so for the "price taker" party whose sign is maybe 50% opposite to the "price makers" sign. The BRP has therefore an incentive to keep the balancing costs at zero and let other parties carry the balancing costs. The BRP can achieve this by using the ensemble minimum as a safe base generation. However, this principle leaves some excess energy at times, that needs to be traded with short notice on average under the sport market price scheme.

4.3.   The "Price Taker" Optimisation Problem

A small pool (in MW) can be traded and optimised with a so called "price taker" policy. This means, that the pool does not necessarily need price predictions, but only needs to keep the generation profiles according to their schedule. The "price taker" can assume a diurnal cycle of the demand and a pricing that follows this pattern. As a refinement, the "price taker" can try to predict the generation profile of competing intermittent generation and from this compute a new price profile of the trading interval. This would in some cases encourage to reschedule generation to achieve the highest possible price. Thus, instead of predicting the price for the bid, the "price taker" predicts the intervals with highest prices and schedules the generation, so that most hours are delivered during the high price time interval. This strategy secures that the "price taker" will get rid of the energy at the highest possible price. However, it also means that the "price taker" needs a rather flexible pool of generating units.

4.4.   The Combi-Pool Optimisation Scheme

The lower the predictability of the energy pool the better. Pumped hydro energy is one of the key storage units along with combined heat and power (CHP) and in smaller portions with biogas (e.g. [7], [8], [9]). In a CHP plant the energy can be stored as heat, if the energy regulations allow to do this on market terms. It is however useful to introduce the concept of scheduled demand, which is a better description of what such a "combi-pool" needs to include. Some markets do not allow direct coupling of generation and demand. It is nevertheless the most efficient way to level out differences between demand and generation.

Generally, the system operators often export imbalances and there seems to be a trend that larger markets work with increased import and export. This is on the one hand levelling prices out and helping non-scheduled generation and it may in many cases even be a better alternative to use scheduled demand. This would mean that heavy industry could benefit from low prices. Using both would allow for more intermittent energy on the grid.

What is going to increase the efficiency of a BRP is therefore a number of inventive solutions. The difficulty in predicting the price and output from a BRP increases with the amount of negative scheduled MWh and MW in the pool. However, information of the pool needs to be kept highly confidential for maximum competitiveness, which is under strong debate in Europe at present [10], if the BRP are TSO's.

Last but not least, the question has to be raised whether it is scientifically correct to optimise a system with a mechanism, where the primary target is to generate confusion for the market participants? The answer is yes, because this is the primary principle of the free market to maintain fair competition. Wind power does not operate under such fair competition, because it is exposed to the world via weather forecasts, which is strictly speaking against market principles.

5.   Wind Power Forecasting Methods

Now that various optimisation strategies have been discussed, it is important to get an understanding of the forecasting methodologies and possibilities to set up optimisation functions for the trading of wind energy. Therefore, the following sections will provide various approaches of wind power forecasting and discuss the error that the forecasting process is subject to. An error decomposition is used to give insight into the forecasting problem, but also into the limitations of forecasting. These limitations are the prerequisite to build up an optimisation system that benefits from this information to predict the required reserve to balance the intermittent energy source.

Following these principles, two different types of optimisation schemes for existing BRP's are described and demonstrated. These are discussed and conclusions are drawn to allow other optimisations scenarios to be set up as described above.

5.1. Different Approaches to Forecast Power Output

The aim of a wind power forecast is to link the wind prediction of the NWP model to the power output of the turbine. Three fundamentally different approaches can be distinguished (e.g. [5], [11], [12]):

- The physical approach aims to describe the physical process of converting wind to power and models all of the steps involved.

- The statistical approach aims at describing the connection between predicted wind and power output directly by statistical analysis of time series from data in the past.

- The learning approach uses artificial intelligence (AI) methods to learn the relation between predicted wind and power output from time series of the past.

In practical applications these methods are sometimes combined or mixed. The different types of wind power forecasting methods and systems currently in use worldwide were summarised by Giebel et al. [11].

It should be noted that, whenever possible, dispersion of wind power over large areas should be performed, as the aggregation of wind power leads to significant reduction of forecast errors as well as short-term fluctuations. In countries with a longer tradition and fixed feed-in tariffs this seems to be the natural way wind power is deployed. However, in countries, where wind power is relatively new and where larger single wind farms have been and are being built, developers, TSO's, authorities and policy makers will have to consider in the future pooling of energy sources not only to it's technical and economic feasibility, but also to allow for and set rules for such approaches, if efficient and environmentally clean deployment of renewables, especially wind energy, is a target.

5.2. Ensemble Prediction Systems

The use of ensembles is intended to provide a set of forecasts which cover the range of possible uncertainty, recognising that it is impossible to obtain a single deterministic forecast which is always correct [13]. An ensemble prediction system (EPS) is one that produces a number of numerical weather forecasts, as opposed to a single, deterministic forecast. Ensemble techniques have been employed for some time in operational medium-range weather forecasting systems [14]. Three approaches dominate the field: (1) Ensemble Kalman Filter (EnKF) approach (e.g. [15], [16], [17]), (2) Singular vector (SV) approach (e.g. [18], [19], [20]) and (3) Breeding approach (e.g. [21], [22]). There are two other ensemble methods, the multi- model approach (e.g. [23], [24], [25], [26], [27], [28]) and the multi-scheme approach. These are discussed and tested in several studies of their feasibility (e.g. [27], [29], [30], [31], [32]). The multi-model method results in independent forecasts, but the exact reason of the independent solutions will never be understood. Similar arguments do not apply to other ensemble prediction methods.

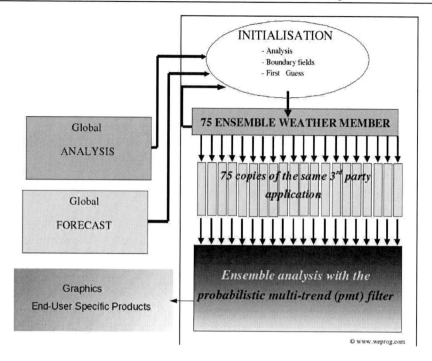

Figure 1. Schematic of the MSEPS ensemble weather & wind power prediction system.

5.3.   The MSEPS Forecasting System

The Multi-Scheme Ensemble Prediction System (MSEPS) that has been used for our studies and which will be described later in this chapter, is a limited area ensemble prediction system using 75 different NWP formulations of various physical processes. These individual "schemes" each mainly differ in their formulation of the fast meteorological processes: dynamical advection, vertical mixing and condensation. The focus is on varying the formulations of those processes in the NWP model that are most relevant for the simulation of fronts and the friction between the atmosphere and earth's surface, and hence critical to short-range numerical weather prediction. Meng and Zhang [31] found that a combination of different parameterisation schemes has the potential to provide better background error covariance estimation and smaller ensemble bias. Using an EPS for wind power prediction is fundamentally different from using one consisting of a few deterministic weather prediction systems, because severe weather and critical wind power events are two different patterns. The severity level increases with the wind speed in weather, while wind power has two different ranges of winds that cause strong ramping, one in the middle range and a narrow one just around the storm level (the cutoff level). Wind power forecasting models therefore have to be adopted to the use of the ensemble data. In general, a wind power prediction model or module, that is directly implemented into the MSEPS is different from traditional power prediction tools, because the ensemble approach is designed to provide an objective uncertainty of the power forecasts due to the weather uncertainty and requires adaptation to make use of the additional information provided by the ensemble. Figure 1 shows the principle of the MSEPS wind power forecasting system. Apart from the direct

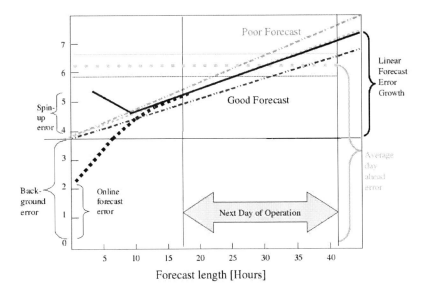

Figure 2. Error decomposition example generated with 1-year of data from the western part of Denmark.

implementation (e.g. [33], [34], [35]), some effort has been made in recent years by adopting traditional wind power prediction tools to ensemble data from the MSEPS system in research projects and studies (e.g. [36], [37], [38]).

6.   Aspects of the Forecasting Error

There is still a prevailing opinion in the wind energy community that the wind power prediction error is primarily generated by wrong weather forecasts (e.g. [11], [39]). From a meteorological perspective, this is a statement that may cause misunderstanding, because part of the error is due a complex mixture of weather related errors. The weather forecast process itself can only be blamed for the linear error growth with forecast length. We have therefore conducted an error decomposition in order to quantify the different error sources with a large ensemble of MSEPS weather forecasts. Traditionally, increased spatial weather prediction model resolution has been said to provide better forecasts (e.g. [40], [41]), but the shortest model waves may anti-correlate with the truth and cause double punishment in the verification (e.g. [32],[42]) and thereby additional model error. An ensemble prediction approach is another way to improve forecasts with fewer anti-correlation hours and the possibility to predict and understand forecast errors.

6.1.   Wind Power Error Decomposition

In order to understand the forecasting error in wind power, we have carried out a decomposition of the influencing components that have let to misunderstanding in the past. The analysis to do this includes forecasts generated in 6 hour frequency. This means that one particular forecast horizon will verify at four different times each day. The forecast error was accumulated in 6 hour bins with centre at 3, 9, ..., 45 hour prediction horizon for additional smoothing. This hides the disturbing diurnal cycle arising from verification once per day. Thereby a linear error growth with prediction horizon is achieved. Figure 2 shows such an error decomposition for a 1-year verification over the forecast length 3-45 hours with data from the western part of Denmark. By generating a decomposition of the forecasting error it can be illustrated, which parts of the error can actually be due to the weather forecast process. In Figure 2, we have split the error into a background error and a prediction system error with a linear error growth. We illustrated the potential improvements from the weather part with "good forecast", "average forecast" and "poor forecast", which differ to a certain degree. However, when looking at the background error, then the differences between a "good forecast" and a "poor forecast" is less significant. The background error is not directly "felt" by end users, because this initial error is in the daily operation either recovered by short-term forecasts with use of online measurements or if this is not available, by extrapolating the online measurements a few hours ahead. In the day-ahead trading, which takes place in most countries approximately 12 hours before the time period at which the bids have to be given on the market, the first few hours of the forecast are also irrelevant. The light gray line represents a typical pattern of an online forecast. This forecast has a steeper error growth, but starts from zero. A short-term forecast 1-2 hours ahead is typically close to the persistence level except in time periods, where the wind power ramps significantly.

The linear error growth indicates that the weather forecast is responsible for about 1/3 of the error in the forecast for the next day and the remainder is a background error originating from different sources. These additional error sources were found to be due to: (i) the initial weather conditions; (ii) sub grid scale weather activity; (iii) coordinate transformations; (iv) the algorithm used to compute the wind power; (v) imperfection of turbines and measurement errors. The question remains, which fraction of the background error is caused by imperfect initial conditions of the weather forecast and which fraction is due to erroneous wind power parameterisations. By extrapolating the linear forecast error growth from 9-45 hours down to the 0 hour forecast, the background mean absolute error (MAE) could be estimated to be just under 4% of installed capacity. Therefore, we added an additional fixed uncertainty band of +/- 4% of the installed capacity to the native MSEPS ensemble uncertainty to account for the background error that exists in addition to the weather forecast generated error. With this band, we achieved that 8120 hours out of 9050 hours or 89.7% of the hours are covered by the predicted uncertainty interval. The remaining 10% have numerically large errors that are only partially covered by the MSEPS uncertainty prediction. Figure 3 shows a scatter plot of this test. The x-axis shows the measured wind power [MWh] and the y-axis the mean absolute error (MAE) in % installed capacity. The black crosses are those forecasts that deviate less than +/- 4% from the measurements. Here 8120 hours are equivalent to 89.7% of the time. The gray crosses towards the top show those errors that are greater than +4% and are measured for 576 hours, equivalent to 6.4%. The gray crosses towards the lower boundary are 354 hours and equivalent to 3.9% of the time.

A 4% constant background error is a poor approximation and probably the explanation why 10% of the error events are unpredicted. Part of the background error is due to the computation of the wind power. The inherent error from the conversion of wind to power of course also has an impact on the error, not only the weather forecast. The difference of different methodologies for different forecast problems can be quite large. We will therefore demonstrate this difference in the next section. The following list shows the methodologies that have been used in this demonstration. The power prediction methods can be distinguished as follows:

- Method 1: Direction and time independent simple sorting algorithm.

- Method 2: Time dependent and direction independent least square algorithm.

- Method 3: Direction dependent least square algorithm.

- Method 4: Direction dependent least square algorithm using combined forecasts.

- Method 5: Same as method 4, but including stability dependent corrections.

- Method 6: 300 member ensemble forecast of method 5 - all farms are handled individually and 6 parameters for each one of the 75 members are used to compute the power.

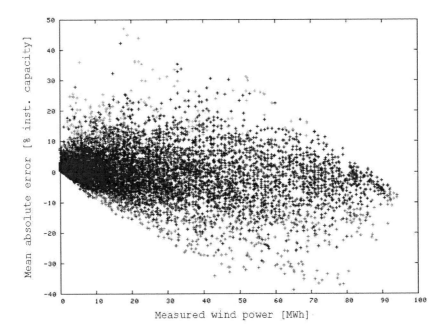

Figure 3. Scatter Plot of the mean absolute error (MAE) with a constant background error of 4% added to the native forecast of the EPS mean. The black crosses are EPS mean forecasts +/- 4% error, while the gray crosses display the errors that lie outside this band.

Note, when considering the results from investigating these different power prediction methods, all values are given as the improvement in % to the mean absolute error (MAE) for the day-ahead forecasts based on only one daily (00UTC) forecast performed with the basic power prediction (method 1).

To verify the impact of the conversion from meteorological parameters to wind power on the forecasting error, we use these 6 different methods to convert wind and other weather parameters into wind power output. Table 1 shows the results of the different power prediction approaches based on the same weather data.

Each result differs only in the level of detail of the statistical computation of wind power. The verification for Ireland took place for a period of 1-year (2005) for 51 individual wind farms of a total capacity of 497 MW, the Danish verification of aggregated wind power was up-scaled from 160 sites and took place for the period 09/2004-10/2005 and in Germany the verification was conducted for Germany as a whole and three individual TSO's for a 10-month period with the up-scaled online measurements valid for approximately 19 GW from estimated measurements published by the three TSO's (01/2006-10/2006). It can be seen from Table 1 that the forecast quality has been improved significantly for all investigated areas from a relatively simple power curve conversion method (i.e. method 1) to a more complex method (e.g. method 5 or 6), when making use of the additional information from the ensemble. The relatively low improvement in Australia is due to the fact that the background error is higher for only 6 wind farms in comparison to the areas with a large number of farms.

Three major results can be drawn from from this investigation:

- The advanced methods, using several parameters and EPS information are providing superior results to using simple methods.

- The forecasting error is dependent on the location, the size and the load of the farms/areas.

- Detailed measurements can significantly reduce the forecast error.

Table 1. Statistics of the investigation of different wind power forecast methodologies in various countries.

Method	Ireland	Australia	Denmark	Germany	E.ON	VE	RWE
Cap [MW]	497	388	1830	19030	7787	7486	3464
Load [%]	33	30	21	16	16	15	16
1	1.00	1.00	1.00	1.00	1.00	1.00	1.00
2	2.18	3.01	8.52	1.40	1.09	2.11	2.26
3	9.34	1.51	9.87	12.32	8.30	12.05	15.44
4	14.44	4.38	12.41	12.32	8.30	12.05	15.44
5	15.29	6.58	13.60	15.41	11.57	14.80	16.95
6	17.23	9.18	n/a	n/a	n/a	n/a	n/a

7. Reserve Prediction and Optimisation

The economic value of wind power is related to the predictability of the weather and hence the wind power. The required forecast horizon depends on the market structure and the inertia of the conventional power plant. An economic benefit from uncertainty predictions can in the most static markets only be achieved, if the prediction ranges up to 48 hours ahead. The trading of up-regulation and down-regulation can then be synchronised with the trading of the total generation and therefore the handling costs of wind power on the grid can be optimised.

This creates not only an economic advantage for the system operator, but also a fair strategy for the intermittent energy source, as it prevents dominant generators with large market shares gaining exclusive contracts caused by the imbalance from wind generation. In addition to the economic benefit, there is a security benefit of trading reserve capacity in advance, as the capacity can then be scheduled with the focus on grid security.

7.1. Optimisation of Reserve Predictions: Example Denmark

Next an example of a reserve prediction scenario is presented, based on a situation in which there is knowledge of the uncertainty of the wind power forecast in advance. This assumption is reflected by the statistical scores.

The up-regulation and down-regulation has been counted separately, because of the asymmetry in the market pricing structure. Note however, that for convenience, the computation of the *optimal forecast* in the example below is for simplicity based on the assumption that the price of up-regulation and down-regulation is the same, which would normally not be the case. It should in most cases be more efficient to trade with a lower prediction of future production of wind power than expected and arrange for down-regulation, when the wind power production increases above the predicted power, because down-regulation is in many markets n average only 25% of the price for up-regulation.

In our example, we compared different forecasting scenarios:

1. Scenario: Use of the optimal forecast and reserve prediction for next day's wind generation.

2. Scenario: Use of the optimal forecast for next day's wind generation, where reserve is only allocated on the next day according to the demand.

3. Scenario: Use of the Ensemble average forecast for next day's wind generation, where reserve is only allocated on the next day according to the demand.

The scenarios in this example were constructed for a fully competitive market, where reserve capacity is traded day-ahead. In this example, we used the same single forecast in both scenario no. 1 and no. 2. In scenario no. 1 we allocated reserve capacity in a market with competition on reserve. In case no. 2 and no. 3 all regulation was traded according to the demand on the short-term spot market.

Table 2 shows the results of the 3 scenarios. Note, that all values are in units percent of installed capacity. The numbers are taken from prioritised production in the western part of Denmark in the period January to April 2005. The forecast horizon is 17 to 41 hours until daylight saving starts, then the forecast horizon changes to 16 to 40 hours. The total prioritised production is rated at 1900 MW in this period, which is almost equivalent to the minimum consumption in this area.

Table 2. Results of the 3 scenarios of using and not using spinning reserve predictions in the wind power production forecasts for the western part of Denmark in the period January to April 2005. All statistical quantities are given as [%] of installed wind power capacity of Western Denmark

Scenario no.	1	2	3
EPS configuration	optimal*	optimal	average
Number of Hours	2331	2331	2331
Forecasted Mean	27.01	26.6	26.7
Bias	-0.77	-1.19	-1.02
Mean Absolute Error	2.72	5.30	5.83
Stdev Error	5.39	7.60	8.32
Correlation	0.977	0.953	0.946

It can be seen in Table 2 that (scenario no. 1) using the optimal forecast and reserve prediction for next day's wind generation almost halves the mean absolute error (MAE) of both scenario no. 2 and scenario no. 3 and hence the costs for expensive spinning reserve for unexpected events (errors). The root mean square scores are slightly lower, but still of the order 30% and 40% better for scenario 1 than for scenario 2 and 3, respectively. The lower improvement in the root mean square (RMSE) is related to the fact that the prediction of the forecast error removes the smaller errors in a band-like way around the optimal forecast. Since the RMSE is more sensitive to the larger errors, the improvement is lower.

Table 3 shows the percentage of reserve of installed capacity that is required, predicted and non-predicted in scenario 1. It can be observed that the predicted up-regulation provided better results than the down-regulation. This is the result of optimising the forecast towards the less costly reserve.

Table 3. Results of the verification of the predicted regulative power on a long-term basis for the Western part of Denmark in the period January to April 2005. The regulation magnitude is given as % of installed wind power capacity. The results correspond to scenario no. 1.

Regulation type	Regulation magnitude	description
Average Up	2.40	required regulation
Average Down	3.43	required regulation
Predicted Up	1.43	traded day ahead
Predicted Down	1.68	traded day ahead
Unpredicted Up	0.98	traded on the day or handled by flexible contracts
Unpredicted Down	1.75	traded on the day or handled by flexible contracts
Unused	0.91	unnecessary regulation capacity

The average of the ensemble has a score of 5.83%. This is approximately the same score as the best single ensemble member. The *optimal forecast* gives the same period 5.35% absolute error. In the above example we found that if we add the uncertainty band and trade the predicted amount of reserve for both up-regulation and down-regulation on the market, we can reduce the error to 2.72%. This means that we have to trade regulation for only 2.72% of the capacity instead of 5.35% in the short-term spot market with high prices.

In this computation, it is assumed that there is no error when the error of the *optimal forecast* lies within the predicted uncertainty band. This increases the correlation significantly, which indicates that a large fraction of the forecast errors within the uncertainty band are completely unpredictable. We suggest that these errors are handled most efficiently by being balanced with pre-allocated constant reserve.

To be able to estimate the impact of such a pre-allocation, we considered 5 cases, with different percentages of pre-allocation and computed the amount of hours, where the pre-allocation accounts for the forecast error. With these results, it becomes feasible to set up a cost analysis of the optimal amount of pre-allocation. Our intention here is however merely to demonstrate how to design an optimised prediction system for a specific problem.

In the following we have therefore constructed 4 cases, where a fixed fraction of the installed capacity is pre-allocated as reserve with a long term contract and additional capacity is assumed to be allocated in a market with competition according to the predicted requirement (see scenario 1 above).

1. Case: Additional long-term contract for reserve of 0.8% of the inst. capacity

2. Case: Additional long-term contract for reserve of 1.6% of the inst. capacity

3. Case: Additional long-term contract for reserve of 3.2% of the inst. capacity

4. Case: Additional long-term contract for reserve of 6.4% of the inst. capacity

5. Case: Additional long-term contract for reserve of 12.8% of the inst. capacity

The following equation has been used to compute the number of hours when the pre-allocated reserve accounts for the forecast error:

$$R = max(R_{pre}, a_{stability}x + b_{stability}) \tag{1}$$

where R_{pre} is the pre-allocated reserve from Table 4 and x is the ensemble spread.

The sum of the hours, where the resulting pre-allocated reserve fully covers the forecast error is shown in Table 4. It can be seen in the table, how well the ensemble spread covers the forecast error for different levels of pre-allocation. The validation also reveals that only 3 hours out of 2331 hours (97 days) had forecast errors that were not covered by the reserve given by equation 1 when using a 12.8% pre-allocation.

A pre-allocation of nearly 13% to reach 96% coverage by using this type of reserve is quite a large amount compared to the mean absolute error (MAE) of the forecast (¡ 6%). It is therefore recommended to study and evaluate from case to case and with real market data, if more than 10% of pre-allocation is a reasonable result of the optimisation. However,

Table 4. Number of hours out of 2331 hours where no additional reserve to the day-ahead and pre-allocated reserve is required for different levels of constant pre-allocation.

case no.	pre-allocated reserve [%]	no reserve required [hours]	percent coverage [%]
1	0.8	1049	45%
2	1.6	1439	62%
3	3.2	1696	73%
4	6.4	1990	85%
5	12.8	2228	96%

since there exists a pronounced price difference between up-regulating and down-regulating reserve, this amount could still proof as the most cost efficient.

It should also be noted, that this procedure does not eliminate the forecast error. Nevertheless, the reserve is traded more competitively than without using the predictions and hence the balancing costs can be reduced significantly.

Another aspect that may be of relevance in other countries is the combination of wind power and large power plants with respect to reserve requirements. In the western part of Denmark for example, the largest plant is 640 MW and corresponds to 33% of the prioritised wind generation. When this plant is in operation, then reserve requirements for wind power will never exceed the reserved 640 MW of up-regulation, which is required for this plant. A combination of wind power forecast failure and a 640 MW plant failure could of course cause a problem for grid security. However, it becomes clear from this example, that in a grid with large plants, it is not the amount of installed wind power that is responsible for the maximum reserve requirement and often also not for the large balancing costs. However, the shown methodology may also only provide a first step in how to combine the reserve allocation of larger power plants and wind power within an acceptable risk management structure and well working interconnections to cover parts of missing power in case of correlated failures.

7.2.   Optimisation of the Reserve Prediction: Example Canada

As described above, the reasoning for using cost functions rather than standard statistical measures is that the uncertainty in the weather is random. Converted to wind power, the weather uncertainty is sometimes found to be very low and sometimes found to be very high. In the long-term and if a reasonable amount of wind power is installed in an area, a constant reserve would consequently be inefficient and expensive. Additionally, full coverage of all forecast errors is expensive, also because there is always a risk of non-weather related accidents during operation. The following is an example of an area, where the weather is highly variable and so is the wind power production. Within Alberta Electric System Operator's AESO's wind power pilot study [43] it has also been found that strong ramping is not seldom. In such areas larger amounts of constant reserve allocation are very expensive. The investigated scenarios are therefore based on this experience and the next step in benchmarking and adopting a forecasting system no longer to a low mean absolute error (MAE), but rather to the conditions under which the wind power integration takes place and optimise the forecasting system with this knowledge.

7.2.1. Optimisation Scenarios

1. Scenario: Static Regulation
    In this scenario the upper limit for reserve allocation from the error statistic of one forecast over 1 year has been determined. The actual reserve allocation is limited by the forecast with the following restriction:

$$FC + R < UB \tag{2}$$

and

$$FC - R > LB \tag{3}$$

with

$$FC - R < G < FC + R \tag{4}$$

Here, FC is the forecast, R is the reserve, G is the actual generation, UB is the upper bound of full generation and LB is the lower bound of no generation. R is chosen to secure that the generation is always lower than the sum of the forecast and reserve and higher than the forecast without reserve, which is supposed to historically always be valid (here: 90%).

2. Scenario: Deterministic Forecast Regulation
    In this scenario the reserve R is chosen as a fixed reserve allocation for upward and downward ramping. We chose +/-11% of installed capacity, which is equivalent to the mean absolute error (MAE).

3. Scenario: Security Regulation
    In this scenario, the reserve R is computed from the difference between minimum and maximum of the ensemble in each hour of the forecast. There may however be areas, where it will be necessary to adopt the difference of minimum and maximum in such a way that single outliers are not increasing the spread unnecessarily.

4. Scenario: Economic Regulation
    In this scenario, an optimisation of scenario 3 is used, where the unused reserve allocation is reduced for economic reasons. As a first approximation the 70% quantile of the level in Scenario 3 was used. In a future optimisation, the skewness of the price of positive and negative reserve also would have to be considered.

The data that were used to simulate different scenarios was 1-year (2006) of forecasting data for two regions in Alberta, the South-West region and the South-Centre region, where 5 wind farms with a combined installed capacity of 251.4 MW were used. The simulation was carried out with 12-18 hour forecasts issued every 6th hour. The forecasting model system was run in 22.5 km horizontal resolution and the wind power forecast was a probabilistic forecast generated with 2 combinations of different power conversion methods each using 300 weather parameters from the 75 member MSEPS ensemble system.

In Table 5 statistics of the aggregated forecasts for the 5 wind farms are displayed.

Table 5. Statistics of wind power forecasts without reserve prediction.

Statistics Parameter	[% rated capacity]
Bias (FC)	-2.10
Mean absolute Error (FC)	11.60
RMSE (FC)	16.80
Correlation (FC)	0.85

In Table 6 all parameters are given in % of installed capacity. The relative cost of various types of reserve was not taken from market reports, but estimated as a percentage relative to urgent reserve on the spot market, which was set to 1.0. This gives the following estimates:

- Urgent reserve = 1.0%

- Unused passive reserve = 0.1%

- Allocated reserve = 0.6%

The "allocated reserve" is the reserve that is allocated according to the forecasts and assumed to be bought day-ahead. The statistical parameter in the first rows of Table 6 are based on the forecasts including a reserve allocation. This means that the estimated error is added or subtracted from the forecast according to the used optimisation scheme. Therefore, these parameters have an index FCR (forecast + reserve), while the statistical parameters in Table 5 are based on the raw forecasts and indexed FC. These FC forecasts were optimised on the mean absolute error (MAE) to the observations.

There are furthermore output results of three types of reserve in the table, the required reserve, the predicted reserve and the unpredicted reserve, respectively. These are named UpReg for up-regulation or DownReg for down-regulation of the power on the electricity grid because of incorrect forecasts. Although the difference in price for up-regulation and down-regulation, where down-regulation is often a factor of 3-5 cheaper than up-regulation, was not accounted for in these simulations, the cost function is more favourable towards down-regulation than up-regulation. This can be seen in scenario 2 and 4, which are more cost efficient and use less up-regulating expensive reserve.

Table 6 also shows the significant difference in effective costs for the first scenario, the purely static reserve allocation, in comparison to the other 3 scenarios. However, when comparing the hours covered by the reserve, then it becomes clear that the coverage and the effective cost are cross-correlated for this type of optimisation. Scenario 2 is for example equally cost efficient than scenario 5, but covers only 64.5% of the hours, while scenario 4 covers 76% of all hours. Although the security scenario (no. 3) is slightly less cost efficient, the covered hours are quite significantly higher than for scenario 2 and also scenario 4 (85% versus 76% and 64,5%). Looking at the mean absolute error (MAE) or the root mean

Table 6. Optimisation Scenarios for the AESO area.

Optimisation	Scenario 1	Scenario 2	Scenario 3	Scenario 4
	Static Reserve [% rated cap.]	Determin. FC Res. [% rated cap.]	Security Reserve [% rated cap.]	Economic Reserve [% rated cap.]
Reserve Predictor	75% reserve	FC+/-11%	max-min	0.7*(max-min)
Bias (FCR)	0.00	-0.85	-0.50	-0.80
MAE (FCR)	0.00	4.54	1.40	2.50
RMSE (FCR)	0.00	10.15	5.40	7.22
Correlation (FCR)	1.00	0.95	0.99	0.97
Required UpReg (7)	4.70	4.70	4.70	4.70
Required DownReg (8)	6.80	6.80	6.80	6.80
Predicted UpReg (9)	4.70	2.84	4.20	3.83
Predicted DownReg (10)	6.80	4.10	5.80	5.13
Unpredicted UpReg (7-9)	0.00	1.80	0.50	0.87
Unpredicted DownReg (8-10)	0.00	2.70	1.00	1.67
Unused Regulation	32.30	3.30	9.30	5.67
Effective cost	15.90	10.80	11.40	10.83
Hours covered by reserve	100.00	64.50	85.10	76.00

square error (RMSE), scenario 3 seems to also outperform scenario 2 and 4. However, when comparing the unused regulation, then the security scenario has almost double the amount of scenario 4, and three times as much unused regulation as scenario 2. Dependent on the pricing structure of the market, which was excluded in this experiment, this could even change the effective cost levels of the scenarios,

' The results of the Canadian example demonstrate once again that the statistical error measures are not capable of providing a complete answer for an optimisation target. However, the results do provide an insight of the complexity of the optimisation of cost functions of reserve to end-users requirements.

8. Summary and Discussion

The ensemble prediction method has a number of applications in wind energy integration and in energy in general. From the discussion in this chapter it can be concluded that energy markets can benefit from using ensemble predictions. Table 7 shows a general description of which forecasts from an ensemble forecasting system should be chosen for minimal costs. We have assumed that the capacity of wind power is sufficient for wind to be the "price maker" when the wind conditions are optimal. It can be seen in Table 7 that the forecast with the lowest mean absolute error (MAE) is not always the forecast that will generate the lowest cost of integration.

Although the original scope of ensemble prediction was to be able to conduct risk analysis of severe weather, it appears that the application in energy is not limited to grid security, but extends to trading and management of weather dependent energy generation systems. This includes all generation methods except nuclear power after the Kyoto protocol has become effective. Nuclear power generators do not have a CO_2 problem and have therefore the least incentive to participate in the balance of wind power. In the future, even nuclear plants may however need forecasts to be able to give bids on the market and to operate efficiently.

Table 7. Summary of the cost optimised forecast selection.

Predicted load factor in [%]	Competition on regulation	Forecast choice	Reserve allocation
0-10	Good on down-regulation	EPS minimum	Downward
70-100	Good on up-regulation	EPS maximum	Upward
20-70	Good for up and down	Best forecast	Down and up

The electricity price has a high volatility level because of the limited storage capacity and the strong relationship to oil prices, political disputes and not to forget the uncertainty in the weather development.

An increasing number of people around the world make their living on trading and because of the automatisation fewer people are required in today's production processes. This means that in the future, increasing volatility of stocks and energy can and have to be expected.

However, the volatility of the energy pricing may increase more than that of stocks for two reasons:

- The amount of intermittent renewables will increase more than the available storage capacity.

- The energy markets are developing slowly with new trading options.

Increased volatility on pricing will result in increased volatility on the generation as well, and consequently lower efficiency and higher costs. Increased volatility can also trigger instabilities on the grid. A typical example could be two competing generators that have to ramp with opposite sign to stay in balance. Increased volatility implies that the frequency of ramps will increase. Such ramps are not dangerous, but certainly do not add to the system security. The generators will bare the loss during the ramp, because of the higher average price.

The optimisation strategies that we have presented here serve to dampen volatility and the intermittent energy price. The main ingredients in this optimisation is ensemble forecasting, which increases the robustness of the decision process. Decisions will be taken on the basis of many results that are generated by some kind of perturbation. The market participants will, with the help of ensemble predictions in the future know in which range competing parties plan to set their bid on the market. There is also more continuity in time by using ensemble forecasting, because the decision process changes slowly hour by hour. This leads to a more stable decision process. Ensemble forecasting makes market participants aware of the risk of any speculation, although it may not be enough to prevent speculations.

A cost efficient way to dampen volatility is to allocate reserve in advance according to the uncertainty forecast for the intermittent energy source. The pre-allocation of reserve secures that a high level of competition can be achieved and last-minute volatility will be reduced.

A next step to reduce volatility is to create energy pools that allow phase shifting in time for intermittent energy. This will cap some of the price off the intermittent energy, because the balancing parties in the pool will demand a higher price than the intermittent generators. The more uncertainty there is on the pool's output for other market participants, the better for the pool. There is an incentive to make the pool members dynamic, so the other market participants cannot guess how much the pool is dependent on the intermittent generation pattern.

An additional step is required to secure fair prices during periods, where the intermittent energy is in excess. The market will often know the periods, where the weather is well predictable and the market price drops. Trading the intermittent energy several days in advance will allow scheduled generators to reduce their emissions and withhold their own generation. Their incentive to do so increases, if they can schedule the production well in advance and thereby cut the marginal costs down.

Trading of intermittent energy requires therefore a special effort and forecasts with an ensemble technique along with temporary pooling with other energy sources will aid in achieving efficiency and stable prices.

9. Conclusion

The major benefit of ensemble prediction methods is that weather dependent energy generation can be classified as certain and uncertain. It has been tradition to not separate between certain and uncertain generation in the trading, because day-ahead markets were used to sell according to one best possible forecast only.

A creative trader could still trade strategically on the basis of a deterministic forecast and pool the deterministic forecasts with ancillary services and bid in the sum of the intermittent and ancillary service generation with a higher price.

Ensemble forecasting however opens possibilities for more creative and efficient trading strategies. Additionally, the trading process is going to become more complex and in fact too complex for a subjective decision process in the future. The objective decision process then has to be done on a computer based on predefined criteria.

The analysis in this chapter suggests that the model for the objective decision process has to be kept small. This secures fast convergence of the solution as well as the possibility for the trained user to redefine boundary conditions and test the solution's sensitivity to various likely and less likely events.

Another result that can be derived from our analysis is that a balance responsible party for intermittent energy should issue regular tenders on pool participation. This will allow an energy pool to deliver power according to a different profile than the weather allows for. Typically the generation should be phase-shifted to match the demand better. The net result is then that the market can no longer force the intermittent generator to bid in with low prices. The intermittent generator thereby gains freedom with the dynamically changing "cocktail" of pooled energy, weather uncertainty and advances trading according to the ensemble minimum. The more degrees of freedom, the higher the price for the intermittent energy, which is critical for the success rate in the future energy markets, because it can not be expected that the required investments in renewable energy will increase or even be kept at today's rate without an economic incentive in addition to the environmental benefits.

References

[1] Global Wind Energy Council (www.gwec.net), Jan. 2008.

[2] B. Lange, M. Wolff, R. Mackensen, R. Jursa, K. Rohrig, D. Braams, B. Valov, L. Hofmann, C. Scholz, and K. Biermann, "Operational control of wind farm clusters for transmission system operators," in *Prod. European Offshore Conference*, Berlin, 2007.

[3] IEA, "Variability of wind power and other renewables: Management options and strategies", IEA/OECD, Paris, 2005.

[4] IEA, "World energy outlook 2004", IEA/OECD, Paris, 2004.

[5] B. Ernst, B. Oakleaf, M. L. Ahlstrom, M. Lange, C. Möhrlen, B. Lange, U. Focken, and K. Rohrig, "Predicting the wind", *IEEE Power and Energy Magazine*, vol. 5, no. 6, pp. 78-89, 2007.

[6] Europressedienst (www.europressedienst.com/news/?newsdate=2-2007), Feb. 2007.

[7] T. Acker, "Characterization of wind and hydropower integration in the USA," in *Proc. AWEA Conference Windpower*, Denver, May 2005.

[8] E. D. Castronuovo and J. A. P. Lopes, "Optimal operation and hydro storage sizing of a wind-hydro power plant," *El. Power Energy Sys.*, vol. 26, pp. 771-778, 2004.

[9] M. Pahlow, L.-E. Langhans, C. Möhrlen, and J. U. Jørgensen, "On the potential of coupling renewables into energy pools," *Z. Energiewirtschaft*, vol. 31, pp. 35-46, 2007.

[10] www.bundesnetzagentur.de, Nov. 2007.

[11] G. Giebel, G. Kariniotakis, and R. Brownsword, "The state- of-the-art in short-term forecasting of wind power - a literature overview," *Position paper ANEMOS Project*, 38 pp., 2005.

[12] G. Hassan, "Short-term wind energy forecasting: Technology and policy," *Report for the Canadian Wind Energy Assocoation*, 2006.

[13] T. P. Legg, K. R. Mylne, and C. Woolcock, "Use of medium-range ensembles at the Met Office I: PREVIN - a system for the production of probabilistic forecast information from the ECMWF EPS," *Met. Appl.*, vol. 9, pp. 255-271, 2002.

[14] R. Buizza, P. L. Houtekamer, Z. Toth, G. Pellerin, M. Wei, and Y. Zhu, "A comparison of the ECMWF, MSC, and NCEP global ensemble prediction systems," *Month. Weather Rev.*, vol. 133, pp. 1076-1097, 2005.

[15] G. Evensen, "Ensemble Kalman Filter: theoretical formulation and practical implementation," *Ocean Dynamics*, vol. 53, pp. 347-367, 2003.

[16] T. M. Hamill and C. Synder, "A hybrid ensemble Kalman Filter-3D variational analysis scheme," *Month. Weath. Rev.*, vol. 128, pp. 2905-2919, 2000.

[17] P. L. Houtekamer, L. Herschel, and L. Mitchell, "A sequential Ensemble Kalman Filter for atmospheric data assimilation," *Month. Weath. Rev.*, vol. 129, pp. 123-137, 2001.

[18] T. N. Palmer, "A non-linear dynamical perspective on model error: A proposal for non-local stochastic-dynamic parameterisation in weather and climate prediction models," *Q. J. R. Meteorol. Soc.*, vol. 127, pp. 279-304, 2001.

[19] R. Buizza, M. Miller, and T. N. Palmer, "Stochastic representation of model uncertainties in the ECMWF Ensemble Prediction System," *Q. J. R. Meteorol. Soc.*, vol. 125, pp. 2887-2908, 1999.

[20] J. Barkmeijer, R. Buizza, T. N. Palmer, K. Puri, and J.-F. Mahfouf, "Tropical Singular Vectors computed with linearized diabatic physics," *Q. J. R. Meteorol. Soc.*, vol. 127, pp. 685-708, 2001.

[21] Z. Toth and E. Kalnay, "Ensemble forecasting at NMC: the generation of perturbations," *Bull. Americ. Meteorol. Soc.*, vol. 74, pp. 2317-2330, 1993.

[22] M. S. Tracton, K. Mo, W. Chen, E. Kalnay, R. Kistler, and G. White, "Dynamic extended range forecasting (DERF) at the National Meteorological Center: practical aspects," *Month. Weath. Rev.*, vol. 117, pp. 1604-1635, 1989.

[23] P. L. Houtekamer and J. Derome, "Prediction experiments with two-member ensembles", *Month. Weath. Rev.*, vol. 122, pp. 2179-2191, 1994.

[24] P. L. Houtekamer and L. Lavaliere, "Using ensemble forecasts for model validation," *Mon. Weath. Rev.*, vol. 125, pp. 796-811, 1997.

[25] R. E. Evans, M. S. J. Harrison, R. J. Graham, and K. R. Mylne, "Joint medium-range ensembles form the Met. Office and ECMWF systems," *Month. Weath. Rev.*, vol. 128, pp. 31043127, 2000.

[26] T. M. Krishnamurti, C. M. Krishtawal, Z. Zhang, T. LaRow, D. Bachiochi, and C. E. Williford, "Multimodel ensemble forecasts for weather and seasonal climate," *J. Climate*, vol. 13, pp. 4196-4216, 2000.

[27] D. J. Stensrud, J. W. Bao, and T. T. Warner, "Using initial condition and model physics perturbations in short-range ensemble simulations of mesoscale convective systems," *Month. Weath. Rev.*, vol. 128, pp. 2077-2107, 2000.

[28] P. J. Roebber, D. M. Schultz, B. A. Colle, and D. J. Stensrud, "Toward improved prediction: High-resolution and ensemble modelling systems in operations," *Weath. Forecasting*, vol. 19, pp. 936-949, 2004.

[29] K. Mylne and K. Robertson, "Poor Man's EPS experiments and LAMEPS plans at the Met Office," *LAM EPS Workshop*, Madrid, October, 2002.

[30] K. Sattler and H. Feddersen, "EFFS Treatment of uncertainties in the prediction of heavy rainfall using different ensemble approaches with DMI-HIRLAM," Scientific Report 03-07, Danish Meteorological Institute, 2003.

[31] Z. Meng and F. Zhang, "Tests of an Ensemble Kalman Filter for Mesoscale and Regional-Scale Data Assimilation. Part II: Imperfect Model Experiments," *Month. Weath. Rev.*, vol. 135, pp. 1403-1423, 2007.

[32] C. Möhrlen, "Uncertainty in wind energy forecasting," Ph.D. dissertation, University College Cork, Ireland, 2004.

[33] C. Möhrlen and J. U. Jørgensen, "Forecasting wind power in high wind penetration markets using multi-scheme ensemble prediction methods," in *Proc. German Wind Energy Conference DEWEK*, Bremen, Nov. 2006.

[34] S. Lang, C. Möhrlen, J. U. Jørgensen, B. P. Ó Gallachóir, and E. J. McKeogh, "Forecasting total wind power generation on the Republic of Ireland grid with a Multi-Scheme Ensemble Prediction System," *Proc. Global Windpower*, Adelaide, Sept. 2006.

[35] S. Lang, C. Möhrlen, J. U. Jørgensen, B. P. Ó Gallachóir, and E. J. McKeogh, "Application of a Multi-Scheme Ensemble Prediction System for wind power forecasting in Ireland and comparison with validation results from Denmark and Germany," in *Proc. European Wind Energy Conference*, Athens, Feb. 2006.

[36] Ü. Cali, K. Rohrig, B. Lange, K. Melih, C. Möhrlen, and J.U. Jørgensen, "Wind power forecasting and confidence interval estimation using multiple numerical weather prediction models," in *Proc. European Wind Energy Conference*, Milan, 2007.

[37] C. Möhrlen, J. U. Jørgensen, P. Pinson, H. Madsen, and J. Runge Kristofferson, "HRENSEMBLEHR - High resolution ensemble for Horns Rev: A project overview," in *Proc. European Offshore Wind Energy Conference*, Berlin, 2007.

[38] P. Pinson, H. Madsen, C. Möhrlen, and J. U. Jørgensen, "Ensemble-based forecasting at Horns Rev," DTU Technical Report, 2008.

[39] J. Jackson, "A cry for better forecasters in Denmark," *Wind Power Monthly*, vol. 12, pp. 40-42, 2003.

[40] C. Möhrlen and J. U. Jørgensen, Verification of Ensemble Prediction Systems for a new market: wind energy. ECMWF Special Project Interim Report, 3, 2005.

[41] C. Keil and M. Hagen, Evaluation of high resolution NWP simulations with radar data," *Physics and Chemistry of the Earth B*, vol.25, pp. 1267-1272, 2000.

[42] R. N. Hoffman, Z. Liu, J.-F. Louis, and C. Grassotti, "Distortion representation of forecast errors," *Mon. Wea. Rev.*, vol. 123, pp. 2758-2770, 2005.

[43] Alberta Electric System Operator (AESO) wind power pilot study (http://www.aeso.ca/gridoperations/13825.html), 2006.

INDEX

D

X

Y

Z